Neutron Spectroscopy

Neutron Spectroscopy

Yu. A. Izyumov

Institute of Metal Physics
Ural Division of the Russian Academy of Sciences
Ekaterinburg (Sverdlovsk), Russia

and

N. A. Chernoplekov

Kurchatov Institute
Moscow, Russia

Translated from Russian by
Joachim Büchner

CONSULTANTS BUREAU • NEW YORK AND LONDON

Library of Congress Cataloging-in-Publication Data

Izi͡umov, I͡U. A. (I͡Uriĭ Aleksandrovich), 1933-
[Neĭtronnai͡a spektroskopii͡a. English]
Neutron spectroscopy / Yu.A. Izyumov and N.A. Chernoplekov ; translated from Russian by Joachim Büchner.
p. cm.
Translation of: Neĭtronnai͡a spektroskopii͡a.
Includes bibliographical references and index.
ISBN 0-306-11033-4
1. Solids--Spectra. 2. Neutrons--Spectra. 3. Neutrons--Scattering. I. Chernoplekov, N. A. (Nikolaĭ Alekseevich) II. Title.
QC176.8.O6I9513 1992
530.4'11--dc20 92-4677
CIP

This translation is published under an agreement with the Russian Copyright Agency (VAAP)

ISBN 0-306-11033-4

A Division of Plenum Publishing Corporation
233 Spring Street, New York, N.Y. 10013

Printed in the United States of America

PREFACE

This book deals with the neutron spectroscopy of solids, i.e., research concerning the dynamics of the crystal state with the aid of inelastic neutron scattering. Thermal neutrons are scattered on a crystal with absorption or emission of elementary excitations, mainly phonons and magnons. Processes in which a single exciting quantum participates are particularly suitable for providing information because in regard to the energy distribution of the coherently scattered neutrons, such processes provide pronounced peaks whose positions directly determine the excitation energy, whereas the width of a peak is indicative of the lifetime. Moreover, the energy distribution of incoherent single-quantum scattering provides information on the density of states in the spectrum of the elementary excitation.

The energy analysis of neutrons scattered in various directions helps to determine the dispersion laws of the phonons and magnons in crystals, and this turns out to be possible for an entire Brillouin zone. No other method offers these possibilities. Thus, acoustic methods make it possible to determine only the initial sections of the dispersion curves of phonons, the sections corresponding to small wave vectors; optical spectroscopy provides only the frequencies of phonons with vanishing wave vectors, similar to magnetic resonances providing the frequencies of spin waves with the wave vector zero. Measurements of the intensities of the peaks of inelastic nuclear or magnetic scattering make it possible to determine the polarization vectors of the phonon and magnon modes, respectively.

Thus, inelastic neutron scattering provides full information on the normal vibration modes of crystal lattices and magnetic lattices of solids.

In the two decades, the techniques of neutron spectroscopy have been used to study the phonon spectra of the basic types of crystal structures. Combined with theoretical models, the mechanisms of interatomic interactions in various crystals could be studied. The same, though to a lesser degree, holds for systems with magnetic order where an experimental determination of the dispersion curves of magnons (spin waves) makes it possible to improve our knowledge of the forces accounting for the development of a magnetic structure in a crystal.

At the present time, investigations of complex crystals with many atoms per unit cell have been initiated. Therefore, the development of theoretical methods of analyzing phonon and magnon spectra and the analysis of neutron scattering cross sections on the basis of the theory of crystal symmetry are of interest. The theory of space group representations facilitates the practical analysis of spectra of crystals of any complexity. The exposition of the theory of harmonic vibrations of crystal lattices and magnetic lattices and of elastic neutron scattering on them using the theory of symmetry is one of the main goals of this book. The method developed allows the reader (without resorting to other sources) to independently perform symmetry analysis of phonons for any crystal and of magnons for magnetic materials that can be described by Shubnikov symmetry.

Another important problem considered in the book is the theory of neutron scattering near a structural transition or a magnetic phase transition, i.e., scattering either at soft modes or at fluctuations of the order parameter with proper regard for the dynamics of the fluctuations. These problems, as well as the dynamics in the harmonic approximation, are treated on the basis of the theory of space group representations.

The third major problem considered in the monograph involves the description of scanning methods on the reciprocal lattice, i.e., the strategies of experimental investigations: determination of the dispersion curves, the spectral density, etc. Thus, the book familiarizes the reader with the contemporary state of neutron spectroscopy from both the theoretical and experimental aspects.

The limited volume of the book does not allow the exposition of all results of research on crystal dynamics by the technique of inelastic neutron scattering and, hence, the authors have analyzed only individual examples. The data on soft modes in crystals undergoing structural phase transitions, the dynamics of hydrogen-containing compounds, and the electron interaction-induced features of the dynamics of metals are fully considered. The selection of experimental results was guided by both the personal interest of the authors in the field of neutron spectroscopy and from the fact that these branches of science have been treated less extensively in reviews and monographs. Excellent review articles (see, e.g., the collection edited by S. Lovesey and T. Springer [27a]) have been published on neutron-aided research on the dynamics of dielectrics and semiconductors, molecular crystals, quantum crystals, and other crystalline objects. With respect to research on the dynamics of substances with magnetic order, the authors tried to provide the fullest possible exposition of the experimental results (to the extent to which this is possible within a single book).

The large number of problems treated using neutron spectroscopy but omitted in the present book due to its space limitations fall into three fields.

The first field deals with research on incommensurate structures of long periodicity, particularly in the region of phase transition from an incommensurate to a commensurate phase.

Collective motions of such a system near a phase transition, the so-called phasons, are at the present time the objects of research conducted with the aid of inelastic neutron scattering.

The second field deals with the dynamics of systems without order, metallic glasses and spin glasses, amorphous semiconductors, dielectrics, etc., which is another problem of current interest.

In both fields, investigations with neutrons have only been started, but the results are of major importance.

The third field deals with crystal field effects; research on this subject has been done on a large scale and is now completed so that a special review or monograph devoted to it is now due.

The authors thank their colleagues O. V. Gurin, M. G. Zemlyanov, P. P. Parshan, S. B. Petrov, A. Yu. Rumyantsev, V. N. Syromyatnikov, and A. A. Chernyshov for their assistance.

NOTATION

$\mathbf{k}_0$	wave vector of the incident neutron
$\mathbf{k}$	wave vector of the scattered neutron
$\varkappa = \mathbf{k} - \mathbf{k}_0$	scattering vector
$\mathbf{e} = \varkappa/\varkappa$	unit scattering vector
$\varepsilon = E_{\mathbf{k}} - E_{\mathbf{k}_0}$	energy transfer
$\mathbf{q}$	wave vector of a quasiparticle (phonon, magnon)
k	wave vector of the superlattice
$\mathbf{p}_0$	polarization vector of the incident beam
$\mathbf{p}$	polarization vector of the scattered beam
$V_{\mathbf{kk}_0}$	amplitude of the interaction of the neutron with the scatterer
$l = \{ni\}$	number of the atom at a lattice point in the crystal
n	number of the cell in the crystal
i	number of the atom in the cell
$\mathbf{x}_i$	radius vector of the ith atom in the unit cell
M_l	mass of the atom
$F_l(\mathbf{q})$	magnetic form factor of the atom
$W_l(\mathbf{q})$	thermal factor
S_l	atomic spin
$\mathbf{u}_l$	displacement of the atom in the crystal
$\varphi_{ij}^{\alpha\beta}(\mathbf{q})$	force matrix

$\Phi_{ij}{}^{\alpha\beta}(\mathbf{q})$	dynamic matrix
$\omega_J(\mathbf{q})$	frequency of the phonon (magnon) branch
$\mathbf{e}_J(\mathbf{q} \mid i)$	vector of phonon polarization
σ	number of atoms in a primitive cell of the crystal
σ_M	number of magnetic atoms in a magnetic cell of the crystal
$\mathbf{t}_n$	arbitrary translation in a lattice
$\mathbf{b}$	vector of the reciprocal lattice
$\mathbf{b}_1, \mathbf{b}_2, \mathbf{b}_3$	basic (shortest) vectors of the reciprocal lattice
G	space group of the crystal
$G_{\mathbf{q}}$	group of the wave vector
$g = \{h \mid \tau_h\}$	element of the space group: h denotes its rotational part; τ_h, the accompanying translation
M	group of the magnetic symmetry of the crystal
$\{\mathbf{q}\}$	star of the wave vector
$\mathbf{K}_L$	arm of the star
$d^{q\nu}$	irreducible representation of the group $G_{\mathbf{q}}$
l_ν	dimension of the irreducible representation of the group $G_{\mathbf{q}}$
τ_ν	irreducible representation (for the Kovalev tables)
$\varepsilon(\mathbf{q}_\lambda{}^\nu \mid i)$	atomic component of the λth basis function of an irreducible representation
$d_m{}^{\mathbf{q}}$	mechanical representation of the group $G_{\mathbf{q}}$
$d_M{}^{\mathbf{q}}$	magnetic representation of the group $G_{\mathbf{q}}$
$d_{SW}{}^q$	spin-wave representation of the group $G_{\mathbf{q}}$
$N(\mathbf{q})$	Bose distribution function
T_c	phase-transition temperature
$\tau = (T - T_c)/T_c$	relative temperature
$\kappa_1 = 1/\xi$	reciprocal correlation length

CONTENTS

Chapter 1. THE THEORY OF NEUTRON SCATTERING AT VIBRATIONS OF THE CRYSTAL LATTICE **1**

1. General Expressions for the Scattering Cross Section . . . 1
2. Single-Phonon Scattering at an Ideal Crystal 12
3. Single-Phonon Scattering at a Crystal Containing Impurities 18
4. Description of Crystal Lattice Vibrations on the Basis of Symmetry Theory 36
5. Selection Rules in Coherent Single-Phonon Scattering . . . 56
6. Phonons and Structural Phase Transitions 65
7. Critical Scattering of Neutrons 77

Chapter 2. DYNAMICS OF THE MAGNETIC SYSTEM OF A CRYSTAL AND INELASTIC SCATTERING OF NEUTRONS **86**

8. General Expressions for the Cross Section of Magnetic Scattering 87

9. Cross Section of Scattering at Spin Waves 90

10. Single-Magnon Scattering of a Polarized Neutron at an Arbitrary Magnetic Structure 100

11. Classification of Spin Waves in Terms of Symmetry 110

12. Scattering of Neutrons at Hybrid Magnon–Phonon Excitations 126

13. Experimental Investigations of the Spin-Wave Spectrum of Magnetic Materials 145

14. Neutron Scattering in a Collectivized Electron System . . . 160

15. Scattering at Soliton Excitations 171

16. Critical Scattering near a Magnetic Phase Transition 182

17. Critical Dynamics of Ferromagnets 195

Chapter 3. STRUCTURAL PHASE TRANSITIONS AND THE DYNAMICS OF CRYSTALS 214

18. Structural Phase Transitions in Perovskites 214

19. Other Structural Transitions of the Displacement Type . . . 223

20. Structural Transitions of the Order–Disorder Type 232

21. Critical Dynamics in a Structural Phase Transition 242

Chapter 4. EXPERIMENTAL METHODS IN NEUTRON SPECTROSCOPY OF PHONONS 253

22. Measurement of the Density Function of Phonon States . . 254

23. Measurement of Dispersion Curves in the Case of Phonons 261

Chapter 5. PHONON SPECTRA OF CRYSTALS 268

24. Phonon Spectra of Non-Transition Metals 268

25. Transition Metals 290

26. Nonmetal Crystals 294

Chapter 6. ELECTRON–PHONON INTERACTION AND ANOMALIES OF THE PHONON SPECTRA OF METALS 301

27. Kohn Anomalies on the Phonon Dispersion Curves 302

28. Anomalies of the Phonon Spectra of Superconductors . . . 318

Chapter 7. THE DYNAMICS OF METAL HYDRIDES . . 322

29. General Remarks on the Dynamics of Hydrides 322

30. Inelastic Scattering of Neutrons at Metal Hydrides 325

31. Hydrides of Transition Metals of the 5th Group 331

32. Metal Hydrides Based on a Face-Centered Cubic Metal Lattice 336

33. Dispersion Curves for Phonons in Hydrides 348

CONCLUSION 359

REFERENCES 361

Neutron Spectroscopy

Chapter 1

THE THEORY OF NEUTRON SCATTERING AT VIBRATIONS OF THE CRYSTAL LATTICE

This chapter is a systematic and detailed exposition of the theory of harmonic vibrations of a crystal and relates to the application of this theory to slow neutron scattering (Sections 1-3). The treatment focuses on single-quantum neutron scattering processes, the analysis of which makes it possible to reconstruct the phonon dispersion curves and their spectral density from experimental data.

The use of symmetry theory for the description and theoretical construction of the phonon modes of a crystal is emphasized. Section 4 presents the mathematical apparatus required for this purpose; the method is based on the theory of space group representations. Section 5 outlines the method of investigating the selection rules in coherent single-phonon scattering.

The scattering of neutrons near a structural phase transition in the form of a displacement is considered in Sections 6 and 7 on the basis of the theory. Formulas are derived for the scattering cross section at a soft phonon mode and at fluctuations of an order parameter. The resulting equations are applicable to crystals of any complexity. The symmetry analysis of a phonon spectrum, the selection rules, and structural phase transitions are illustrated by the example of cubic perovskites.

1. The General Expressions for the Scattering Cross Section

The physics of the scattering of slow neutrons at solids has been expounded in many textbooks (see, e.g., [24]); therefore, assuming that the reader is familiar with the elementary scattering events of neutrons at atoms, we consider the scattering at a crystal as a whole. Inelastic scattering processes are conveniently described within the framework of standard perturbation theory making use of the so-called time formalism which can provide a general expression for the differential scattering cross section through correlators of quantities characterizing a scattering crystal.

The van Hove time formalism [348]. The Born approximation for the probability $W_{n\mathbf{k}|n_0\mathbf{k}_0}$ of a transition from an initial state $n_0\mathbf{k}_0$ to a final state $n\mathbf{k}$ is used in the theory of scattering of slow neutrons:

$$W_{n\mathbf{k}|n_0\mathbf{k}_0} = \frac{2\pi}{\hbar} |\langle n | V_{\mathbf{k}\mathbf{k}_0} | n_0 \rangle|^2 \delta(E_{n_0} + E_{\mathbf{k}_0} - E_n - E_{\mathbf{k}}). \quad (1.1)$$

where $\mathbf{k}_0$ and $E_{\mathbf{k}_0}$ denote the wave vector and the energy of an incident neutron; $\mathbf{k}$ and $E_{\mathbf{k}}$ denote the wave vector and the energy of a scattered neutron; $|\, n_0\rangle$ and $|\, n\rangle$ denote the states of the scatterer (crystal) before and after a scattering event; E_n and E_{n_0} denote the energies of the scatterer in these states; and V denotes the operator of the interaction of a neutron with the scatterer.

The double-differential neutron scattering cross section per unit solid angle Ω and unit energy interval $E_{\mathbf{k}}$ is obtained by averaging the probability $W_{n\mathbf{k}|n_0\mathbf{k}_0}$ over the initial states and by summation over the final states of the scatterer:

$$\frac{d^2\sigma}{d\Omega dE_{\mathbf{k}}} = \frac{m^2}{(2\pi\hbar)^3} \frac{k}{k_0} \sum_{nn_0} \rho_{n_0} \overline{W}_{n\mathbf{k}|n_0\mathbf{k}_0}, \quad (1.2)$$

where ρ_{n_0} denotes the distribution function of the states n_0; and m denotes the mass of the neutron. The bar on top denotes averaging over the spin polarizations in the neutron beam.

The latter expression can be easily rewritten in integral form when one uses in Eq. (1.1) the well-known representation of the δ function:

$$\delta(E) = \frac{1}{2\pi\hbar} \int_{-\infty}^{\infty} e^{-\frac{1}{\hbar}Et} dt. \quad (1.3)$$

Then the final result for the double-differential cross section is conveniently represented in operator form:

$$\frac{d^2\sigma}{d\Omega d\varepsilon} = \frac{m^2}{(2\pi)^3 \hbar^5} \frac{k}{k_0} \int_{-\infty}^{\infty} e^{\frac{i}{\hbar}\varepsilon t} \overline{\langle V^+_{\mathbf{kk}_0}(0)\, V_{\mathbf{kk}_0}(t)\rangle}\, dt. \tag{1.4}$$

where $V_{\mathbf{kk}_0}(t)$ denotes the Heisenberg representation of the operator $V_{\mathbf{kk}_0}$ at the time t with the Hamiltonian $\mathcal{H}$ of the scatterer:

$$V_{\mathbf{kk}_0}(t) = e^{i\frac{\mathcal{H}t}{\hbar}}\, V_{\mathbf{kk}_0}\, e^{-i\frac{\mathcal{H}t}{\hbar}}. \tag{1.5}$$

The matrix element

$$V_{\mathbf{kk}_0} = \int e^{-i\mathbf{kr}}\, V(\mathbf{r})\, e^{i\mathbf{k}_0\mathbf{r}}\, d\mathbf{r} \tag{1.6}$$

is an operator for the dynamic variables describing the state of the scatterer, and the statistical averaging is performed in Eq. (1.4) in accordance with the well-known rule

$$\langle\ldots\rangle = \mathrm{Sp}\,(e^{-\mathcal{H}/kT}\ldots)/\mathrm{Sp}\, e^{-\mathcal{H}/kT}. \tag{1.7}$$

The quantity

$$\varepsilon = E_{\mathbf{k}} - E_{\mathbf{k}_0}, \tag{1.8}$$

was introduced in expression (1.4) for the cross section, and this quantity is termed the transfer energy; the matrix element $V_{\mathbf{kk}_0}$ depends upon the quantity

$$\varkappa = \mathbf{k} - \mathbf{k}_0, \tag{1.9}$$

which is termed the transfer momentum or, more frequently, the scattering vector, so that the cross section defined by Eq. (1.4) depends upon the two variables $\varkappa$ and ε which, indeed, are measured in experiments. It follows from Eq. (1.4) that $d^2\sigma/d\Omega d\varepsilon$ is the Fourier component of some correlation functions for the dynamic variables of the scattering, with the dynamic variables referring to the two instants 0 and t. The calculation of these functions is the problem underlying the theory of scattering.

By replacing the variable t by $-t$ in the integral of Eq. (1.4), we obtain another equally popular expression for the cross section:

$$\frac{d^2\sigma}{d\Omega d\varepsilon}=\frac{m^2}{(2\pi)^3\hbar^5}\frac{k}{k_0}\int\limits_{-\infty}^{\infty} e^{-i\frac{\varepsilon}{\hbar}t}\,\overline{\langle V^+_{\mathbf{kk}_0}(t)\,V_{\mathbf{kk}_0}(0)\rangle}\,dt. \qquad (1.10)$$

Equations (1.4) and (1.10) reflect the basic mathematical tools in research on the cross section of inelastic neutron scattering at crystals. Let us note that in the two equations, V^+ denotes the Hermitian conjugate of V. Finally, according to what has been said above [see Eq. (1.2)], the bar over the operator means

$$\bar{L} = \mathrm{Sp}\ (\rho_\sigma L), \qquad (1.11)$$

where ρ_σ denotes the operator of the spin density in the initial beam.

Nuclear scattering at a crystal. The energy of the interaction of a slow neutron with an individual nucleus is described by the Fermi pseudopotential

$$V(\mathbf{r}) = \alpha\delta(\mathbf{r} - \mathbf{R}_l), \qquad (1.12)$$

where $\mathbf{r}$ and $\mathbf{R}_l$ denote the coordinates of the neutron and of the nucleus, respectively, and α defines the scattering amplitude. In the case of an interaction with a system of N nuclei, the operator

$$V=\sum_{l=1}^{N}\alpha_l\,\delta(\mathbf{r}-\mathbf{R}_l), \qquad (1.13)$$

and the corresponding matrix element $V_{\mathbf{kk}_0}$ in accordance with Eqs. (1.6) and (1.9) are equal to

$$V_{\mathbf{kk}_0}=\sum_l \alpha_l\, e^{-i\varkappa\mathbf{R}_l}. \qquad (1.14)$$

By substituting this equation into Eq. (1.10), we observe that the cross section of nuclear scattering at a crystal is given by

$$\frac{d^2\sigma}{d\Omega d\varepsilon}=\frac{m^2}{4\pi^2\hbar^4}\frac{k}{k_0}\sum_{ll'}\overline{\alpha_l\alpha_{l'}}\,\frac{1}{2\pi\hbar}\int\limits_{-\infty}^{\infty} e^{-\frac{i}{\hbar}\varepsilon t}\langle e^{i\varkappa\mathbf{R}_l(t)}\,e^{-i\varkappa\mathbf{R}_{l'}(0)}\rangle\,dt, \qquad (1.15)$$

in which the averaging over the spin polarization of the incident beam still has to be performed. One must take into account that when a spin of magnitude I is present in the nucleus, the amplitude α_l has the following structure:

$$\alpha_l = A_l + B_l\,(\mathbf{S}_n \mathbf{I}), \tag{1.16}$$

where $\mathbf{S}_n$ denotes the neutron spin and A and B denote constants which do not depend upon the energy and the spin. In the case of a non-polarized neutron beam, the spin density is $\rho_\sigma = 1/2$ (the two orientations of the neutron spin are equally likely) and therefore

$$\overline{\alpha_l \alpha_{l'}} = A_l A_{l'} + (1/4) B_l^2 I_l\,(I_l + 1)\,\delta_{ll'}. \tag{1.17}$$

Equation (1.15) makes it possible to average also over the isotope distribution because the amplitudes α_l differ for the various isotopes of a particular element. This averaging is conveniently performed by introducing the population numbers P_l^s which are equal to unity if the crystal site l is occupied by an isotope of the variety s, and which are equal to zero otherwise. Then the averaging of the physical quantities over the isotope composition is reduced to averaging over these population numbers, e.g.,

$$\widetilde{A_l A_{l'}} = \sum_{ss'} A^s A^{s'}\,\widetilde{P_l^s P_{l'}^{s'}}, \tag{1.18}$$

where the tilde denotes averaging over the isotopes and A^s denotes the spin-independent part of the amplitude for the isotope of the variety s. Since the isotope distribution in a crystal is stochastic, we obtain

$$\widetilde{P_l^s P_{l'}^{s'}} = c^s c^{s'} + (c^s \delta_{ss'} - c^s c^{s'})\,\delta_{ll'}, \tag{1.19}$$

where c^s denotes the relative concentration of a particular isotope. Making use of this equation, we write Eq. (1.18) as

$$\widetilde{A_l A_{l'}} = (\tilde{A})^2 + \{\widetilde{A^2} - (\tilde{A})^2\}\,\delta_{ll'}, \tag{1.20}$$

where $\tilde{A}$ and $\widetilde{A^2}$ denote the quantities

$$\tilde{A} = \sum_s c^s A^s, \quad \widetilde{A^2} = \sum_s c^s\,(A^s)^2, \tag{1.21}$$

averaged over the isotope composition. Thus, by combining Eqs. (1.17) and (1.20), we obtain

$$\widetilde{\alpha_l \alpha_{l'}} = (\tilde{A})^2 + \{\widetilde{A^2} - (\tilde{A})^2 + (1/4)\,\widetilde{B^2(I+1)I}\}\,\delta_{ll'}, \tag{1.22}$$

which makes it possible to take into account the isotope-averaging effect in Eq. (1.15).

In Eq. (1.15), $\mathbf{R}_l$ and $\mathbf{R}_{l'}$ denote the instantaneous positions of the crystal nuclei which perform thermal vibrations. Let us represent $\mathbf{R}_l$ in the form

$$\mathbf{R}_l(t) = \mathbf{R}^\circ_l + \mathbf{u}_l(t), \tag{1.23}$$

where $\mathbf{R}_l^\circ$ denotes the equilibrium position of atom l in the crystal, and $\mathbf{u}_l(t)$ denotes the atom's displacement. By substituting Eqs. (1.23) and (1.22) in Eq. (1.15), we can write the cross section in its final form

$$\frac{d^2\sigma}{d\Omega d\varepsilon} = \frac{k}{k_0}\sum_{ll'}\{\sigma^{\mathrm{coh}} + \delta_{ll'}\,\sigma^{\mathrm{incoh}}\}\,e^{i\varkappa(\mathbf{R}_l^\circ - \mathbf{R}_{l'}^\circ)} \times$$
$$\times \frac{1}{2\pi\hbar}\int_{-\infty}^{\infty} e^{-\frac{i}{\hbar}\varepsilon t} K_{ll'}(t)\,dt, \tag{1.24}$$

where

$$K_{ll'}(t) = \langle e^{i\varkappa\mathbf{u}_l(t)}\,e^{-i\varkappa\mathbf{u}_{l'}(0)}\rangle \tag{1.25}$$

represents the correlator of the atomic displacements and σ^{coh} and σ^{incoh} denote the effective cross sections of coherent and incoherent scattering of neutrons at an isolated nucleus:

$$\sigma^{\mathrm{coh}} = \frac{m^2}{4\pi^2\hbar^4}(\tilde{A})^2, \quad \sigma^{\mathrm{incoh}} = \frac{m^2}{4\pi^2\hbar^4}\left\{\widetilde{A^2} - (\tilde{A})^2 + \frac{1}{4}\widetilde{B^2 I(I+1)}\right\}. \tag{1.26}$$

Equation (1.24) together with Eq. (1.25) are the starting formulas for calculations of the cross section of inelastic neutron scattering at vibrating crystals. The theory of the normal vibration modes of a crystal lattice must be employed to calculate the correlator of Eq. (1.25).

The normal vibration modes. In the harmonic approximation, the vibrations of a crystal are described by the Hamiltonian

$$\mathcal{H} = \sum_{l\alpha} \frac{(P_l^\alpha)^2}{2M_l} + \frac{1}{2} \sum_{ll'} \sum_{\alpha\beta} \varphi_{ll'}^{\alpha\beta} u_l^\alpha u_{l'}^\alpha, \tag{1.27}$$

where the momentum operator P_l^α and the atomic displacement u_l^α satisfy the quantum-mechanical commutation relation

$$[u_l^\alpha, P_{l'}^\beta] = i\hbar\delta_{\alpha\beta}\,\delta_{ll'}. \tag{1.28}$$

Regardless of whether or not the masses M_l of all the atoms of the crystal are equal, the quadratic form can always be converted to its diagonal form, and the eigenvectors of this quadratic form (1.27), which correspond to the eigenvalues, are termed the normal vibration modes. The transition to the normal modes is accomplished with the following equations:

$$u_l^\alpha = \sum_\lambda \sqrt{\frac{\hbar}{2M_l\,\omega_\lambda}}\,\{B_{l\lambda}^\alpha b_\lambda + B_{l\lambda}^{*\alpha} b_\lambda^+\}; \tag{1.29}$$

$$P_l^\alpha = -i\sum_\lambda \sqrt{\frac{\hbar M_l\,\omega_\lambda}{2}}\,\{B_{l\lambda}^\alpha b_\lambda - B_{l\lambda}^{*\alpha} b_\lambda^+\}. \tag{1.30}$$

The amplitudes b_λ and b_λ^+ of the normal vibrations satisfy the Bose commutation rules

$$[b_\lambda, b_{\lambda'}^+] = \delta_{\lambda\lambda'},\ [b_\lambda, b_{\lambda'}] = [b_\lambda^+, b_{\lambda'}^+] = 0, \tag{1.31}$$

and the vector quantities $B_{l\lambda}^\alpha$ satisfy the orthogonality and completeness equations

$$\left.\begin{aligned} &\sum_{l\lambda} \overset{*}{B}{}_{l\lambda}^\alpha B_{l\lambda'}^\alpha = \delta_{\lambda\lambda'}; \\ &\sum_\lambda \overset{*}{B}{}_{l\lambda}^\alpha B_{l'\lambda}^\beta = \delta_{\alpha\beta}\,\delta_{ll'}. \end{aligned}\right\} \tag{1.32}$$

It is easy to verify that the initial commutational relation (1.28) is satisfied if one inserts into it Eqs. (1.29) and (1.30) and makes use of Eqs. (1.31) and (1.32). It is also easy to verify that if the quantities $B_{l\lambda}^\alpha$ and ω_λ^2 denote, respectively, the eigenvectors and the eigenvalues of the dynamic matrix

$$\Phi_{ll'}^{\alpha\beta} = \frac{1}{\sqrt{M_l M_{l'}}} \varphi_{ll'}^{\alpha\beta}, \tag{1.33}$$

the Hamiltonian of Eq. (1.27) is diagonal in terms of the Bose amplitudes b_λ, namely

$$\mathcal{H} = \sum_\lambda \hbar\omega_\lambda \left(b_\lambda^+ b_\lambda + \frac{1}{2} \right). \tag{1.34}$$

As a consequence of what has been indicated above, $B_{l\lambda}{}^\alpha$ and $\omega_\lambda{}^2$ are obtained from the equation

$$\sum_{l'\beta} \Phi_{ll'}^{\alpha\beta} B_{l'\lambda}^{\beta} = \omega_\lambda^2 B_{l\lambda}^{\alpha}, \tag{1.35}$$

in which the dynamic matrix is assumed to be given. Thus, using the transformations of Eqs. (1.29) and (1.30) under the assumption that the quantities $B_{l\lambda}{}^\alpha$ and ω_λ satisfy Eq. (1.35) and, hence, Eqs. (1.32), means transition to the normal vibration modes. In terms of the normal modes, Eq. (1.24) for the cross section can be fully calculated.

Calculation of the cross section. First let us transform the correlator of (1.25) which appears in the expression for the cross section. For this purpose we use the expansion of Eq. (1.29) which is conveniently stated in simplified form:

$$\mathbf{u}_l = \sum_\lambda (\mathbf{S}_{l\lambda} b_\lambda + \overset{*}{\mathbf{S}}_{l\lambda} b_\lambda^+). \tag{1.36}$$

According to Eq. (1.34), the time dependence of this quantity is given by the expression

$$\mathbf{u}_l(t) = e^{\frac{i}{\hbar}\mathcal{H}t} \mathbf{u}_l e^{-\frac{i}{\hbar}\mathcal{H}t} = \sum_\lambda (\mathbf{S}_{l\lambda} e^{-i\omega_\lambda t} b_\lambda + \overset{*}{\mathbf{S}}_{l\lambda} e^{i\omega_\lambda t} b_\lambda^+). \tag{1.37}$$

Since the commutator of the quantities $i\boldsymbol{\varkappa}\mathbf{u}_l(t)$ and $i\boldsymbol{\varkappa}\mathbf{u}_{l'}(0)$, which appear in the exponentials of Eq. (1.25), is a C number, the product of the exponentials can be represented as a single exponential (see, e.g., [24]) so that

$$K_{ll'}(t) = e^{\frac{1}{2}[i\boldsymbol{\varkappa}\mathbf{u}_l(t),\, -i\boldsymbol{\varkappa}\mathbf{u}_{l'}(0)]} \langle e^{i\boldsymbol{\varkappa}\mathbf{u}_l(t) - i\boldsymbol{\varkappa}\mathbf{u}_{l'}(0)} \rangle. \tag{1.38}$$

This representation is advantageous because one can use the cumulative expansion method, i.e., one can use the identity

$$\langle e^{ax} \rangle = \exp \sum_{n=1}^{\infty} \frac{a^n}{n!} M_n(x). \tag{1.39}$$

The cumulative factors $M_n(x)$ of the first orders are expressed through the moments of the averaged quantity with the aid of relations which can be easily verified by comparing the expansions of the exponentials on the left and on the right side of Eq. (1.39)

$$\langle x \rangle = M_1$$

$$\langle x^2 \rangle = M_1^2 + M_2$$

$$\langle x^3 \rangle = M_1^3 + 3\, M_1 M_2 + M_3$$

$$\langle x^4 \rangle = M_1^4 + 6\, M_1^2 M_2 + 4\, M_1 M_3 + 3\, M_2^2 + M_4$$

. . . .

In the case of Eq. (1.38), the averages over the odd powers cancel so that the cumulative factors of odd order vanish, whereas the first even cumulative factors are

$$M_2 = \langle x^2 \rangle, \quad M_4 = \langle x^4 \rangle - 3\,\langle x^2 \rangle^2, \ldots \tag{1.40}$$

By using Eqs. (1.36) and (1.37), we can rewrite the average in Eq. (1.38) in the following form:

$$\langle \exp\,[i\varkappa \mathbf{u}_l(t) - i\varkappa \mathbf{u}_{l'}(0)] \rangle = \prod_{\lambda} \langle \exp\,[i\,(\eta_{ll'}^{\lambda}\, b_\lambda + \overset{*}{\eta}{}_{ll'}^{\lambda}\, b_\lambda^{+})] \rangle = \tag{1.41}$$

$$= \prod_{\lambda} \exp\left[-\frac{1}{2} \langle (\eta_{ll'}^{\lambda}\, b_\lambda + \overset{*}{\eta}{}_{ll'}^{\lambda}\, b_\lambda^{+})^2 \rangle \right] = \exp\left[-\sum_{\lambda} |\eta_{ll'}^{\lambda}|^2 \left(N_\lambda + \frac{1}{2} \right) \right].$$

For the sake of simplicity the following notation was introduced:

$$\eta_{ll'}^{\lambda} = (\varkappa \mathbf{S}_{l\lambda})\, e^{-i\omega_\lambda t} - (\varkappa \mathbf{S}_{l'\lambda})$$

and

$$N_\lambda = \langle b_\lambda^+ b_\lambda \rangle = \frac{1}{e^{\hbar\omega_\lambda/kT} - 1} \tag{1.42}$$

for the Bose function.

In the system of equations (1.41), we took into account that only the first cumulative factor M_2 must be considered in the averaged exponential because the quantity $\mathbf{S}_{l\lambda}$, which defines the amplitude of the normal mode (and this means also $\eta_{ll'}{}^\lambda$), is of the order of $1/\sqrt{N}$ (this will be shown in the next section). The cumulative factors of the next orders provide corrections $1/N$ relative to 1 and can be omitted with this degree of accuracy. After substituting the expression for $\eta_{ll'}{}^\lambda$ into Eq. (1.41) and calculating the commutator in Eq. (1.38), we obtain the final form

$$K_{ll'}(t) = e^{-W_l - W_{l'}} \exp\left\{\sum_\lambda (\varkappa \mathbf{S}_{l\lambda})(\varkappa \mathbf{S}_{l'\lambda})^* e^{-i\omega_\lambda t}(N_\lambda + 1) + \right. \\ \left. + \sum_\lambda (\varkappa \mathbf{S}_{l\lambda})^* (\varkappa \mathbf{S}_{l'\lambda}) e^{i\omega_\lambda t} N_\lambda \right\}, \tag{1.43}$$

where

$$W_l = \frac{1}{2} \sum_\lambda |\varkappa \mathbf{S}_{l\lambda}|^2 \operatorname{cth} \frac{\hbar\omega_\lambda}{2kT} \tag{1.44}$$

is the Debye–Waller thermal factor.

Because of Eq. (1.24), the scattering cross section is defined by the Fourier image of Eq. (1.43). It can be calculated simply by expanding the exponential of Eq. (1.43) in a series. Prior to that, Eq. (1.43) is conveniently rewritten in the shorter form

$$K_{ll'}(t) = e^{-W_l - W_{l'}} \exp\left[\sum_\lambda \left(A_\lambda e^{-i\omega_\lambda t} + B_\lambda e^{i\omega_\lambda t}\right)\right]. \tag{1.45}$$

Let us consider the nth-order contribution which contains p_1 times the factor $A_\lambda e^{-i\omega_\lambda t}$ and p_2 times the factor $B_\lambda e^{-i\omega_\lambda t}$ (under the condition that $p_1 + p_2 = n$). This contribution can be stated in the form

$$K_{ll'}(t)\Big|_{p_2}^{p_1} = e^{-W_l - W_{l'}} \frac{1}{p_1!\, p_2!} \sum_{\lambda_1} \ldots \sum_{\lambda_{p_1}} A_{\lambda_1} \ldots A_{\lambda_{p_1}} e^{-i(\omega_{\lambda_1} + \ldots + \omega_{\lambda_{p_1}})t} \times \\ \times \sum_{\mu_1} \ldots \sum_{\mu_{p_2}} B_{\mu_1} \ldots B_{\mu_{p_2}} e^{i(\omega_{\mu_1} + \ldots + \omega_{\mu_{p_2}})t}.$$

After inserting this expression in Eq. (1.24) and integrating over t, we obtain the corresponding contribution to the scattering cross section:

$$\frac{d^2\sigma}{d\Omega d\varepsilon}\Big|_{p_2}^{p_1} = \frac{k}{k_0}\sum_{ll'}\{\sigma^{\text{coh}}\, e^{-W_l-W_{l'}}\, e^{i\varkappa(\mathbf{R}_l^\circ-\mathbf{R}_{l'}^\circ)} + \sigma^{\text{incoh}}\, e^{-2W_l}\delta_{ll'}\}\times$$

$$\times\frac{1}{p_1!\,p_2!}\prod_{i=1}^{p_1}\sum_{\lambda_i}(\varkappa\mathbf{S}_{l\lambda_i})(\varkappa\mathbf{S}_{l'\lambda_i})^*(N_{\lambda_i}+1)\times$$

$$\times\prod_{j=1}^{p_2}\sum_{\lambda_j}(\varkappa\mathbf{S}_{l\lambda_j})^*(\varkappa\mathbf{S}_{l'\lambda_j})N_{\lambda_j}\,\delta\left(-\frac{\varepsilon}{\hbar}-\sum_{i=1}^{p_1}\omega_{\lambda_i}+\sum_{j=1}^{p_2}\omega_{\lambda_j}\right); \qquad (1.46)$$

The δ function accounts for energy conservation in the interaction of a neutron with the crystal; the conservation law can be stated in the form

$$E_{\mathbf{k}_0}+\sum_{j=1}^{p_2}\hbar\omega_{\lambda_j}=E_{\mathbf{k}}+\sum_{i=1}^{p_1}\hbar\omega_{\lambda_i}. \qquad (1.47)$$

This expression explains the meaning of the scattering process which is described by $\dfrac{d^2\sigma}{d\Omega\, d\varepsilon}\Big|_{p_2}^{p_1}$: the expression describes the scattering cross section in which p_2 phonons are absorbed and p_1 other phonons are emitted. Thus, the expansion of the exponential in Eq. (1.45) creates a multiphonon expansion of the cross section of neutron scattering at the crystal. Let us note that including multiphonon processes in accordance with Eq. (1.46) is not very consistent because when a neutron interacts simultaneously with several phonons, anharmonic terms in the Hamiltonian of the crystal, terms completely disregarded by us, become important [see the Hamiltonian of Eq. (1.34)]. But the single-phonon processes defined by Eq. (1.46) are quite accurately described. The following equations, which describe scattering under emission of a single phonon, $d^2\sigma^+/d\Omega d\varepsilon$, and under absorption of a single phonon, $d^2\sigma^-/d\Omega d\varepsilon$, result from Eq. (1.46):

$$\frac{d^2\sigma+}{d\Omega d\varepsilon} \equiv \frac{d^2\sigma}{d\Omega d\varepsilon}\Bigg|_{p_2=0}^{p_1=1} = \frac{k}{k_0}\sum_{ll'}\{\sigma^{\mathrm{coh}}\, e^{-W_l-W_{l'}+i\varkappa(\mathbf{R}_l^\circ-\mathbf{R}_{l'}^\circ)} +$$

$$+\sigma^{\mathrm{incoh}}\, e^{-2W_l}\,\delta_{ll'}\}\sum_\lambda(\varkappa\mathbf{S}_{l\lambda})(\varkappa\mathbf{S}_{l'\lambda})^*(N_\lambda+1)\left(-\frac{\varepsilon}{\hbar}-\omega_\lambda\right); \qquad (1.48)$$

$$\frac{d^2\sigma-}{d\Omega d\varepsilon} = \frac{d^2\sigma}{d\Omega d\varepsilon}\Bigg|_{p_2=1}^{p_1=0} = \frac{k}{k_0}\sum_{ll'}\{\sigma^{\mathrm{coh}}\, e^{-W_l-W_{l'}+i\varkappa(\mathbf{R}_l^\circ-\mathbf{R}_{l'}^\circ)} +$$

$$+\sigma^{\mathrm{incoh}}\, e^{-2W_l}\,\delta_{ll'}\}\sum_\lambda(\varkappa\mathbf{S}_{l\lambda})^*(\varkappa\mathbf{S}_{l'\lambda})\,N_\lambda\,\delta\left(-\frac{\varepsilon}{\hbar}+\omega_\lambda\right). \qquad (1.49)$$

The cross section comprises in both cases two contributions, namely a coherent and a noncoherent term, and their meaning will be explained in the next section when the scattering of neutrons at an ideal crystal is discussed. Equations (1.46), (1.48), and (1.49) formally describe the scattering at some crystal having various defects, but this description implies the harmonic approximation. The cross section is expressed through the quantities $\mathbf{S}_{l\lambda}$ which are given by the expansion of Eq. (1.36) of the atomic displacements in terms of the amplitudes of the normal-vibration modes. Hence, the system of normal modes given by Eq. (1.35) is assumed to be known. In the case of an ideal crystal, the normal modes (phonons) are known and therefore Eqs. (1.48) and (1.49) obtained in this case for the cross section have a very specific meaning.

2. Single-Phonon Scattering at an Ideal Crystal

Phonons in the harmonic approximation. Let us consider a crystal containing N primitive cells with σ atoms in each of the cells. The crystal cells are numbered by the subscript n, and the atoms inside each of the cells are numbered by the subscript i. In the harmonic approximation, the energy of such a crystal can be stated in the following form [cf. Eq. (1.27)]:

$$\mathcal{H} = \sum_{ni\alpha}\frac{(p_{ni}^\alpha)^2}{2M_i} + \frac{1}{2}\sum_{ni\alpha}\sum_{mj\beta}\varphi_{ni,\,mj}^{\alpha\beta}\,u_{ni}^\alpha\,u_{mj}^\beta. \qquad (2.1)$$

Let us introduce the system of eigenvectors and eigenvalues for the dynamic matrix

$$\Phi_{ij}^{\alpha\beta}(\mathbf{q}) = \sum_m \frac{1}{\sqrt{M_i M_j}} \varphi_{ni,\, mj}^{\alpha\beta} \, e^{-i\mathbf{q}(\mathbf{R}_n^\circ - \mathbf{R}_m^\circ)}, \tag{2.2}$$

where the eigenvectors and eigenvalues are determined from the equation

$$\sum_{j\beta} \Phi_{ij}^{\alpha\beta}(\mathbf{q}) \, e_J^\beta(\mathbf{q} \mid j) = \omega_J^2(\mathbf{q}) \, e_J^\alpha(\mathbf{q} \mid i). \tag{2.3}$$

The vector quantities $e_J^\alpha(\mathbf{q} \mid i)$ ($\alpha = x, y, z$) are termed the polarization vectors, and the $\omega_J^2(\mathbf{q})$ are termed the eigenvalues of the dynamic matrix. The subscript J numbers the eigenvalues, whose total number is obviously equal to the dimensionality of the dynamic matrix, i.e., to 3σ. Thus, J indicates the number of the branch of the phonon spectrum. Since the dynamic matrix is Hermitian, the orthogonality and completeness relations apply to the polarization vectors:

$$\sum_i \mathbf{e}_J^*(\mathbf{q} \mid i) \, \mathbf{e}_{J'}(\mathbf{q} \mid i) = \delta_{JJ'}, \quad \sum_J \overset{*}{e}{}_J^\alpha(\mathbf{q} \mid i) \, e_J^\beta(\mathbf{q} \mid j) = \delta_{\alpha\beta} \, \delta_{ij}. \tag{2.4}$$

Also the other relations hold:

$$\omega_J^2(-\mathbf{q}) = \omega_J^2(\mathbf{q}), \quad \mathbf{e}(-\mathbf{q} \mid i) = \mathbf{e}^*(\mathbf{q} \mid i). \tag{2.5}$$

It is easy to verify that the expansion of the atomic displacement $\mathbf{u}_{ni}$ and of the momentum $\mathbf{P}_{ni}$ in a series of Bose amplitudes $b_{\mathbf{q}J}$ and $b_{\mathbf{q}J}{}^+$ of the form [see Eqs. (1.29) and (1.30)]

$$\mathbf{u}_{ni} = \sum_{\mathbf{q}J} \sqrt{\frac{\hbar}{2NM_i \omega_J(\mathbf{q})}} \left\{ \mathbf{e}_J(\mathbf{q} \mid i) \, e^{i\mathbf{q}\mathbf{R}_n^\circ} b_{\mathbf{q}J} + \mathbf{e}_J^*(\mathbf{q} \mid i) \, e^{-i\mathbf{q}\mathbf{R}_n^\circ} b_{\mathbf{q}J}^+ \right\}; \tag{2.6}$$

$$\mathbf{P}_{ni} = -\sum_{\mathbf{q}J} \sqrt{\frac{\hbar M_i \omega_J(\mathbf{q})}{2N}} \left\{ \mathbf{e}_J(\mathbf{q} \mid i) \, e^{i\mathbf{q}\mathbf{R}_n^\circ} b_{\mathbf{q}J} - \mathbf{e}_J^*(\mathbf{q} \mid i) \, e^{-i\mathbf{q}\mathbf{R}_n^\circ} b_{\mathbf{q}J}^+ \right\} \tag{2.7}$$

preserves the commutation relations of Eq. (1.28), and the Hamiltonian of Eq. (2.1) is then reduced to its diagonal form. Its parts relating to the kinetic energy and the potential energy in terms of the Bose amplitudes comprise nondiagonal terms which obviously cancel in the sum. Indeed, we have

$$\sum_{ni\alpha} \frac{(P_{ni}^\alpha)^2}{2M_i} = \frac{1}{2} \sum_{\mathbf{q}J} \hbar\omega_J(\mathbf{q}) \left(b_{\mathbf{q}J}^+ b_{\mathbf{q}J} + \frac{1}{2} \right) -$$

$$-\frac{1}{4}\sum_{\mathbf{q}J}\hbar\omega_J(\mathbf{q})\,(b_{\mathbf{q}J}\,b_{-\mathbf{q}J}+b^{+}_{-\mathbf{q}J}\,b^{+}_{\mathbf{q}J});$$

$$\frac{1}{2}\sum_{ni\alpha}\sum_{mj\beta}\varphi^{\alpha\beta}_{ni,\,mj}\,u^{\alpha}_{ni}\,u^{\beta}_{mj}=\frac{1}{2}\sum_{\mathbf{q}J}\hbar\omega_J(\mathbf{q})\left(b^{+}_{\mathbf{q}J}\,b_{\mathbf{q}J}+\frac{1}{2}\right)+$$

$$+\frac{1}{4}\sum_{\mathbf{q}J}\hbar\omega_J(\mathbf{q})\,(b_{\mathbf{q}J}\,b_{-\mathbf{q}J}+b^{+}_{-\mathbf{q}J}\,b^{+}_{\mathbf{q}J}),$$

so that

$$\mathcal{H}=\sum_{\mathbf{q}J}\hbar\omega_J(\mathbf{q})\left(b^{+}_{\mathbf{q}J}\,b_{\mathbf{q}J}+\frac{1}{2}\right). \tag{2.8}$$

Thus, in the case of an ideal crystal, a normal vibration mode, i.e., a phonon, is characterized by the wave vector **q** and by the branch number J.

General formulas for the cross section of single-phonon scattering. We can use the results of the preceding section, i.e., Eqs. (1.48) and (1.49), in which the subscript l must be replaced by ni, the subscript λ by $\mathbf{q}J$, and the vector $\mathbf{S}_{l\lambda}$ by

$$\mathbf{S}_{ni,\,\mathbf{q}J}=\sqrt{\frac{\hbar}{2NM_i\,\omega_J(\mathbf{q})}}\;\mathbf{e}_J(\mathbf{q}\,|\,i)\,e^{i\mathbf{q}\mathbf{R}^{0}_{n}}. \tag{2.9}$$

The latter equation was obtained from a comparison of Eqs. (2.6) and (1.36).

Before we rewrite in these terms Eqs. (1.48) and (1.49) for the cross section of single-phonon scattering, we generalize Eqs. (1.18)-(1.22) which describe averaging over the isotopes; we relate these equations to the case in which the crystal is supposed to comprise several varieties of atoms numbered by the subscript i and having the mass M_i. We obtain in place of Eqs. (1.18)-(1.22):

$$\widetilde{A_{ni}A_{n'\,i'}}=\sum_{ss'}A^{s}_{i}\,A^{s'}_{i'}\,\widetilde{p^{s}_{ni}\,p^{s'}_{n'\,i'}}; \tag{2.10}$$

$$\widetilde{p^{s}_{ni}\,p^{s'}_{n'\,i'}}=c^{s}_{i}\,c^{s'}_{i'}+\delta_{nn'}\,\delta_{ii'}\,(c^{s}_{i}\,\delta_{ss'}-c^{s}_{i}\,c^{s'}_{i}); \tag{2.11}$$

$$\widetilde{\alpha_{ni}\,\alpha_{n'\,i'}}=\tilde{A}_i\,\tilde{A}_{i'}+\delta_{nn'}\,\delta_{ii'}\left\{\widetilde{A^{2}_{i}}-(\tilde{A}_i)^2+\frac{1}{4}\,\widetilde{B^{2}_{i}I_i(I_i+1)}\right\}. \tag{2.12}$$

Taking into account the latter equation, Eqs. (1.48) and (1.49) for the ideal crystal assume the very specific form

$$\frac{d^2\sigma^{\pm}}{d\Omega d\varepsilon}=\frac{k}{k_0}\sum_{ii'}\sum_{nn'}\{b_i b_{i'}\,e^{-W_i-W_{i'}}\,e^{i\varkappa(\mathbf{R}_n^\circ-\mathbf{R}_{n'}^\circ)+i\varkappa(x_i-x_{i'})}+$$
$$+e^{-2W_i}\sigma_i^{\text{incoh}}\,\delta_{nn'}\,\delta_{ii'}\}\frac{\hbar}{2\sqrt{M_i M_{i'}}}\frac{1}{N}\sum_{\mathbf{q}J}e^{\pm i\mathbf{q}(\mathbf{R}_n^\circ-\mathbf{R}_{n'}^\circ)}\frac{1}{\omega_J(\mathbf{q})}\times$$
$$\times(\varkappa\mathbf{e}_J(\mathbf{q}\,|\,i))(\varkappa\mathbf{e}_J(\mathbf{q}\,|\,i'))^*\left(N_{\mathbf{q}J}+\frac{1}{2}\pm\frac{1}{2}\right)\delta(\varepsilon\pm\hbar\omega_J(\mathbf{q}))\tag{2.13}$$

($\mathbf{x}_i$ denotes the radius vector of the *i*th atom in the unit cell). The following quantities averaged over the isotopes were introduced for the scattering at an atom of variety *i*:

the amplitude of coherent scattering

$$b_i=\sqrt{\frac{m^2}{4\pi^2\hbar^4}}\,\widetilde{A}_i\tag{2.14}$$

the cross section of incoherent scattering

$$\sigma_i^{\text{incoh}}=\frac{m^2}{4\pi^2\hbar^4}\left\{\widetilde{A_i^2}-(\widetilde{A}_i)^2+\frac{1}{4}\,\widetilde{B_i^2 I_i(I_i+1)}\right\}.\tag{2.15}$$

The upper sign in Eq. (2.13) corresponds to the emission of a phonon, the lower sign, to absorption of a phonon.

We must also state the expression for the thermal factor; based on Eq. (1.44), this expression can be rewritten in the form

$$W_i=\frac{\hbar}{4M_i N}\sum_{\mathbf{q}J}\frac{|\varkappa\mathbf{e}_J(\mathbf{q}\,|\,i)|^2}{\omega_J(\mathbf{q})}\,\mathrm{cth}\frac{\hbar\omega_J(\mathbf{q})}{2kT}\,.\tag{2.16}$$

Now we have a closed expression for the cross section of single-phonon scattering. In what follows, we will consider in greater detail the coherent and incoherent parts.

Coherent scattering. In the coherent part of Eq. (2.13), the summation of the exponential phase factors over the cell numbers of the crystal leads to the δ symbol which accounts for the conservation of the wave vectors during single-phonon scattering at the crystal:

$$\frac{1}{N}\sum_{n} e^{i(\varkappa \pm q) R_n^0} = \sum_{b} \delta_{\varkappa \pm q, b}, \tag{2.17}$$

where **b** denotes the vector of the reciprocal lattice. Thus, the cross section of coherent scattering can be stated in the form

$$\frac{1}{N}\frac{d^2\sigma_{coh}^{\pm}}{d\Omega d\varepsilon} = \frac{k}{k_0}\sum_{ii'} b_i b_{i'} e^{-W_i - W_{i'}} e^{i\varkappa(x_i - x_{i'})} \frac{\hbar}{2\sqrt{M_i M_{i'}}} \times$$
$$\times \sum_{qJ} \frac{(\varkappa e_J(\varkappa | i))(\varkappa e_J(\varkappa | i'))^*}{\omega_J(\varkappa)} \left(N_{\varkappa J} + \frac{1}{2} \pm \frac{1}{2}\right) \sum_{b} \delta_{\varkappa \pm q, b}\, \delta(\varepsilon \pm \hbar\omega_J(\varkappa)). \tag{2.18}$$

The periodicity of the functions $\omega_J(\mathbf{q})$ and $\mathbf{e}_J(\mathbf{q}\,|\,i)$ in the reciprocal lattice space was taken into consideration:

$$\omega_J(\mathbf{q} + \mathbf{b}) = \omega_J(\mathbf{q}), \quad \mathbf{e}_J(\mathbf{q} + \mathbf{b} \mid i) = \mathbf{e}_J(\mathbf{q} \mid i). \tag{2.19}$$

It follows from Eq. (2.18) that the cross section of coherent scattering is characterized by sharp peaks, whose positions in regard to angles and energies are given by the momentum and energy conservation laws:

$$\left.\begin{aligned} \varkappa &= \mp \mathbf{q} + \mathbf{b}; \\ \varepsilon &= \mp \hbar\omega_J(\mathbf{q}). \end{aligned}\right\} \tag{2.20}$$

When the scattering vector $\varkappa$, which can be determined from Eq. (2.20), is fixed, the scattering cross section is obviously given by the equation

$$\left.\frac{1}{N}\frac{d\sigma_{coh}^{\pm}}{d\Omega d\varepsilon}\right|_{\varkappa = \mp q + b} = \frac{\hbar}{2}\frac{k}{k_0}\sum_{ii'} b_i b_{i'} e^{-W_i - W_{i'}} e^{i\varkappa(x_i - x_{i'})} \times$$
$$\times \sum_{J} \frac{(\varkappa e_J(\varkappa | i))(\varkappa e_J(\varkappa | i'))^*}{\omega_J(\varkappa)} \left(N_{\varkappa J} + \frac{1}{2} \pm \frac{1}{2}\right) \delta(\varepsilon \pm \hbar\omega_J(\varkappa)). \tag{2.21}$$

This expression can be converted into a more compact form by introducing the vector

$$\xi_J(\varkappa) = \sum_{i} b_i \frac{e^{-W_i}}{\sqrt{M_i}} e^{i\varkappa x_i} e_J(\varkappa | i): \tag{2.22}$$

$$\frac{1}{N}\frac{d^2\sigma^{\pm}_{\rm coh}}{d\Omega d\varepsilon}\bigg|_{\varkappa=\mp\mathbf{q}+\mathbf{b}}=\frac{\hbar}{2}\frac{k}{k_0}\sum_J\frac{|\varkappa\xi_J(\varkappa)|^2}{\omega_J(\varkappa)}\left(N_{\varkappa J}+\frac{1}{2}\pm\frac{1}{2}\right)\delta(\varepsilon\pm\hbar\omega_j(\varkappa)). \quad (2.23)$$

The vector $\xi_J(\varkappa)$ must be termed the structure amplitude of inelastic scattering. The square of the projection of this vector upon the scattering vector $\varkappa$ defines the intensity of a single-phonon peak in accordance with Eq. (2.23).

Equations (2.23) and (2.20) show the principle of the method of determining the dispersion curves of the phonons from inelastic neutron scattering. When at a given geometry, a peak is found in the energy distribution of the scattered neutrons; Eq. (2.20) can be used to determine the phonon energy $\hbar\omega_J(\mathbf{q})$ and the phonon's quasimomentum $\mathbf{q}$. The intensity of the inelastic peak must be measured to determine the polarization vector $\mathbf{e}_J(\mathbf{q}\,|\,i)$. This provides only a projection of the vector $\xi_J(\varkappa)$ upon $\varkappa$. In order to determine the polarization vector proper at each atom of a primitive cell, the intensities of the peaks having the same wave vector $\mathbf{q}$ near various lattice points of the reciprocal lattice must be measured. When the number of such peaks is large enough, Eq. (2.22) can be used to determine all σ vector $\mathbf{e}_J(\mathbf{q}\,|\,i)$, i.e., one can obtain full information on the phonon.

Incoherent scattering. The incoherent part of the cross section of Eq. (2.13) does not have phase factors depending upon the cell coordinates and therefore δ symbols relating to the momentum are not encountered in its expression and the sum over the wave vectors of the phonons does not cancel. We obtain from Eq. (2.13)

$$\frac{1}{N}\frac{d^2\sigma^{\pm}_{\rm incoh}}{d\Omega d\varepsilon}=\frac{\hbar}{2}\frac{k}{k_0}\sum_i\sigma_i e^{-2W_i}\frac{1}{M_i}\frac{1}{N}\sum_{\mathbf{q}J}\frac{|\varkappa\mathbf{e}_J(\mathbf{q}\,|\,i)|^2}{\omega_J(\mathbf{q})}\times$$
$$\times\left(N_{\mathbf{q}J}+\frac{1}{2}\pm\frac{1}{2}\right)\delta(\varepsilon\pm\hbar\omega_J(\mathbf{q})). \quad (2.24)$$

In the integration over $\mathbf{q}$, one obtains a smooth expression for the cross section as a function of the energy transferred, and this expression is without sharp peaks as in the case of coherent scattering.

In crystals of cubic symmetry, the cross section of incoherent scattering can be expressed through the frequency distribution function $g(\omega)$, with the distribution function defined by

$$\sum_{\mathbf{q}j} \ldots = \int d\omega g(\omega) \ldots \tag{2.25}$$

Indeed, in a cubic crystal the second-rank tensor is always diagonal and therefore we obtain from Eq. (2.24)

$$\frac{1}{N}\frac{d^2\sigma^{\pm}_{\text{incoh}}}{d\Omega d\varepsilon} = \frac{\hbar k}{6k_0}\varkappa^2\left(\sum_i \sigma_i e^{-2W_i}\frac{1}{M_i}\right)\frac{1}{2}\left(\text{cth}\,\frac{\varepsilon}{2kT} \pm 1\right) g(\varepsilon/\hbar). \tag{2.26}$$

In a noncubic crystal there is no direct proportionality between the cross section of incoherent scattering and the function $g(\omega)$ but, by measuring the coherent cross section of mutually orthogonal positions of a single crystal, one can derive a simple expression which is proportional to the function $g(\omega)$ [24]. Thus, the problem of deriving the frequency distribution function of the phonon spectrum can be solved in principle for crystals of any symmetry.

We considered coherent and incoherent single-phonon scattering cross sections of neutrons in the case of an ideal crystal. In the following section we will develop the theory of inelastic neutron scattering at a crystal containing an impurity atom. This makes it possible to consider inelastic scattering at a crystal with a random distribution and to treat the dynamics of irregular systems.

3. Single-Phonon Scattering at a Crystal Containing Impurities

Owing to its translational symmetry, the eigenstates of an oscillating lattice and their energies are indexed with the quasimomentum in an ideal crystal. When the symmetry is disturbed by the introduction of impurity atoms, new types of states which are localized near these atoms develop. The nature of such local [64] and quasilocal [45, 46] states was explained using various models and described in detail in the literature (see, e.g., [76]). Inelastic scattering of neutrons at impurity-containing crystals is characterized by specific features resulting from localized excitations in the vibration spectrum. The theoretical aspects of this problem have also been described [76, 41]. These problems will be expounded in the present section in a more succinct and general form making use of Green's functions. In the ensuing sections, this method will be used when various problems of the theory of scattering are discussed; at the beginning of this section we therefore provide a general outline of the Green's function method so that subsequently we can refer to the corresponding passages of this section.

The phonon Green's function. Many problems of the dynamics of crystals are most conveniently solved with the aid of the phonon Green's function. We define the retarded Green's function $G_{ll'}{}^{\alpha\beta}(t)$ of the phonons as the correlator of the atomic displacements at the points l and l' at the time t and 0:

$$G_{ll'}^{\alpha\beta}(t) = -\frac{i}{\hbar}\Theta(t)\langle[u_l^\alpha(t), u_{l'}^\beta(0)]\rangle. \tag{3.1}$$

where

$$\Theta(t) = \begin{cases} 1, & t > 0; \\ 0, & t < 0. \end{cases} \tag{3.2}$$

The expression in the square brackets is the commutator, whereas the parentheses enclose the statistical average of Eq. (1.7) with the Hamiltonian $\mathcal{H}$. In the harmonic approximation, the Hamiltonian $\mathcal{H}$ is given by Eq. (1.27).

The Green's function of Eq. (3.1) can be easily calculated by substituting the equations of motion for it. First one conveniently introduces a more general definition of the retarded Green's function composed of the two operators $A(t)$ and $B(0)$ [7]:

$$\langle\langle A(t) \mid B(0)\rangle\rangle \equiv -\frac{i}{\hbar}\Theta(t)\langle[A(t), B(0)]\rangle, \tag{3.3}$$

where

$$A(t) = e^{\frac{i}{\hbar}\mathcal{H}t} A e^{-\frac{i}{\hbar}\mathcal{H}t}. \tag{3.4}$$

The equation of motion for this function is obtained by differentiating the left and right sides of Eq. (3.3) and by making use of the equation of motion for the operator $A(t)$:

$$i\hbar\frac{dA(t)}{dt} = [A(t), \mathcal{H}]. \tag{3.5}$$

Thus, we obtain the equation

$$i\hbar\frac{d}{dt}\langle\langle A(t) \mid B(0)\rangle\rangle = \delta(t)\langle[A, B]\rangle + \langle\langle[A(t), \mathcal{H}] \mid B(0)\rangle\rangle, \tag{3.6}$$

relating the initial phonon function with a more complicated function for which an analogous equation of motion can be stated, etc.

The Green's function makes it possible to determine the correlators of the quantities $A(t)$ and $B(0)$ of which it is composed. For this purpose, we expand the Green's function and the corresponding correlators in Fourier integrals:

$$\langle\langle A(t) \mid B(0)\rangle\rangle = \frac{1}{2\pi}\int_{-\infty}^{\infty} G(\omega)\, e^{-i\omega t}\, d\omega; \tag{3.7}$$

$$\langle A(t)\, B(0)\rangle = \frac{1}{2\pi}\int_{-\infty}^{\infty} J(\omega)\, e^{-i\omega t}\, d\omega; \tag{3.8}$$

$$\langle B(0)\, A(t)\rangle = \frac{1}{2\pi}\int_{-\infty}^{\infty} J(\omega)\, e^{-\beta\hbar\omega}\, e^{-i\omega t}\, d\omega. \tag{3.9}$$

One can show [7] that the spectral density $J(\omega)$ of the correlator and the Fourier component $G(\omega)$ of the retarded Green's function are related by the simple expression

$$J(\omega) = i\hbar\, \frac{G(\omega + i\delta) - G(\omega - i\delta)}{1 - e^{-\beta\hbar\omega}}. \tag{3.10}$$

Below we give examples of retarded Green's functions composed of quantities A and B of various types; the operators of the atomic displacements, of the spin, or of the electron density will appear as such quantities A and B. We consider in the present section the phonon Green's function $\langle\langle u_l^{\alpha}(t) \mid u_{l'}^{\beta}(0)\rangle\rangle$ which is defined by Eq. (3.1).

For the phonon Green's function of Eq. (3.1), the equation of motion (3.6) is reduced to the following form:

$$i\hbar \frac{d}{dt}\langle\langle u_l^{\alpha}(t) \mid u_{l'}^{\beta}(0)\rangle\rangle = \frac{i\hbar}{M_l}\langle\langle P_l^{\alpha}(t) \mid u_{l'}^{\beta}(0)\rangle\rangle, \tag{3.11}$$

wherein we used Eq. (3.5) for the operator $u_l^{\alpha}(t)$ and the commutation relations of Eq. (1.28). The equation of motion for the Green's function on the right side of the equation can be composed in a similar fashion. This Green's function has the form

$$i\hbar \frac{d}{dt}\langle\langle P_l^{\alpha}(t) \mid u_{l'}^{\beta}(0)\rangle\rangle = -i\hbar\delta(t)\,\delta_{\alpha\beta}\,\delta_{ll'} - i\hbar \sum_{m\gamma} \varphi_{lm}^{\alpha\gamma}\langle\langle u_m^{\gamma}(t) \mid u_{l'}^{\beta}(0)\rangle\rangle. \tag{3.12}$$

By combining these two equations, we obtain a closed equation for the initial phonon Green's function (3.1)

$$-\frac{d^2}{dt^2}G_{ll'}^{\alpha\beta}(t)-\frac{1}{M_l}\sum_{m\gamma}\varphi_{lm}^{\alpha\gamma}G_{ml'}^{\gamma\beta}(t)=\frac{1}{M_l}\delta(t)\,\delta_{\alpha\beta}\,\delta_{ll'}, \tag{3.13}$$

which, after Fourier transformation, can be stated as

$$\sum_{m\gamma}\{\omega^2 M_l\,\delta_{\alpha\gamma}\,\delta_{lm}-\varphi_{lm}^{\alpha\gamma}\}\,G_{ml'}^{\gamma\beta}(\omega)=\delta_{\alpha\beta}\,\delta_{ll'}. \tag{3.14}$$

The resulting equation can also be stated in symmetrized form:

$$\sum_{m\gamma}\left\{\omega^2\,\delta_{\alpha\gamma}\,\delta_{lm}-\frac{1}{\sqrt{M_lM_m}}\,\varphi_{lm}^{\alpha\gamma}\right\}\left(\sqrt{M_m\,M_{l'}}\,G_{ml'}^{\gamma\beta}\right)=\delta_{\alpha\beta}\,\delta_{ll'}. \tag{3.15}$$

For an ideal crystal with σ atoms per unit cell ($l = ni$, $l' = n'i'$), Eq. (3.15) is solved by expansion in a Fourier series. Using Eq. (2.3) for the eigenfrequencies of the matrix and the polarization vectors, we obtain the solution of Eq. (3.15):

$$G^{\circ\alpha\beta}_{ni,\,n'i'}(\omega)=\frac{1}{\sqrt{M_iM_{i'}}}\,\frac{1}{N}\sum_{\mathbf{q}J}\frac{e_J^{*\alpha}(\mathbf{q}\,|\,i)\,e_J^{\beta}(\mathbf{q}\,|\,i')}{\omega^2-\omega_J^2(\mathbf{q})+i\delta}\,e^{-i\mathbf{q}(\mathbf{R}_n^\circ-\mathbf{R}_{n'}^\circ)}\equiv$$

$$\equiv\frac{1}{\sqrt{M_iM_{i'}}}\,\mathcal{G}^{\circ\alpha\beta}_{ni,\,n'i'}(\omega). \tag{3.16}$$

An infinitely small imaginary addition with $\delta > 0$ ensures the correct behavior of the retarded Green's function at large times.

The phonon Green's function contains all information on the crystal dynamics (the entire phonon spectrum and the corresponding polarization vectors) and therefore the function can be used to calculate the various crystal characteristics which are associated with the dynamic properties of the crystal. More particularly, the Green's function method which we introduced facilitates research on the dynamics of a crystal containing impurity atoms.

The dynamics of a crystal with an impurity atom. First let us consider a very simple case: a multiatomic crystal contains a single impurity atom with a mass M' distinguished from the basic atoms of mass M. We assume that the force constants do not change so that the point defect in the crystal is fully localized at the lattice point occupied by the impurity atom. We term this lattice point the zeroth lattice point. The problem of the vibrations of such a crystal with an impurity was solved for the first time solved in [64] and was treated with the Green's function in [174].

We will calculate the phonon Green's functions and thereafter the vibration spectrum of the system under consideration. For some lattice point l of the crystal, the mass of the atom can be conveniently stated with the aid of the expression

$$M_l = M - (M - M')\delta_{l,0} \tag{3.17}$$

and can be inserted into Eq. (3.14) for the Green's function. The resulting equation can be rewritten in the form

$$\sum_{m\gamma}\{\omega^2 M\delta_{\alpha\gamma}\,\delta_{lm} - \varphi_{lm}^{\alpha\gamma}\}\,G_{ml'}^{\gamma\beta} = \delta_{\alpha\beta}\,\delta_{ll'} + \sum_{m\gamma} V_{lm}^{\alpha\gamma}\,G_{ml'}^{\gamma\beta}, \tag{3.18}$$

where the matrix $V_{lm}{}^{\alpha\gamma}$ determines the localized perturbation introduced by the impurity:

$$V_{lm}^{\alpha\gamma} = M\omega^2\rho\delta_{\alpha\gamma}\,\delta_{l0}\,\delta_{l'0} \equiv Mv_0\,\delta_{\alpha\gamma}\,\delta_{l0}\,\delta_{l'0}, \tag{3.19}$$

where $\rho = (M - M')/M$. Assuming that in Eq. (3.18) the perturbation v_0 is equal to zero, we can also state the equation for the Green's function of the ideal crystal:

$$\sum_{m\gamma}\{\omega^2 M\delta_{\alpha\gamma}\,\delta_{lm} - \varphi_{lm}^{\alpha\gamma}\}\,G_{ml'}^{\circ\gamma\beta} = \delta_{\alpha\beta}\,\delta_{ll'}. \tag{3.20}$$

The combination of Eqs. (3.18) and (3.20) results in the standard equation for the Green's function of a crystal with an impurity; this equation can be written in terms of the $\mathcal{G}$ function [$G = (1/M)\,\mathcal{G}$]:

$$\mathcal{G}_{ll'}^{\alpha\beta} = \mathcal{G}_{ll'}^{\circ\alpha\beta} + \sum_{mp}\sum_{\gamma\gamma'}\mathcal{G}_{lm}^{\circ\alpha\gamma}\,V_{mp}^{\gamma\gamma'}\,\mathcal{G}_{pl'}^{\gamma'\beta}, \tag{3.21}$$

where

$$V_{mp}^{\gamma\gamma'} = v_0\delta_{m0}\delta_{p0}\delta_{\gamma\gamma'},\quad v_0 = \rho\omega^2. \tag{3.22}$$

Since the perturbation introduced by the impurity is only local, the solution of Eq. (3.21) can be stated in explicit form:

$$\mathcal{G}_{ll'}^{\alpha\beta} = \mathcal{G}_{ll'}^{\circ\alpha\beta} + \sum_{\gamma\gamma'} \mathcal{G}_{l0}^{\circ\alpha\gamma} \left(\frac{v_0}{1 - v_0 \mathcal{G}_{00}^{\circ}} \right)^{\gamma\gamma'} \mathcal{G}_{0l'}^{\circ\gamma'\beta}. \tag{3.23}$$

where $\mathcal{G}_{00}{}^{\circ}$ denotes the diagonal Green's function of the ideal crystal [see Eq. (3.16)]:

$$\mathcal{G}_{00}^{\circ\alpha\beta}(\omega) = \frac{1}{N} \sum_{\mathbf{q}J} \frac{\overset{*}{e}{}_J^{\alpha}(\mathbf{q})\, e_J^{\beta}(\mathbf{q})}{\omega^2 - \omega_J^2(\mathbf{q}) + i\delta}. \tag{3.24}$$

Now let us generalize the results to the case of a crystal having several atoms per primitive cell. Equation (3.14) for the Green's function is in this case restated in the form ($l = ni$)

$$\sum_{mj\gamma} \{\omega^2 M_{ni}\, \delta_{\alpha\gamma}\, \delta_{nm}\, \delta_{ij} - \varphi_{ni,\, mj}^{\alpha\gamma}\}\, G_{mj,\, n'i'}^{\gamma\beta} = \delta_{\alpha\beta}\, \delta_{nn'}\, \delta_{ii'}. \tag{3.25}$$

Assume that the replacing impurity atom is in sublattice No. 1; then we obtain in place of Eq. (3.17)

$$M_{ni} = M_i - (M_1' - M_1)\, \delta_{n0} \delta_{i1}. \tag{3.26}$$

Equations (3.21)–(3.24) in this case are replaced by the following equations:

$$\mathcal{G}_{ni,\, n'i'}^{\alpha\beta} = \mathcal{G}_{ni,\, n'i'}^{\circ\alpha\beta} + \sum_{\substack{mm'\\jj'}} \sum_{\gamma\gamma'} \mathcal{G}_{ni,\, mj}^{\circ\alpha\gamma}\, V_{mj,\, m'j'}^{\gamma\gamma'}\, \mathcal{G}_{m'j',\, n'i'}^{\gamma'\beta}; \tag{3.27}$$

$$V_{mj,\, m'j'}^{\gamma\gamma'} = v_{01}\, \delta_{m0}\, \delta_{m'0}\, \delta_{j1}\, \delta_{j'1}\, \delta_{\gamma\gamma'}, \tag{3.28}$$

where $v_{01} = \rho_1 \omega^2$, and $\rho_1 = (M_1 - M_1')/M_1$:

$$\mathcal{G}_{ni,\, n'i'}^{\alpha\beta} = \mathcal{G}_{ni,\, n'i'}^{\circ\alpha\beta} + \sum_{\gamma\gamma'} \mathcal{G}_{ni,\, 01}^{\circ\alpha\gamma} \left(\frac{v_{01}}{1 - v_{01} \mathcal{G}_{01,01}^{\circ}} \right)^{\gamma\gamma'} \mathcal{G}_{01,\, n'i'}^{\circ\gamma'\beta}, \tag{3.29}$$

where

$$\mathcal{G}_{01,01}^{\circ\alpha\beta}(\omega) = \frac{1}{N} \sum_{\mathbf{q}J} \frac{\overset{*}{e}{}_J^{\alpha}(\mathbf{q}\,|\,1)\, e_J^{\beta}(\mathbf{q}\,|\,1)}{\omega^2 - \omega_J^2(\mathbf{q}) + i\delta}. \tag{3.30}$$

Our equations (3.23) or (3.29) make it possible to study the vibration spectrum of a crystal with an impurity atom. Since in the presence of an impurity, translational symmetry of a crystal is not preserved, the normal vibration modes cannot be characterized by a wave vector, and in such a situation one conveniently studies not the individual vibration modes and their energies but rather the vibration spectrum as a whole, with the vibration spectrum defined by the state density.

First let us introduce the distribution function $G(\omega^2)$ of the square of the frequencies; in the case of an ideal crystal, this function is

$$G_0(\omega^2)=\frac{1}{N}\sum_{\mathbf{q}J}\delta(\omega^2-\omega_J^2(\mathbf{q})). \tag{3.31}$$

Using the well-known identity

$$\frac{1}{x+i\delta}=\mathcal{P}\frac{1}{x}-i\pi\delta(x) \tag{3.32}$$

we can easily relate $G_0(\omega^2)$ to the phonon Green's function:

$$G_0(\omega^2)=-\frac{1}{\pi N}\operatorname{Im}\operatorname{Sp}\mathcal{G}^\circ(\omega), \tag{3.33}$$

where the trace of $\mathcal{G}_{ni,n'i'}{}^{\circ\alpha\beta}$ is taken over both the lattice-point subscript and the vector subscripts.

A similar relation exists also for the distribution function $G(\omega^2)$ of the square of the frequencies in the vibration spectrum of the crystal with impurities:

$$G(\omega^2)=-\frac{1}{\pi N}\operatorname{Im}\operatorname{Sp}\mathcal{G}(\omega^2). \tag{3.34}$$

Here we introduce Eq. (3.29) for the $\mathcal{G}$ function and make use of the equation

$$\sum_{mj\gamma}\mathcal{G}^{\circ\alpha\gamma}_{ni,mj}\mathcal{G}^{\circ\gamma\beta}_{mj,n'i'}=-\frac{d}{d\omega^2}\mathcal{G}^{\circ\alpha\beta}_{ni,n'i'}, \tag{3.35}$$

so that $G(\omega^2)$ can be represented in the compact form

$$G(\omega^2)=G_0(\omega^2)+\frac{1}{\pi N}\operatorname{Im}\sum_{\gamma\gamma'}v_{01}\left(\frac{1}{1-v_{01}\mathcal{G}^\circ_{01,01}}\right)^{\gamma\gamma'}\left\{\frac{d}{d\omega^2}\mathcal{G}^{\circ\gamma'\gamma}_{01,01}\right\}. \tag{3.36}$$

In the case of a cubic crystal, the first matrix element is proportional to $\delta_{\gamma\gamma'}$ because the tensor $\mathcal{G}_{01,01}{}^{\circ\alpha\beta}$ is diagonal with respect to the indices

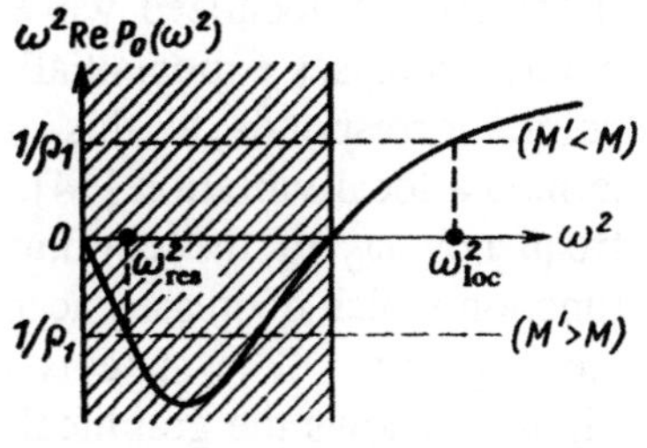

Fig. 1. Local and resonance levels in the vibration spectrum of a crystal with one replacing impurity atom.

α and β so that

$$\mathcal{G}^{\circ\alpha\beta}_{01,01}=\frac{1}{3N}\sum_{\mathbf{q}J}\frac{1}{\omega^2-\omega_J^2(\mathbf{q})+i\delta}\,\delta_{\alpha\beta}\equiv P_0(\omega)\,\delta_{\alpha\beta}. \tag{3.37}$$

In this case, Eq. (3.36) assumes the form

$$G(\omega^2)=G_0(\omega^2)+\frac{3}{\pi N}v_{01}\,\mathrm{Im}\left(\frac{\frac{d}{d\omega^2}P_0(\omega)}{1-v_{01}P_0(\omega)}\right)-G_0(\omega^2)+\frac{3}{\pi N}\rho_1\omega^2\times$$

$$\times\left[\frac{\rho_1\omega^2\,\mathrm{Im}\,P_0(\omega)\frac{d}{d\omega^2}\,\mathrm{Re}\,P_0(\omega)+(1-v_{01}\,\mathrm{Re}P_0(\omega))\frac{d}{d\omega^2}\,\mathrm{Im}\,P_0(\omega)}{[1-\rho_1\omega^2\,\mathrm{Re}\,P_0(\omega)]^2+[\rho_1\omega^2\,\mathrm{Im}\,P_0(\omega)]^2}\right]. \tag{3.38}$$

Thus, the impurity implies a distortion of the density of states of the ideal crystal, and this distortion depends greatly upon the frequency ω. The correction is proportional to $1/N$, and in the case of a finite concentration of the impurities, c must be replaced by c [sic]. The addition to the density of states has a sharp peak at the frequency ω_0, satisfying the equation

$$1-\rho_1\omega_0^2\,\mathrm{Re}\,P_0(\omega_0)=0. \tag{3.39}$$

Depending upon the perturbation parameter ρ_1, this equation can have a solution which falls into the band of the continuous frequency spectrum of the vibrations of the initial crystal or above this band. In the first case, $\mathrm{Im}\,P_0(\omega_0)$ differs from zero and the correction term in Eq. (3.38) has a maximum with finite width $\sim\rho_1\omega_0{}^i\,\mathrm{Im}\,P_0(\omega_0)$ at the frequency ω_0. In the second case, $\mathrm{Im}\,P_0(\omega)=0$ so that an infinite peak of width zero develops in Eq. (3.38), indicating the appearance of a discrete frequency in the spectrum. A localized wave function, which in the space of the lattice points decreases exponentially with the distance from the impurity lattice point, corresponds to this frequency. The corresponding vibration is termed a local vibration [64]. The crystal state, which corresponds to the frequency ω_0 of the continuous spectrum, is characterized by a wave function which decreases according to a power law with the distance from the impurity. The state is termed quasilocal or resonance state [45]. Figure 1 shows the graphical solution of Eq. (3.39) and demonstrates the appearance of local and quasilocal frequencies. For local frequencies to exist, positive (and sufficiently large) ρ_1 values must exist, i.e., local vibrations develop in a crystal containing a light impurity. Quasilocal vibrations develop at negative ρ_1, which correspond to a heavy impurity. In the case $M_1' >> M_1$, the frequency ω_0 falls into the low-frequency region of the continuous spectrum where the density of states $G_0(\omega^2)$ (and, hence, the level width) is small. This means that there appears a sharp resonance peak which substantially changes the thermodynamics of such a crystal at low temperatures [46]:

$$T \lesssim \hbar\omega_0/k. \tag{3.40}$$

In the case of noncubic crystals, one must use the general equation (3.36) from which one can infer that the condition for the local frequency is

$$\det \mid 1 - \rho_1\omega_0^2 \, \mathcal{G}^{\circ}_{01,01}(\omega_0) \mid = 0. \tag{3.41}$$

The quantity $\mathcal{G}_{01,01}{}^{\circ}$ is a 3×3 matrix to be taken over the vector subscripts; we obtain according to Eq. (3.16)

$$\mathcal{G}^{\circ\alpha\beta}_{01,01}(\omega) = \frac{1}{N} \sum_{\mathbf{q}J} \frac{\overset{*}{e}{}^{\alpha}_{J}(\mathbf{q}\mid 1)\, e^{\beta}_{J}(\mathbf{q}\mid 1)}{\omega^2 - \omega_J^2(\mathbf{q}) + i\delta}\,. \tag{3.42}$$

Thus, in the general case Eq. (3.41) defines three different frequencies. In a cubic crystal, these frequencies turn out to be degenerate and this leads to the coefficient 3 in the correction term of Eq. (3.38) for the density of states. It follows from Eq. (3.42) that the local frequencies are given not only by the relative change in the mass upon replacement of an atom of a single sublattice but also by the type of motion of this sublattice in the initial crystal (via the polarization vectors for this sublattice).

Averaging over impurities. Now let us consider a crystal containing a number of impurity atoms at lattice points s. For the sake of simplicity, we restrict our considerations to a single-atom lattice. The Green's function of the crystal with the impurity satisfies the general equation (3.21) in which, however, the perturbation accounts for all impurity atoms so that

$$V_{mp}^{\gamma\gamma'} = v_0\, \delta_{\gamma\gamma'} \sum_s \delta_{ms}\, \delta_{ps}. \tag{3.43}$$

The Green's function averaged over all impurity configurations must be calculated. To this end, we switch from Eq. (3.21) stated in symbolic form

$$\mathcal{G} = \mathcal{G}^\circ + \mathcal{G}^\circ V \mathcal{G}, \tag{3.44}$$

to its other notation

$$\mathcal{G} = \mathcal{G}^\circ + \mathcal{G}^\circ R \mathcal{G}^\circ; \tag{3.45}$$

$$R = V + R\mathcal{G}^\circ V. \tag{3.46}$$

Owing to Eq. (3.45), the averaging of the Green's function is reduced to averaging over R. We state the last equation in explicit form, taking into account the structure of the perturbation matrix of Eq. (3.43). Obviously, R_{mn} is nonvanishing only for lattice points n occupied by the impurity, and therefore Eq. (3.46) assumes the form

$$R_{ms} = v_0\, \delta_{ms} + \sum_{s'} R_{ms'}\, \mathcal{G}^\circ_{s's}\, v_0. \tag{3.47}$$

All quantities are matrices with respect to the vector subscripts. The term with $s' = s$ is conveniently separated in the sum over s' and the equation is rewritten in the form

$$R_{ms} = \Lambda \delta_{ms} + {\sum_{s'}}' R_{ms'}\, \mathcal{G}^\circ_{s's}\, \Lambda, \tag{3.48}$$

where

$$\Lambda = v_0/(1 - v_0\, \mathcal{G}^\circ_{00}) \tag{3.49}$$

is also a 3×3 matrix.

The resulting Eq. (3.48) must also be averaged over the impurities. As in the averaging over the isotope distribution, it is convenient to introduce quantities p_n which are equal to 1 when the lattice point n is occupied by an impurity atom and which are equal to zero otherwise. Then Eq. (3.48) can be given the form

$$R_{mn}\, p_n = \Lambda p_n\, \delta_{mn} + \sum_{n'}{}' R_{mn'}\, p_{n'}\, \mathcal{G}^{\circ}_{n'n}\, p_n\, \Lambda \tag{3.50}$$

(the prime in the sum means that the term with $n' = n$ is excluded). This equation makes it possible to average over the impurities after integration so that

$$\overline{R_{mn}\, p_n} = \Lambda \bar{p}_n\, \delta_{mn} + \sum_{n_1}{}' \delta_{mn_1} \Lambda \mathcal{G}^{\circ}_{n_1 n} \Lambda\, \overline{p_{n_1}\, p_n} + \\ + \sum_{n_2 n_1}{}' \delta_{mn_2} \Lambda \mathcal{G}^{\circ}_{n_2 n_1} \Lambda \mathcal{G}^{\circ}_{n_1 n} \Lambda \overline{p_{n_2}\, p_{n_1}\, p_n} + \dots \tag{3.51}$$

The average value of a product of any number of random quantities can be expressed through the concentration (see, e.g., [41]), and the expansion of Eq. (3.51) makes it possible to determine the quantity $R_{mn}p_n$ in any approximation with respect to the concentration. In a first approximation with respect to c, the result can be obtained by direct averaging of Eq. (3.50) with the aid of the approximation

$$\overline{R_{mn'}\, p_{n'}\, p_n} = \overline{R_{mn'}\, p_{n'}} \cdot \bar{p}_n \quad (n' \neq n). \tag{3.52}$$

Thus, we obtain an equation for the averaged quantity $R_{mn}p_n$:

$$\overline{R_{mn}\, p_n} = c\Lambda \delta_{mn} + c \sum_{n'} \overline{R_{mn'}\, p_{n'}} \cdot \mathcal{G}^{\circ}_{n'n}\, \Lambda, \tag{3.53}$$

and the averaged Green's function according to Eq. (3.45) is expressed through $R_{mn}p_n$ in the following manner:

$$\overline{\mathcal{G}}_{ll'} = \mathcal{G}^{\circ}_{ll'} + \sum_{mn} \mathcal{G}^{\circ}_{lm}\, \overline{R_{mn}\, p_n} \mathcal{G}^{\circ}_{nl'}. \tag{3.54}$$

After averaging over the impurities, the translational symmetry in the crystal is restored, and therefore the solution of the last two equations is obtained by expanding all quantities in Fourier series:

$$\overline{R_{mn}\,p_n} = \frac{1}{N}\sum_{\mathbf{q}} R(\mathbf{q})\,e^{-i\mathbf{q}(\mathbf{R}_n-\mathbf{R}_m)}, \quad \overline{\mathcal{G}}_{ll'} = \frac{1}{N}\sum_{\mathbf{q}} \overline{\mathcal{G}}(\mathbf{q})\,e^{-i\mathbf{q}(\mathbf{R}_l-\mathbf{R}_{l'})}. \tag{3.55}$$

We obtain from Eq. (3.53) a matrix equation for $R(q)$:

$$R(\mathbf{q}) = c\Lambda + R(\mathbf{q})\,\mathcal{G}^{\circ}(\mathbf{q})\,c\Lambda,$$

where

$$R(\mathbf{q}) = c\Lambda/(1 - \mathcal{G}^{\circ}(\mathbf{q})\,c\,\Lambda). \tag{3.56}$$

By substituting this result in Eq. (3.54), we can easily state the equation for the Green's function $\overline{\mathcal{G}}(\mathbf{q})$ in the form of the standard Dyson equation:

$$\overline{\mathcal{G}}(\mathbf{q}) = \mathcal{G}^{\circ}(\mathbf{q}) + \mathcal{G}^{\circ}(\mathbf{q})\,c\Lambda\overline{\mathcal{G}}(\mathbf{q}). \tag{3.57}$$

In the case of a cubic crystal, in which $\Lambda^{\alpha\beta} \propto \delta_{\alpha\beta}$, the solution to this equation must be determined in the form

$$\overline{\mathcal{G}}^{\alpha\beta}(\mathbf{q}) = \sum_J g_J \frac{\overset{*}{e}{}_J^{\alpha}(\mathbf{q})\,e_J^{\beta}(\mathbf{q})}{\omega^2 - \omega_J^2(\mathbf{q})}. \tag{3.58}$$

Taking into account the structure of $\mathcal{G}^{\circ}(\mathbf{q})$, we obtain for the coefficients g_J the following equation:

$$g_J = 1 + \frac{cv}{1 - v\mathcal{G}^{\circ}_{00}} \cdot \frac{1}{\omega^2 - \omega_J^2(\mathbf{q})}\,g_J.$$

Substituting the solution of this equation into Eq. (3.58) renders the final expression for the averaged Green's function of the cubic crystal:

$$\overline{\mathcal{G}}^{\alpha\beta}_{ll'} = \frac{1}{N}\sum_{\mathbf{q}J} \frac{\overset{*}{e}{}_J^{\alpha}(\mathbf{q})\,e_J^{\beta}(\mathbf{q})}{\omega^2 - \omega_J^2(\mathbf{q}) - c\Lambda}\,e^{-i\mathbf{q}(\mathbf{R}_l-\mathbf{R}_{l'})}. \tag{3.59}$$

By contrast to the Green's function of the initial crystal, the denominator of Eq. (3.59) has the self-energy part $c\Lambda$ with a singularity at the same frequency ω_0 satisfying Eq. (3.39) as the density of states of the vibration spectrum. Let us note that the averaging over the impurity configurations [(averaging resulting in Eq. (3.59)] is valid only for the frequencies of the continuous spectrum [76], and therefore singularities of the quantity $c\Lambda$ can be caused only by the resonance frequencies of the spectrum. Equation (3.59) does not adequately describe the lattice dynamics for frequencies above the continuous spectrum, and a separate treatment is required at these frequencies.

Neutron scattering at the vibrations of the continuous spectrum. The double differential cross section of neutron scattering in which a single vibration mode of the crystal participates is in the general case given by the expression

$$\frac{d^2\sigma}{d\Omega d\varepsilon}=\frac{k}{k_0}\sum_{ll'}(b_l b_{l'}+\delta_{ll'}\sigma^{\text{incoh}})\,e^{-W_l-W_{l'}}\,e^{i\varkappa(\mathbf{R}_l^\circ-\mathbf{R}_{l'}^\circ)}\times$$
$$\times\varkappa^\alpha\varkappa^\beta\frac{1}{2\pi\hbar}\int_{-\infty}^{\infty}e^{-\frac{1}{\hbar}\varepsilon t}\langle u_l^\alpha(t)\,u_{l'}^\beta(0)\rangle\,dt, \tag{3.60}$$

which is obtained from the general equation (1.15) when averaging over the isotopes. Equations (3.7)-(3.10) make it possible to express the cross section through the retarded Green's function of Eq. (3.1)

$$\frac{d^2\sigma}{d\Omega d\varepsilon}=\frac{k}{k_0}\sum_{ll'}(b_l b_{l'}+\delta_{ll'}\sigma^{\text{incoh}})\,e^{-W_l-W_{l'}}\,e^{i\varkappa(\mathbf{R}_l^\circ-\mathbf{R}_{l'}^\circ)}\times$$
$$\times\varkappa^\alpha\varkappa^\beta\frac{1}{\pi}\,\mathrm{Im}\,G_{ll'}^{\alpha\beta}\left(-\frac{\varepsilon}{\hbar}\right)\frac{1}{e^{\beta\varepsilon}-1}. \tag{3.61}$$

When the Green's function of Eq. (3.16) of an ideal crystal is taken as $G_{ll'}{}^{\alpha\beta}$, Eq. (3.61) leads to our well-known expression which is the sum of the two terms of Eq. (2.13) describing the cross section under emission or absorption of a phonon.

The expression for the cross section in the form of Eq. (3.61) is convenient for calculating the scattering cross section at a crystal containing impurities. The observed cross section represents the value of Eq. (3.61) averaged over the distribution of the impurities; not only the Green's function $G_{ll'}{}^{\alpha\beta}$, but also the amplitude of coherent scattering, must be averaged. Let us write the random amplitude at the atom l

through the population number p_l*:

$$b_l = b + (b_0 - b)\, p_l. \tag{3.62}$$

The averaged expression in Eq. (3.61) can be split into a sum of three terms [let us recall that $G_{ll'} = (1/M)\mathcal{G}_{ll'}$] in the case of a multiatomic crystal:

$$\overline{\frac{1}{N}\sum_{ll'} b_l b_{l'} e^{i\varkappa(\mathbf{R}_l-\mathbf{R}_{l'})}\mathcal{G}_{ll'}} = b^2 \sum_{ll'} e^{i\varkappa(\mathbf{R}_l-\mathbf{R}_{l'})}\overline{\mathcal{G}_{ll'}} +$$
$$+ 2b(b-b_0)\frac{1}{N}\sum_{ll'} e^{i\varkappa(\mathbf{R}_l-\mathbf{R}_{l'})}\overline{p_l\mathcal{G}_{ll'}} +$$
$$+ (b_0-b)^2\frac{1}{N}\sum_{ll'} e^{i\varkappa(\mathbf{R}_l-\mathbf{R}_{l'})}\overline{p_l p_{l'}\mathcal{G}_{ll'}}. \tag{3.63}$$

The expression for $\overline{\mathcal{G}}_{ll'}$ was calculated above and is given by Eq. (3.59). Now the mean values $\overline{p_l\mathcal{G}_{ll'}}$ and $\overline{p_l p_{l'}\mathcal{G}_{ll'}}$ must be calculated. The first of these values can be easily expressed through the mean values $\overline{R_{ml}p_l}$ and $T_{l,mn} = \overline{p_l R_{mn} p_n}$ (with $l \neq n$), making use of the basic Eq. (3.45) for the Green's function:

$$\overline{p_l\mathcal{G}_{ll'}} = c\mathcal{G}^{\circ}_{ll'} + \sum_m \mathcal{G}^{\circ}_{lm}\overline{R_{ml}p_l}\mathcal{G}^{\circ}_{ll'} + \sum_m \sum_{n(\neq l)}{}' \mathcal{G}^{\circ}_{lm} T_{l,mn}\mathcal{G}^{\circ}_{nl'}. \tag{3.64}$$

The quantity $\overline{R_{ml}p_l}$ was calculated above, and $T_{l,mn}$ can be obtained as a series, provided that Eq. (3.50) is used. One can show (for details, see [41]) that in the lowest approximation with respect to the concentration, the equation

$$T_{l,mn} = c\overline{R_{mn}p_n} + c\delta_{ml}\Lambda\sum_{n'}\mathcal{G}^{\circ}_{ln'}\overline{R_{n'n}p_n} \tag{3.65}$$

holds.

---

*The thermal factor e^{-W_l} can be included in the scattering amplitude b_l. In the ensuing discussion of the present section, b_l will be understood as the expression $b_l e^{-W_l}$. We will assume that W_l in a crystal with an impurity assumes only two values, one of which corresponds to the impurity atom and the other, to some matrix atom. The following discussion will show that this approach is only an approximation.

As to the quantity $\overline{p_l \mathcal{G}_{ll'} p_{l'}}$, obviously at $l \neq l'$ its expansion with respect to the concentration begins with a quadratic term, and therefore we have in the lowest approximation with respect to c

$$\overline{p_l \mathcal{G}_{ll'} p_{l'}} = \delta_{ll'} \overline{p_l \mathcal{G}_{ll}}. \tag{3.66}$$

After a Fourier transformation, Eqs. (3.64) and (3.66) become

$$\overline{p_l \mathcal{G}_{ll'}} = c\,(1 + \mathcal{G}^{\circ}_{00}\,\Lambda)\,\overline{\mathcal{G}_{ll'}};$$

$$\overline{p_l \mathcal{G}_{ll'} p_{l'}} = c\,(\mathcal{G}^{\circ}_{00} + \mathcal{G}^{\circ}_{00}\,\Lambda \mathcal{G}^{\circ}_{00})\,\delta_{ll'} = c\mathcal{G}^{(1)}_{00}\,\delta_{ll'},$$

where $\mathcal{G}_{00}{}^{(1)}$ denotes the Green's function at the impurity lattice point of the crystal which has a single impurity atom.

After substituting these expressions in Eq. (3.63), we state the final expression for the averaged sum of Eq. (3.63):

$$\frac{1}{N} \sum_{ll'} b_l b_{l'} \mathrm{e}^{i\varkappa (\mathbf{R}_l - \mathbf{R}_{l'})} \mathcal{G}_{ll'} =$$

$$= b^2 \overline{\mathcal{G}}(\varkappa) + 2b\,(b_0 - b)\,c\,(1 + \mathcal{G}^{\circ}_{00}\,\Lambda)\,\overline{\mathcal{G}}(\varkappa) + (b_0 - b)^2 \mathcal{G}^{(1)}_{00}. \tag{3.67}$$

All quantities are matrices with the vector subscripts. The expression for the cross section is most simply transformed further in the case of a cubic crystal. In this case, Eq. (3.59) can be used for the quantity $\overline{\mathcal{G}}(\varkappa)$, and the cross section of Eq. (3.61) can be stated in the final form [46, 54, 55, 76]:

$$\frac{d^2\sigma}{d\Omega d\varepsilon} = \frac{k}{k_0 M} \frac{1}{e^{\beta\varepsilon} - 1} \sum_J |\varkappa \mathbf{e}_J(\varkappa)|^2 \times$$

$$\times \left\{ \left[b^2 + 2cb\,(b_0 - b)\,\frac{\mathrm{Re}\,\Lambda(-\varepsilon/\hbar)}{\rho\,(\varepsilon/\hbar)^2} \right] \frac{1}{\pi} \times \right.$$

$$\times \frac{c\,\mathrm{Im}\,\Lambda(-\varepsilon/\hbar)}{[(\varepsilon/\hbar)^2 - \omega_J^2(\varkappa) - c\,\mathrm{Re}\,\Lambda(-\varepsilon/\hbar)]^2 + [c\,\mathrm{Im}\,\Lambda(-\varepsilon/\hbar)]^2} +$$

$$\left. + c\left[(b_0 - b)^2 + 2b\,(b_0 - b)\,\frac{\rho\,(\varepsilon/\hbar)^2}{(\varepsilon/\hbar)^2 - \omega_J^2(\varkappa)} \right] \frac{1}{\pi}\,\mathrm{Im}\,\Lambda(-\varepsilon/\hbar) \right\}. \tag{3.68}$$

The real and imaginary parts are given by

$$\Lambda(\omega) = \rho\omega^2 \frac{1 - \rho\omega^2\,\mathrm{Re}\,P_0(\omega) + i\rho\omega^2\,\mathrm{Im}\,P_0(\omega)}{[1 - \rho\omega^2\,\mathrm{Re}\,P_0(\omega)]^2 + [\rho\omega^2\,\mathrm{Im}\,P_0(\omega)]^2}, \tag{3.69}$$

where

$$P_0(\omega) = \frac{1}{3N} \sum_{\mathbf{q}J} \frac{1}{\omega^2 - \omega_J^2(\mathbf{q}) + i\delta} \,. \tag{3.70}$$

In order to establish the structure of Eq. (3.68), we take the limit $c \to 0$. Then the last term in the braces cancels and the first term in the limit yields the expression $b^2 \delta (\varepsilon^2/\hbar^2 - \omega_J^2(\mathbf{q}))$; this implies in Eq. (3.68) the sum of two terms one of which coincides with the expression for the coherent cross section of neutron scattering with emission of a phonon, whereas the other term means absorption. At a finite, though low, concentration of the impurity, the first term of Eq. (3.68) describes the coherent scattering at phonons with a frequency defined by the condition

$$\omega^2 = \omega_J^2(\mathbf{q}) + c \operatorname{Re} \Lambda (\omega_J(\mathbf{q})), \tag{3.71}$$

and with an attenuation $\sim c \operatorname{Im} \Lambda(\omega_J(\mathbf{q}))$. This is a resonance-type attenuation when the solution of Eq. (3.39) falls into a band of the continuous spectrum. The second term in Eq. (3.68) is proportional to the concentration and describes incoherent scattering resulting from the difference in the amplitudes of scattering of the atoms of the impurity and the atoms of the initial crystal. Also this expression has a sharp maximum when Eq. (3.39) has a resonance solution. But strictly speaking, this division of the scattering into a coherent and an incoherent part is rather arbitrary as the terms associated with the second component in the square brackets of Eq. (3.68) are mixed terms, i.e., they reflect features of both coherent and incoherent scattering.

Scattering at local vibrations. At a transfer energy ε exceeding the maximum frequency ω_{max} of the continuous spectrum, single-quantum scattering of a neutron can occur only at local vibrations. A local vibration involves atomic displacements localized near an impurity atom, and therefore the scattering at the individual local vibrations is incoherent. The scattering cross section at the crystal as a whole can be obtained after calculating the scattering cross section at a crystal containing a single impurity atom and by multiplying thereafter the result with the total number of impurity atoms.

Let us denote the Green's function of a crystal with a single impurity atom by $\mathcal{G}_{ll'}^{(1)}$. Using the relation $b_l = b + (b_0 - b)\,\delta_{l0}$, we split the sum over the lattice points in Eq. (3.61) for the cross section in the following form:

$$\frac{1}{N}\sum_{ll'} b_l b_{l'} e^{i\varkappa(\mathbf{R}_l-\mathbf{R}_{l'})} \mathcal{G}_{ll'}^{(1)} = b^2 \sum_{ll'} e^{i\varkappa(\mathbf{R}_l-\mathbf{R}_{l'})} \mathcal{G}_{ll'}^{(1)} +$$

$$+ b(b_0-b)\left[\sum_l e^{i\varkappa\mathbf{R}_l}\mathcal{G}_{l0}^{(1)} + \sum_{l'} e^{-i\varkappa\mathbf{R}_{l'}}\mathcal{G}_{0l'}^{(1)}\right] + (b_0-b)^2\mathcal{G}_{00}^{(1)} \quad (3.72)$$

and insert the expression for $\mathcal{G}_{ll'}^{(1)}$:

$$\mathcal{G}_{ll'}^{(1)} = \mathcal{G}_{ll'}^{\circ} + \mathcal{G}_{l0}^{\circ} v \frac{1}{1-\mathcal{G}_{00}^{\circ} v}\mathcal{G}_{0l'}^{\circ}. \quad (3.73)$$

As a consequence, the sum of interest to us is converted to the form

$$\frac{1}{N}\sum_{ll'} b_l b_{l'} e^{i\varkappa(\mathbf{R}_l-\mathbf{R}_{l'})}\mathcal{G}_{ll'}^{(1)} = b^2\,\mathcal{G}^{\circ}(\varkappa) + \frac{1}{N} b(b_0-b)\,\mathcal{G}^{\circ}(\varkappa) +$$

$$+\frac{1}{N}\left\{b^2\mathcal{G}^{\circ}(\varkappa)\, v\frac{1}{1-\mathcal{G}_{00}^{\circ} v}\mathcal{G}^{\circ}(\varkappa) + b(b_0-b)\,\times\right.$$

$$\times\left[v\frac{1}{1-\mathcal{G}_{00}^{\circ} v}\mathcal{G}_{00}^{\circ} + \mathcal{G}_{00}^{\circ} v\frac{1}{1-\mathcal{G}_{00}^{\circ} v}\right] + \left.(b_0-b)^2\frac{1}{1-\mathcal{G}_{00}^{\circ} v}\mathcal{G}_{00}^{\circ}\right\}, \quad (3.74)$$

where $\mathcal{G}^{\circ}(\varkappa)$ denotes a Fourier component $\mathcal{G}_{ll'}^{\circ}$:

$$\mathcal{G}^{\circ\alpha\beta}(\varkappa) = \sum_J \frac{\overset{*}{e}{}_J^{\alpha}(\varkappa)\, e_J^{\beta}(\varkappa)}{\omega^2-\omega_J^2(\varkappa)+i\delta} \quad (3.75)$$

[we recall that the quantity obtained in Eq. (3.74) is a matrix over the vector subscripts].

According to Eq. (3.61), the cross section is given by the imaginary part of Eq. (3.74). For $|\omega| > \hbar\omega_{max}$, Im $\mathcal{G}^{\circ}(\varkappa)$ vanishes, and therefore only terms with the factor $1/(1 - \mathcal{G}_{00}v^{\circ})$ can be nonvanishing in Eq. (3.74) since the imaginary part thereof does not vanish when a local frequency exists. Taking into account what has been stated, it is easy to show that the imaginary part of Eq. (3.74) for a cubic crystal can be transformed to the form

$$\operatorname{Im}\frac{1}{N}\sum_{ll'} b_l b_{l'} e^{i\varkappa(\mathbf{R}_l-\mathbf{R}_{l'})}\mathcal{G}_{ll'}^{(1)\alpha\beta}(\omega) = \frac{1}{N}\sum_J \overset{*}{e}{}_J^{\alpha}(\varkappa)\, e_J^{\beta}(\varkappa)\,\times$$

$$\times \left[\frac{b}{\omega^2 - \omega_J^2(\varkappa)} + (b_0 - b)\,\mathrm{Re}\,P_0(\omega) \right]^2 \times$$

$$\times \frac{\rho^2 \omega^4 \,\mathrm{Im}\, P_0(\omega)}{[1 - \rho\omega^2 \,\mathrm{Re}\, P_0(\omega)]^2 + [\rho\omega^2 \,\mathrm{Im}\, P_0(\omega)]^2} \, . \tag{3.76}$$

For $\omega > \omega_{max}$, we have $\mathrm{Im}\, P_0(\omega) = 0$, and therefore the last factor in Eq. (3.76) vanishes at all ω values except for $\omega = \omega_0$ where Eq. (3.39) holds. At $\omega = \omega_0$, a 0/0 singularity arises and must be determined with the aid of the following method. We expand of the expression $1 - \rho\omega^2 \,\mathrm{Re}\, P_0(\omega)$ in the neighborhood of the point ω_0:

$$1 - \rho\omega^2 \,\mathrm{Re}\, P_0(\omega) = \frac{d}{d\omega} [1 - \rho\omega^2 \,\mathrm{Re}\, P_0(\omega)]_{\omega=\omega_0} (\omega - \omega_0) + \ldots \equiv$$

$$\equiv 2\rho\omega_0 B(\omega_0^2)(\omega - \omega_0), \tag{3.77}$$

where

$$B(\omega^2) = \frac{1}{N} \sum_{\mathbf{q}J} \frac{\omega_J^2(\mathbf{q})}{\omega_0^2 - \omega_J^2(\mathbf{q})} \tag{3.78}$$

[we used Eq. (3.39) when we derived Eq. (3.77)]. With the well-known representation of the δ function

$$\delta(x) = \frac{1}{\pi} \frac{\delta}{x^2 + \delta^2}, \tag{3.79}$$

we find that the last factor in Eq. (3.76) is equal to $(\omega_0/2B(\omega_0^2)\,\delta(\omega - \omega_0)$. Now let us insert the result obtained for Eq. (3.76) into Eq. (3.61) and determine the final expression for the scattering cross section at a local vibration in a cubic crystal [172]:

$$\frac{d^2\sigma}{d\Omega d\varepsilon} = \frac{k}{k_0} \frac{c\hbar\omega_0}{2MB(\omega_0^2)} \sum_J |\varkappa \mathbf{e}_J(\varkappa)|^2 \times$$

$$\times \left[\frac{b_0 - b}{\rho\omega_0^2} + \frac{b}{\omega_0^2 - \omega_J^2(\varkappa)} \right]^2 \{(N_0 + 1)\,\delta(\varepsilon + \hbar\omega_0) + N_0\,\delta(\varepsilon - \hbar\omega_0)\}, \tag{3.80}$$

where N_0 denotes the equilibrium number of local vibrations at a given temperature:

$$N_0 = \left(e^{\frac{\hbar\omega_0}{kT}} - 1\right)^{-1} . \tag{3.81}$$

The expression in the brackets in Eq. (3.80) defines the structure factor of the local vibration; the structure factor is related to the wave function of the local state by the Fourier transform.

We observe that scattering at local vibrations manifests itself in the form of sharp peaks in the inelastic cross section for energies $|\varepsilon| > \hbar\omega_{\max}$ of transitions developing in processes with emission or with absorption. Equation (3.80) was obtained in [172], where also the results of numerical calculations of the cross section were presented for various values of the parameter ρ.

Summarizing our considerations of inelastic scattering at a crystal containing a certain concentration of randomly distributed impurities, we conclude that despite missing translational symmetry in such a system, the scattering of neutrons at the vibrations of the continuous spectrum ($|\varepsilon| < \hbar\omega_{\max}$) is essentially incoherent. Strictly speaking, these results are valid only at low impurity concentrations. When the concentrations are increased, the interaction of the impurities leads to a broadening of isolated local levels and to their transition into a band and also causes fluctuation-induced smearing of the upper boundary of the continuous spectrum which, under certain conditions, may overlap with the impurity-induced band. These effects have been treated theoretically in [65].

4. Description of the Crystal Lattice Vibrations on the Basis of Symmetry Theory

Phonons and irreducible representations of space groups. The normal vibration modes of a crystal (phonons) are characterized by the frequencies $\omega_j(\mathbf{q})$ and by the polarization vectors $\mathbf{e}_j(\mathbf{q} \mid i)$ which are the eigenfrequencies and the eigenvectors, respectively, of the dynamic matrix $\Phi_{ij}^{\alpha\beta}(\mathbf{q})$. By virtue of the well-known Wigner theorem, the energy levels of the system are characterized by irreducible representations of its symmetry group. The symmetry of a crystal is described by the space group G and therefore the normal vibration modes must be characterized by irreducible representations of this group.

The theory of representations of space groups has been expounded in detail in the second volume of the present work [42], and therefore we state now only the basic definitions and conclusions required for the ensuing considerations. The most important points are:

1. An irreducible representation $D^{\{\mathbf{q}\}\nu}$ of a space group G is given by the star of the wave vector $\{\mathbf{q}\}$ and the number ν of the representation.

2. The star of the wave vector is by definition the set of nonequivalent vectors $\mathbf{q}_1, \mathbf{q}_2, \ldots, \mathbf{q}_{l_q}$ (termed the arms of the star) which can be obtained from a particular are, e.g., $\mathbf{q}_1$, by the effect of all elements g of the space group. Two arms differing only by some vector of the reciprocal lattice are termed equivalent arms. Since equivalent vectors can be obtained when various elements g act upon the vector $\mathbf{q}$, the number of arms in a star is smaller than the number of elements in the group. We denote the elements of the group G, which generate the arms of the star, $g_1, g_2, \ldots, g_L, \ldots, g_{l_\mathbf{q}}$, and term them representative elements; then

$$\mathbf{q}_L = g_L \mathbf{q}. \tag{4.1}$$

3. The elements g of the group G which leave the vector $\mathbf{q}$ unchanged or convert it into an equivalent vector form a subgroup $G_\mathbf{q}$ of the group G, the so-called group of the wave vector. The equation

$$g\mathbf{q} = \mathbf{q} + \mathbf{b}. \tag{4.2}$$

expresses the condition for an element g to be included in the group $G_\mathbf{q}$. One can show that the representative elements g_L are given by the expansion of the group G in terms of the subgroup $G_\mathbf{q}$ in left-hand mixed classes:

$$G = g_1 G_\mathbf{q} + g_2 G_\mathbf{q} + \ldots + g_{l_\mathbf{q}} G_\mathbf{q} = \sum_L g_L G_\mathbf{q}. \tag{4.3}$$

4. The representation $D^{\{\mathbf{q}\}\nu}$ of the space group G is derived from the irreducible representations $d^{\mathbf{q}\nu}$ of the group $G_\mathbf{q}$ of the wave vector, where one irreducible representation $D^{\{\mathbf{q}\}\nu}$ of the dimension $l_\mathbf{q} \times l_\nu$ corresponds to each irreducible representation $d^{\mathbf{q}\nu}$, with l_ν denoting the dimensionality of the representation $d^{\mathbf{q}\nu}$. All stars of the 230 space groups and the matrices of all irreducible representations $d^{\mathbf{q}\nu}$ have been tabulated in handbooks, e.g., in Kovalev's book [50].

5. The elements g of the space group can be given in the Seitz notation: $g = \{h \mid \tau_h + \mathbf{t}\}$, where h denotes the rotation operation; τ_h denotes the concurrent nontrivial translation, and $\mathbf{t}$ denotes a whole translation of the group G. The following relations which explain the meaning o the symbols introduced are applicable:

$$g\mathbf{x} = h\mathbf{x} + \tau_h; \tag{4.4}$$

$$gg' = \{h \mid \mathbf{t}\}\{h' \mid \mathbf{t}'\} = \{hh' \mid h\mathbf{t}' + \mathbf{t}\}; \tag{4.5}$$

$$g^{-1} = \{h \mid \mathbf{t}\}^{-1} = \{h^{-1} \mid -h^{-1}\mathbf{t}\}. \tag{4.6}$$

6. For the matrix $d_{\lambda\mu}{}^{\mathbf{q}\nu}(g)$ of some element of the group $G_{\mathbf{q}}$ corresponding to the irreducible representation $d^{\mathbf{q}\nu}$, the equation

$$d_{\lambda\mu}^{\mathbf{q}\nu}(\{h \mid \tau_h + \mathbf{t}\}) = e^{-i\mathbf{q}\mathbf{t}}\, d_{\lambda\mu}^{\mathbf{q}\nu}(\{h \mid \tau_h\}) \tag{4.7}$$

holds [67]; this equation expresses the matrices of the representation of some element, $\{h \mid \tau_h + \mathbf{t}\}$, through the matrices $\{h \mid \tau_h\}$ of the elements not containing whole translations. The set of all these elements is termed the zeroth block of the group $G_{\mathbf{q}}$. Their number coincides with the number of elements of the point group $G_{\mathbf{q}}{}^{\circ}$ corresponding to the group $G_{\mathbf{q}}$ of the wave vector, but these elements by themselves do not form a group.

What has been outlined above provides a grouptheoretical classification of the normal modes of the crystal vibrations. By virtue of the Wigner theorem, the frequency $\omega_J(\mathbf{q})$ belongs to some irreducible representation $D^{\{\mathbf{q}\}\nu}$ of the space group of the crystal with the star of the wave vector $\mathbf{q}$, and the subscript J (number of the phonon branch) is the subscript of the set and indicates the number of this representation and the number λ of the basis function. The dimensionality of the irreducible representation $D^{\{\mathbf{q}\}\nu}$ indicates the degeneracy of the phonon frequency of multiplicity $l_{\mathbf{q}} \times l_{\nu}$; the frequencies corresponding to the various arms $\mathbf{q}_L$ of the star $\{\mathbf{q}\}$ naturally coincide:

$$\omega_J(q_L) = \omega_J(\mathbf{q}_1). \tag{4.8}$$

As to the polarization vectors $\mathbf{e}_J(\mathbf{q}_L \mid i)$, they depend upon the selected arm of the star and the number of the basis function. Our problem is to develop a grouptheoretical scheme which facilitates the determination of the multiplicity of phonon frequency degeneracy at a given wave vector

q and also to determine the polarization vectors corresponding to this frequency. The latter problem has mainly a grouptheoretical aspect, i.e., it is defined by the symmetry properties of the crystal under consideration, whereas the calculation of the phonon frequencies proper requires specific knowledge of the dynamic matrix, i.e., of the interaction forces.

The mechanical representation. Information on the symmetry properties of the vibrations of crystal atoms near their equilibrium positions can be found in the so-called mechanical representation of a space group [67, 246]. This representation can be introduced by considering the set of displacements $\mathbf{u}_{ni}$ for each atom occupying an equilibrium position $\mathbf{R}_{ni}{}^{\circ}$ (n denotes the number of the cell, and i, the number of the atom in the cell).

The $3\sigma N$-component column vector composed of the quantities $u_{ni}{}^{\alpha}$ (N denotes the number of cells in the crystal, and σ denotes the number of the atoms in the cell) is by the element g of the group G converted into another $3\sigma N$-dimensional vector. Such a linear transformation defines a $3\sigma N \times 3\sigma N$ matrix which provides a representation of the group G termed the mechanical representation. Obviously, this representation encompasses all information not only about the space group of the crystal but also on its structure because the representation encompasses the equilibrium positions $\mathbf{R}_{ni}{}^{\circ}$ of all atoms of the crystal. For this reason this representation is the starting point for developing a classification of the phonons and for calculating the polarization vectors.

The system of basis vectors in the $3\sigma N$-dimensional space is conveniently introduced for the actual derivation of the mechanical representation. As basis functions we choose the function

$$\psi^{\mathbf{q}}_{j\beta} = \sum_{n} \oplus\, \varepsilon^{\circ}_{j\beta}\, e^{i\mathbf{q}\mathbf{t}_n}, \tag{4.9}$$

which can be characterized by some wave vector **q**. The direct summation over the primitive cells of the crystal means that there exists an N-component column vector composed of the quantities $\varepsilon_{j\beta}{}^{\circ}$ which are repeated in each cell of the crystal; $\varepsilon_{j\beta}{}^{\circ}$ represents a 3σ-dimensional column in which all components except for one (which is equal to 1) are equal to zero, corresponding to the displacement of the atom with the number j in the β direction.

The function $\psi_{j\beta}{}^{\mathrm{q}}$ are the eigenfunctions of the translation operator $\hat{T}(\mathbf{t})$, and therefore the self-evident relation

$$\hat{T}(\mathbf{t})\,\psi_{j\beta}^{\mathbf{q}} = e^{-i\mathbf{q}\mathbf{t}}\,\psi_{j\beta}^{\mathbf{q}} \tag{4.10}$$

is valid. Let us find out how the functions $\psi_{j\beta}{}^{\mathbf{q}}$ transform under the influence of the element $g = \{h \mid \tau_h\}$ of the group $G_{\mathbf{q}}$ of the wave vector. Assume that the atom with the number j and the coordinate $\mathbf{x}_j$ from the zeroth cell in the general case is transformed into the atom with the number i of another cell:

$$g\mathbf{x}_j = h\mathbf{x}_j + \boldsymbol{\tau}_h \equiv \mathbf{x}_j + \mathbf{a}_p(g, j), \tag{4.11}$$

where $\mathbf{a}_p(g, j)$ denotes the translation connecting the zeroth cell with this new cell; this can be symbolically expressed as

$$g\,(j0) \rightarrow (i\mathbf{a}_p). \tag{4.12}$$

In addition, the vector of the atomic displacement is rotated by the rotational operation h included in the element g so that the results of the application of the operator $\hat{T}(g)$ to the function $\psi_{j\beta}{}^{\mathbf{q}}$ can be stated in the form

$$\hat{T}(g)\,\psi_{j\beta}^{\mathbf{q}} = e^{-i\mathbf{q}\mathbf{a}_p(g,\,j)} \sum_{\alpha} R_{\alpha\beta}^{h}\,\psi_{i\alpha}^{\mathbf{q}}, \tag{4.13}$$

where $R_{\alpha\beta}{}^{h}$ denotes the rotation matrix corresponding to the rotational operation h. The numbers of the atoms are changed in accordance with Eq. (4.12). The resulting equation (4.13) can be stated in the more general form

$$\hat{T}(g)\,\psi_{j\beta}^{\mathbf{q}} = \sum_{i\alpha} \{d_m^{\mathbf{q}}(g)\}_{i\alpha,\,j\beta}\,\psi_{i\alpha}^{\mathbf{q}}, \tag{4.14}$$

where the matrix $\{d_m{}^{\mathbf{q}}(g)\}_{i\alpha,j\beta}$ is by definition the mechanical representation of the group $G_{\mathbf{q}}$ of the wave vector. By comparing Eqs. (4.14) and (4.13), the explicit form of the matrices of the mechanical representation can be obtained:

$$\{d_m^{\mathbf{q}}(g)\}_{i\alpha,\,j\beta} = e^{-i\mathbf{q}\mathbf{a}_p(g,\,j)}\,\delta_{i,\,gj}\,R_{\alpha\beta}^{h}. \tag{4.15}$$

In this expression, the δ symbol accounts for condition (4.12) for the transposition of the atoms under the influence of the element g.

Obviously, the dimensionality of the mechanical representation can be expanded in irreducible representation $d^{\mathbf{q}\nu}$ of the group $G_{\mathbf{q}}$. The expansion is made in accordance with the standard formulas of group theory [14, 67]:

$$d_m^{\mathbf{q}} = \sum_{\nu} n_{\nu} d^{\mathbf{q}\nu}; \tag{4.16}$$

$$n_{\nu} = \frac{1}{n(G_{\mathbf{q}}^{\circ})} \sum_{g \in G_{\mathbf{q}}^{\circ}} \chi_m^{\mathbf{q}}(g) \overset{*}{\chi}{}^{\mathbf{q}\nu}(g) \tag{4.17}$$

and expresses the coefficient n_{ν} of the appearance of an irreducible representation through the characters $\chi^{\mathbf{q}\nu}(g)$ of this representation and the characters $\chi_m{}^{\mathbf{q}}(g)$ of the mechanical representation. According to Eq. (4.15)

$$\chi_m^{\mathbf{q}}(g) = \sum_{j} e^{-i\mathbf{q}\mathbf{a}_p(g,j)} \delta_{j,gj} \operatorname{Sp} R^h. \tag{4.18}$$

In Eq. (4.17), the summation is extended only over those elements g of the group $G_{\mathbf{q}}$ which are from the zeroth block. The number of these elements is small and coincides with the number $n(G_{\mathbf{q}}{}^{\circ})$ of elements of the point group $G_{\mathbf{q}}{}^{\circ}$ corresponding to the group $G_{\mathbf{q}}$.

Let us note that the representation obtained with the aid of Eq. (4.25) is the mechanical representation of the group $G_{\mathbf{q}}$ of the wave vector but is not the representation of the entire space group G. Since a unique correspondence exists between the irreducible representations of the groups $G_{\mathbf{q}}$ and G, the problem of the composition of the mechanical representation of the space group is reduced to calculations based on Eqs. (4.16)-(4.18) for the group of the wave vector.

The basis functions and polarization vectors. The mechanical representation $d_m{}^{\mathbf{q}}$ is that set of irreducible representations $d^{\mathbf{q}\nu}$ of the group G whose basis makes it possible to describe any, even the most general, type of atomic displacements corresponding to a particular wave vector for the crystal. The normal mode of crystal vibrations (phonons), which is characterized by some wave vector and by the number of the irreducible representation, is described by the basis functions of this irreducible representation. Thus, the calculation of the polarization vectors $\mathbf{e}_J(\mathbf{q} \mid i)$, where the subscript J denotes the number of some irreducible representation $d^{\mathbf{q}\nu}$, is reduced to the calculation of the basis functions of this irreducible representation.

The basis functions $\psi_\lambda{}^{\mathbf{q}\nu}$ of the irreducible representation $d^{\mathbf{q}\nu}$ of the group $G_\mathbf{q}$ can be derived with the general formula known from group theory [67]:

$$\psi_\lambda^{\mathbf{q}\nu} = \frac{1}{N} \sum_{g \in G_\mathbf{q}} \overset{*}{d}{}^{\mathbf{q}\nu}_{\lambda\mu}(g)\, \widehat{T}(g)\, \psi, \tag{4.19}$$

where ψ denotes some initial function, and $d_{\lambda\mu}{}^{\mathbf{q}\nu}$ denotes the matrices of the representation selected. When the subscript μ is fixed, Eq. (4.19) determines the l_ν basis functions of the l_ν-dimensional irreducible representation $d^{\mathbf{q}\nu}$. In order to derive a basis capable of describing the displacements of the crystal atoms, we use the basis vector of the mechanical representation [Eq. (4.9)] as an initial function $\psi_{j\beta}{}^{\mathbf{q}}$. Taking into account the effect of an element of the group $G_\mathbf{q}$ upon $\psi_{j\beta}{}^{\mathbf{q}}$ in accordance with Eq. (4.13), we state the expression for $\psi_\lambda{}^{\mathbf{q}\nu}$ in the form

$$\psi_\lambda^{\mathbf{q}\nu} = \sum_{g \in G_q^\circ} \overset{*}{d}{}^{\mathbf{q}\nu}_{\lambda\mu}(g) \sum_{i\alpha} e^{-i\mathbf{q}\mathbf{a}_p(g,\, j)}\, \delta_{i,gj}\, R^h_{\alpha\beta}\, \psi^{\mathbf{q}}_{i\alpha}. \tag{4.20}$$

Taking into account the structure defined by Eq. (4.9) for the function $\psi_{i\alpha}{}^{\mathbf{q}}$, we state Eq. (4.20) in the same form:

$$\psi_\lambda^{\mathbf{q}\nu} = \sum_n \oplus E_\lambda^{\mathbf{q}\nu}\, e^{i\mathbf{q}\mathbf{t}_n}, \tag{4.21}$$

i.e., in the form of a straightforward sum of 3σ-dimensional columns:

$$E_\lambda^{\mathbf{q}\nu} = \sum_{g \in G_\mathbf{q}^\circ} \overset{*}{d}{}^{\mathbf{q}\nu}_{\lambda\mu}(g) \sum_{i\alpha} e^{-i\mathbf{q}\mathbf{a}_p(g,\, j)}\, \delta_{i,gj}\, R^h_{\alpha\beta}\, \varepsilon_{i\alpha}.$$

Such a multi-dimensional vector, in turn, is conveniently represented in the form of a straightforward sum of σ three-dimensional vectors $\varepsilon\left(\begin{smallmatrix}\mathbf{q}\nu \\ \lambda\end{smallmatrix} \mid i\right)$:

$$E_\lambda^{\mathbf{q}\nu} = \sum_{i=1}^{\sigma} \oplus \varepsilon \left(\begin{matrix} \mathbf{q}\nu \\ \lambda \end{matrix} \middle| i \right),$$

where

$$\varepsilon^\alpha \left(\begin{matrix} \mathbf{q}\nu \\ \lambda \end{matrix} \middle| i \right) = \sum_{g \in G_\mathbf{q}^\circ} \overset{*}{d}{}^{\mathbf{q}\nu}_{\lambda[\mu]}(g)\, e^{-i\mathbf{q}\mathbf{a}_p(g,\, j)}\, \delta_{i,g[j]}\, R^h_{\alpha[\beta]}. \tag{4.22}$$

The vector $\varepsilon\left(\begin{smallmatrix}\mathbf{q}\nu \\ \lambda\end{smallmatrix} \mid i\right)$ is the atomic component of the basis function of the irreducible representation $d^{\mathbf{q}\nu}$ for atoms in the zeroth cell. For atoms from the other cells, we obtain from Eq. (4.21) the following expression:

$$\boldsymbol{\varepsilon}\begin{pmatrix}\mathbf{q}\nu \\ \lambda\end{pmatrix}\Bigg| ni\Bigg) = e^{i\mathbf{q}\mathbf{t}_n}\,\boldsymbol{\varepsilon}\begin{pmatrix}\mathbf{q}\nu \\ \lambda\end{pmatrix}\Bigg| i\Bigg), \tag{4.23}$$

which reduces the calculation of an entire basis function to the calculation of the σ atomic components in the zeroth cell of the crystal.

Equation (4.22) is very convenient for practical calculations. The subscripts μ, j, and β in the square brackets must be fixed. The transition to other subscripts leads either to a new system of basis functions (in cases in which the given irreducible representation is contained not even once in the composition of the mechanical representation) or to a system identical to zero.

We state in conclusion the expressions for the atomic components of the basis functions which correspond to some arm $\mathbf{q}_L$ of the star $\{\mathbf{q}\}$ of the wave vector and which are expressed through the basis functions of the particular arm:

$$\varepsilon^{\alpha}\left(\begin{matrix}\mathbf{q}_L\nu \\ \lambda\end{matrix}\Bigg| i'\right) = e^{-i\mathbf{q}_L \mathbf{a}_p(g,\,i)} \sum_{\beta} R^{h_L}_{\alpha\beta}\, \varepsilon^{\beta}\left(\begin{matrix}\mathbf{q}\nu \\ \lambda\end{matrix}\Bigg| i\right). \tag{4.24}$$

The numbers i and i' of the atoms are related by the expression

$$g_L \mathbf{x}_i = h_L \mathbf{x}_i + \boldsymbol{\tau}_{h_L} \equiv \mathbf{x}_{i'} + \mathbf{a}_p(g,\, i), \tag{4.25}$$

which is analogous to Eq. (4.11). Thus, the transition from one arm to another is, in essence, reduced to a rotational transformation of the atomic vector $\boldsymbol{\varepsilon}\left(\begin{smallmatrix}\mathbf{q}\nu \\ \lambda\end{smallmatrix} \mid i\right)$.

The atomic components of the basis functions of the irreducible representations of the group $G_{\mathbf{q}}$ satisfy the orthogonality and completeness relations:

$$\sum_{i\alpha} \overset{*}{\varepsilon}{}^{\alpha}\left(\begin{matrix}\mathbf{q}\nu \\ \lambda\end{matrix}\Bigg| i\right) \varepsilon^{\alpha}\left(\begin{matrix}\mathbf{q}\nu' \\ \lambda'\end{matrix}\Bigg| i\right) = \delta_{\nu\nu'}\,\delta_{\lambda\lambda'}; \tag{4.26}$$

$$\sum_{\nu\lambda} \overset{*}{\varepsilon}{}^{\alpha}\left(\begin{matrix}\mathbf{q}\nu \\ \lambda\end{matrix}\Bigg| i\right) \varepsilon^{\beta}\left(\begin{matrix}\mathbf{q}\nu \\ \lambda\end{matrix}\Bigg| j\right) = \delta_{\alpha\beta}\,\delta_{ij}. \tag{4.27}$$

If some representation appears several times (assume r times) in the composition of the mechanical representation, one can obtain r sets of basis functions $\varepsilon_s{}^\alpha({}^{\mathbf{q}\nu}_{\lambda} \mid i)$ which are numbered by the subscript $s = 1, \ldots, r$. These functions, which were calculated with Eq. (4.24), in general need not be orthogonal to each other within the particular set but can be made orthogonal by an appropriate unitary transformation. Thus, in the case of the repeated representations, Eqs. (4.26) and (4.27) can be stated in the form

$$\sum_{i\alpha} \overset{*}{\varepsilon}{}_s^\alpha \begin{pmatrix} \mathbf{q}\nu \\ \lambda \end{pmatrix} i \Big) \varepsilon_{s'}^\alpha \begin{pmatrix} \mathbf{q}\nu' \\ \lambda' \end{pmatrix} i \Big) = \delta_{\nu\nu'}\, \delta_{ss'}\, \delta_{\lambda\lambda'}; \tag{4.28}$$

$$\sum_{\nu\lambda s} \overset{*}{\varepsilon}{}_s^\alpha \begin{pmatrix} \mathbf{q}\nu \\ \lambda \end{pmatrix} i \Big) \varepsilon_s^\beta \begin{pmatrix} \mathbf{q}\nu \\ \lambda \end{pmatrix} j \Big) = \delta_{\alpha\beta}\, \delta_{ij}. \tag{4.29}$$

Since the subscript s numbers the repeated representations, Eqs. (4.28) and (4.29) can be stated in the usual forms (4.26) and (4.27) when we assume that the subscript ν passes through all irreducible representations, including all the repeated representations. The relation between the two forms of stating orthogonality and completeness conditions can be interpreted in another way. All r repeated irreducible representations can be considered a single representation of an r-times greater dimensionality, and then the composite subscript $s\lambda$ numbers the basis functions of this representation and Eqs. (4.28) and (4.29) again can be stated in the form of Eqs. (4.26) and (4.27).

Polarization vectors. The components $\varepsilon({}^{\mathbf{q}\nu}_{\lambda} \mid i)$ of a basis function are closely related with the polarization vectors $\mathbf{e}_J(\mathbf{q} \mid i)$ for the phonon modes characterized by the subscript $J = \nu\lambda$. According to the definition (see Section 2), the polarization vectors are the eigenvectors of the dynamic matrix and are associated with the eigenfrequency $\omega_J(\mathbf{q})$. The multiplicity of the degeneracy of the frequencies $\omega_J(\mathbf{q})$ is for a particular wave vector given by the crystal symmetry which implies certain relations for the elements of the dynamic matrix. As we will see, these relations can be determined with the aid of the matrices of the mechanical representation introduced above.

Let us consider the general type of the displacements of the crystal atoms, the wave being characterized by the wave vector $\mathbf{q}$. The displacements can be fully determined by the set of the 3σ quantities $\{u_j{}^\beta\}$ which are the projections of the displacements of the atoms in the zeroth cell. The $3\sigma N$-dimensional vector u of the atomic displacements of the entire crystal obviously can be expressed through the unit vectors of Eq. (4.9) in $3\sigma N$-dimensional space:

$$u = \sum_{\beta j} u_j^\beta \psi_{j\beta}^{\mathbf{q}}. \tag{4.30}$$

The transformation properties of this $3\sigma N$-dimensional vector under the influence of an element of the group $G_\mathbf{q}$ can be obtained with the aid of the transformation relations (4.14) for the unit vectors $\psi_{j\beta}{}^\mathbf{q}$. The result is the relation

$$\widehat{T}(g)\, u = \sum_{i\alpha} \left[\sum_{j\beta} \{d_m^\mathbf{q}(g)\}_{i\alpha,\, j\beta}\, u_j^\beta \right] \psi_{i\alpha}^\mathbf{q} \equiv \sum_{i\alpha} u_i'^\alpha\, \psi_{i\alpha}^\mathbf{q},$$

which can be interpreted as the transformation law of the quantities $u_i{}^\alpha$:

$$\widehat{T}(g)\, u_i^\alpha = \sum_{j\beta} \{d_m^\mathbf{q}(g)\}_{i\alpha,\, j\beta}\, u_j^\beta. \tag{4.31}$$

Thus, the components of the vectors of the atomic displacements of the zeroth cell are transformed with the matrices of the mechanical representation.

From the condition that the potential energy of the crystal must be invariant,

$$g \sum_{ij\alpha\beta} u_i^\alpha\, \Phi_{ij}^{\alpha\beta}(\mathbf{q})\, u_j^\beta = \sum_{ij\alpha\beta} u_i'^\alpha\, \Phi_{ij}^{\alpha\beta}(\mathbf{q})\, u_j'^\beta = \sum_{ij\alpha\beta} u_i^\alpha\, \Phi_{ij}^{\alpha\beta}(\mathbf{q})\, u_j^\beta$$

we obtain a relation which must be satisfied by the dynamic matrix. In matrix notation, this relation has the form

$$(d_m^\mathbf{q}(g))^{-1}\, \Phi(\mathbf{q})\, (d_m^\mathbf{q}(g)) = \Phi(\mathbf{q}). \tag{4.32}$$

It suffices to take in this equation the generators of the group $G_\mathbf{q}$ as g in order to obtain all relations between the matrix elements of the dynamic matrix $\Phi(\mathbf{q})$, with the relations resulting from crystal symmetry. The explicit form of the matrices of the mechanical representation can be found with Eq. (4.15).

Having brought into account the symmetry of the crystal in this way, the dynamic matrix Φ can be reduced to diagonal form by a unitary transformation U derived with the aid of the basis functions of the irreducible representations $d^{\mathbf{q}\nu}$ from the composition of the mechanical representation $d_m{}^\mathbf{q}$. This possibility of diagonalizing follows from group theory, and we give here only instructions for deriving the required unitary matrix. The matrix $U_{s\Lambda}$ has $3\sigma \times 3\sigma$ size; its columns are numbered by the subscript Λ of the set consisting of the numbers ν of the irreducible representations $d^{\mathbf{q}\nu}$ of the composition of the mechanical representation and the numbers λ of the basis functions of these representations. The set's subscript s, which numbers the lines, consists of the numbers i of the atoms in the zeroth cell and of the vector subscripts α. The atomic component of the function $\psi_{\lambda}{}^{\mathbf{q}\nu}$ of the irreducible representation for the atoms from the zeroth cell turns out to be an element of the matrix U. These relations are expressed by the precise mathematical formula

$$U^{\mathbf{q}}_{s\Lambda} \equiv U^{\mathbf{q}}_{i\alpha,\, \nu\lambda} = \varepsilon^{\alpha}\left(\begin{matrix}\mathbf{q}\nu \\ \lambda\end{matrix}\middle| i\right), \tag{4.33}$$

where the quantities $\varepsilon^{\alpha}(^{\mathbf{q}\nu}_{\lambda} \mid i)$ are given by Eq. (4.24) under the condition that all these quantities, which belong to a single column $U_{s\Lambda}$, were normalized to unity. The matrix

$$\Phi'(\mathbf{q}) = \overset{+}{U}{}^{\mathbf{q}}\, \Phi(\mathbf{q})\, U^{\mathbf{q}} \tag{4.34}$$

must be composed of blocks along the diagonal. Each of the blocks corresponds to an irreducible representation which is contained in the mechanical representation. When an irreducible representation appears only once in the composition of the mechanical representation, the corresponding block in Eq. (4.34) is diagonal and its multiplicity coincides with the dimension of the irreducible representation. When a certain representation appears r times ($r > 1$) in $d_m{}^{\mathbf{q}}$, the corresponding block is non-diagonal and has the dimensions $l_\nu r \times l_\nu r$. Its diagonalization cannot be accomplished on the basis of the theory of symmetry and must be additionally performed.

Let us now consider the relation of the eigenfunctions of the dynamic matrix with the basis functions of the irreducible representation. It is a well-known fact that if some matrix is composed so that its columns are the eigenvectors of a given matrix, the first matrix is a unitary matrix which diagonalizes the given matrix. When we recall how the unitary matrix of Eq. (4.33) was derived and that this unitary matrix brings the dynamic matrix to diagonal form, we arrive at the conclusion that the quantities $\varepsilon^{\alpha}(^{\mathbf{q}\nu}_{\lambda} \mid i)$ – the atomic components of the basis function of an irreducible representation – are also the polarization vectors of the corresponding mode of vibration, provided that its representation appears only once in the mechanical representation:

$$e_J^{\alpha}(\mathbf{q}\,|\,i) = \varepsilon^{\alpha}\begin{pmatrix}\mathbf{q}\nu \\ \lambda\end{pmatrix}\!\!\left|\,i\right) \quad \text{for} \quad r_\nu = 1 \; (J = \nu\lambda). \tag{4.35}$$

In the event that some representation $d^{\mathbf{q}\nu}$ appears r times in the mechanical representation, the polarization vectors are the superposition of the sets of basis functions of this irreducible representation so that

$$e_J^{\alpha}(\mathbf{q}\,|\,i) \equiv e_{\nu n}^{\alpha}(\mathbf{q}\,|\,i) = \sum_{s\lambda} U_{s\lambda,\,n}\, \varepsilon_s^{\alpha}\begin{pmatrix}\mathbf{q}\nu \\ \lambda\end{pmatrix}\!\!\left|\,i\right), \tag{4.36}$$

where $U_{s\lambda,n}$ denotes an $rl_\gamma \times rl_\gamma$ matrix satisfying the usual conditions of unitarity:

$$\left.\begin{aligned} &\sum_n \overset{*}{U}_{s\lambda,\,n} U_{s'\lambda',\,n} = \delta_{ss'}\,\delta_{\lambda\lambda'}; \\ &\sum_{s\lambda} \overset{*}{U}_{s\lambda,\,n}\, U_{s\lambda,\,n'} = \delta_{nn'}. \end{aligned}\right\} \tag{4.37}$$

In this case, rl_ν polarization vectors $e_{\nu n}{}^{\alpha}(\mathbf{q}\,|\,i)$ correspond to the r eigenfrequencies $\omega_{\nu p}(\mathbf{q})$ ($p = 1,\ldots,r$) with the multiplicity l_ν of the degeneracy.

Making use of Eqs. (4.28) and (4.29) for the basis functions, we can use Eqs. (4.35) and (4.36) to state the conditions of orthogonality and completeness for the polarization vectors:

$$\left.\begin{aligned} &\sum_{i\alpha} \overset{*}{e}{}_{\nu n}^{\alpha}(\mathbf{q}\,|\,i)\, e_{\nu' n'}^{\alpha}(\mathbf{q}\,|\,i) = \delta_{\nu\nu'}\,\delta_{nn'}; \\ &\sum_{\nu n} \overset{*}{e}{}_{\nu n}^{\alpha}(\mathbf{q}\,|\,i)\, e_{\nu n}^{\beta}(\mathbf{q}\,|\,j) = \delta_{\alpha\beta}\,\delta_{ij}. \end{aligned}\right\} \tag{4.38}$$

This is the grouptheoretical way of stating the previously derived Eqs. (2.4). Let us recall that if an irreducible representation appears only once in the mechanical representation, the subscript n in Eq. (4.38) coincides with ν.

Let us summarize. Symmetry analysis of the vibration spectrum of a crystal is fully based on the concept of the mechanical representation. The composition of this representation determines the multiplicity of the degeneracy of the phonon frequencies with a particular wave vector and

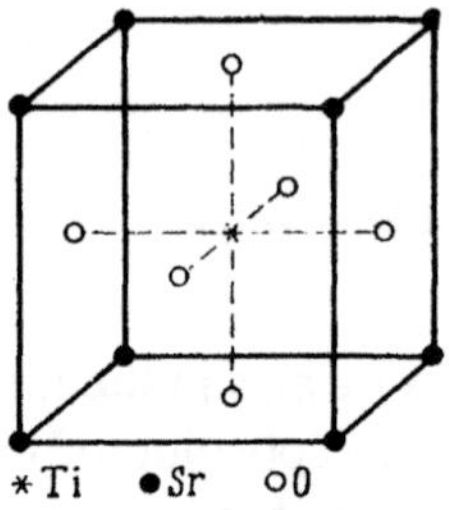

Fig. 2. Unit cell of the $SrTiO_3$ crystal with perovskite structure.

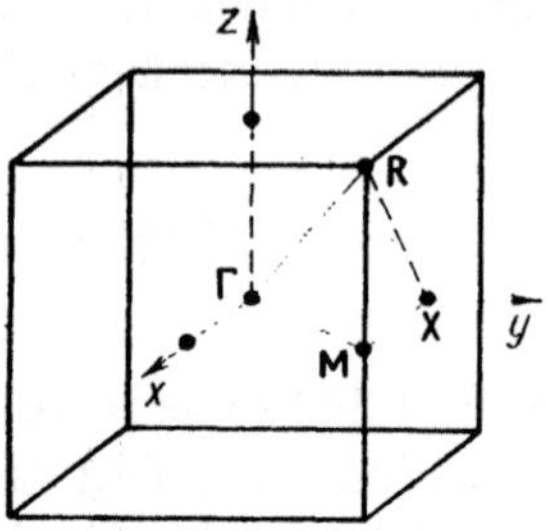

Fig. 3. Symmetry points and symmetry directions of the Brillouin zone of a simple cubic lattice.

determines also the type of symmetry of the vibrations. The basis functions of the mechanical representation determine the polarization vectors, i.e., the possible modes of vibrations, and also facilitate diagonalization of the dynamic matrix, i.e., in principle, make it possible to determine the spectrum of the phonon frequencies.

Example. Let us illustrate the above-described symmetry analysis of the vibration spectrum of a crystal on the example of a cubic crystal with perovskite structure ($SrTiO_3$). Crystals with this structure will be repeatedly discussed below in relation to the structural phase transitions observed in them.

Their space group is $O_h{}^1$ and the atoms occupy the following positions:

$$\mathrm{Sr} - 1\,(b),\quad \mathrm{Ti} - 1\,(a),\quad \mathrm{O} - 3\,(c). \tag{4.39}$$

The atoms belonging to an unit cell have the following coordinates (Fig. 2):

$$1\,(a) - (000),\ 1\,(b) - \left(\frac{1}{2}\,\frac{1}{2}\,\frac{1}{2}\right),$$
$$3\,(c) = 1\left(0\,\frac{1}{2}\,\frac{1}{2}\right),\ 2\left(\frac{1}{2}\,0\,\frac{1}{2}\right),\ 3\left(\frac{1}{2}\,\frac{1}{2}\,0\right). \tag{4.40}$$

Let us state the results of calculations of the basis functions of the mechanical representation of the group $O_h{}^1$ for some wave vector stars

TABLE 1. The Basis Functions of the Mechanical Representation for $\mathbf{k} = 0$ of the Group O_h^1 in the Case of a Perovskite Crystal*

Representation	Atoms				
	1 (*a*)	1 (*b*)	3 (*c*)		
			1	2	3
τ_8			 010 $00\bar{1}$	$\bar{1}00$ 001	100 $0\bar{1}0$
τ_{10}	100 010 001	100 010 001	100 	 010 	 001
τ'_{10}			 010 001	100 001	100 010
τ_{10}	100 010 001	100 010 001	100 010 001	100 010 001	100 010 001
τ'_{10}			200 $0\bar{1}0$ $00\bar{1}$	$\bar{1}00$ 020 $00\bar{1}$	$\bar{1}00$ $0\bar{1}0$ 002

*The unitary-equivalent basis functions of the repeated representation τ_{10}, in which the acoustic mode was separated, are stated in the lower part of the table.

which were taken from the first Brillouin zone. Figure 3 illustrates the Brillouin zone in a simple cubic lattice and also shows the standard notation of the symmetry points and symmetry directions.

The point Γ. The single-arm $\mathbf{k} = 0$ corresponds to this point. Making use of the Kovalev tables [50] of representations of space groups, we find that the group O_h^1 for this star has 10 irreducible representations: $\tau_1, \ldots, \tau_4$ are one-dimensional, τ_5 and τ_6 are two-dimensional, and $\tau_7, \ldots, \tau_{10}$ are three-dimensional representations (the Kovalev notation is used here and everywhere in what follows). With Eqs. (4.15)-(4.18) and the tables of the irreducible representations, we can determine the composition of the mechanical representation $d_m^{\mathbf{k}}$ for $\mathbf{k} = 0$ and all positions occupied in perovskite:

$$d_m^{\mathbf{k}}(1a) = \tau_{10},\ d_m^{\mathbf{k}}(1b) = \tau_{10},\ d_m^{\mathbf{k}}(3c) = \tau_8 + 2\tau_{10}. \tag{4.41}$$

the basis functions which were calculated with Eq. (4.22) are listed in Table 1. Each triple of numbers indicates the components of the displacement of a particular atom in the oscillatory mode. The atomic components of the basis function for the representation τ_8 are the polarization vectors of the particular phonon mode. The polarization vectors of the phonons corresponding to the repeated representation τ_{10} are a superposition of all six basis functions of this representation.

Let us note that the basis functions can be subjected to a unitary transformation. Thus, in place of the two sets of functions for the representation τ_{10}, which were directly obtained from Eq. (4.22), one can use the unitary transformation to obtain the two sets which are shown in the lower part of Table 1. The first of these sets obviously corresponds to the acoustic branch of the vibrations when all the atoms of a cell are displaced together. The second set of the representation (denoted by a prime) describes some optical mode.

It is easy to verify that the transition to a new set of functions of the representation τ_{10} is obtained with the aid of the unitary matrix

$$U = \frac{1}{\sqrt{3}} \begin{pmatrix} 1 & 0 & 0 & \sqrt{2} & 0 & 0 \\ 0 & 1 & 0 & 0 & \sqrt{2} & 0 \\ 0 & 0 & 1 & 0 & 0 & \sqrt{2} \\ \sqrt{2} & 0 & 0 & -1 & 0 & 0 \\ 0 & \sqrt{2} & 0 & 0 & -1 & 0 \\ 0 & 0 & \sqrt{2} & 0 & 0 & -1 \end{pmatrix}. \tag{4.42}$$

When a line composed of the original six basis functions of the representations τ_{10} and τ_{10}' is multiplied by the matrix U, one obtains a line consisting of the set of functions indicated in the lower part of Table 1. It is necessary to normalize all basis functions to the same magnitude.

TABLE 2. The Basis Functions of the Mechanical Representation of the Wave Vector $\mathbf{k} = \frac{1}{2}(\mathbf{b}_1 + \mathbf{b}_2 + \mathbf{b}_3)$ of the Star $\{\mathbf{k}_{13}\}$ ($\varepsilon = e^{i(2\pi/3)}$)

Representation	Atoms				
	1 (*a*)	1 (*b*)	3 (*c*)		
			1	2	3
τ_4			100	010	001
τ_6			100 $\bar{\varepsilon}00$	$0\varepsilon0$ $0\bar{1}0$	$00\varepsilon^2$ $00\overline{\varepsilon^2}$
τ_7		100 010 001			
τ_8			001 $0\bar{1}0$	$00\bar{1}$ 100	010 $\bar{1}00$
τ_{10}	100 010 001		001 010	001 100	010 100

The point R. The star [$\{\mathbf{k}_{13}\}$ in the Kovalev notation] of the wave vector $\mathbf{k} = 1/2(\mathbf{b}_1 + \mathbf{b}_2 + \mathbf{b}_3)$ corresponds to this point. For it the group $O_h{}^1$ has ten irreducible representations: $\tau_1, \ldots, \tau_4$ are one-dimensional, τ_5 and τ_6 are two-dimensional, and $\tau_7, \ldots, \tau_{10}$ are three-dimensional representations. The composition of the mechanical representation is

$$d_m^{\mathbf{k}}(1a) = \tau_{10},\ d_m^{\mathbf{k}}(1b) = \tau_7,\ d_m^{\mathbf{k}}(3c) = \tau_4 + \tau_6 + \tau_8 + \tau_{10}. \tag{4.43}$$

The basis functions are listed in Table 2.

The point M. This point belongs to the three-dimensional star $\{\mathbf{k}_{11}\}$ with the wave vectors

$$\kappa_1 = \frac{1}{2}(\mathbf{b}_1 + \mathbf{b}_2),\ \kappa_2 = \frac{1}{2}(\mathbf{b}_1 + \mathbf{b}_3),\ \kappa_3 = \frac{1}{2}(\mathbf{b}_2 + \mathbf{b}_3). \tag{4.44}$$

The group $G_{\mathbf{k}}$ of the wave vector has in its zeroth block 16 rather than 48

TABLE 3. The Basis Functions of the Mechanical Representation for the Three-Arm Star $\{\mathbf{k}_{11}\}$ of the Wave Vector $\mathbf{k} = \frac{1}{2}(\mathbf{b}_1 + \mathbf{b}_2)$

Representation	Arm	Atoms 1 (a)	1 (b)	3 (c) 1	2	3
τ_1	κ_1			010	100	
	κ_2			001		100
	κ_3				001	010
τ_3	κ_1			100	$0\bar{1}0$	
	κ_2			$\bar{1}00$		001
	κ_3				010	$00\bar{1}$
τ_4	κ_1	001				
	κ_2	010				
	κ_3	100				
τ_5	κ_1			010	$\bar{1}00$	
	κ_2			$00\bar{1}$		100
	κ_3				001	$0\bar{1}0$
τ_8	κ_1		001			001
	κ_2		010		010	
	κ_3		100	100		

Representation	Arm	Atoms 1 (a)	1 (b)	3 (c) 1	2	3
τ_7	κ_1			100	010	
	κ_2			100		001
	κ_3				010	001
τ_9	κ_1			001 001	$00\bar{1}$ 001	
	κ_2			$0\bar{1}0$ 010		010 010
	κ_3				100 100	$\bar{1}00$ 100
τ_{10}	κ_1	100 010	100 010			110 $1\bar{1}0$
	κ_2	001 100	001 100		101 101	
	κ_3	010 001	010 001	011 $01\bar{1}$		

elements and ten irreducible representations: $\tau_1, \ldots, \tau_8$ are one-dimensional, and τ_9 and τ_{10} are two-dimensional. The composition of the mechanical representation of the group $G_\mathbf{k}$ is

$$\left.\begin{aligned} d_m^{\kappa}(1a) &= \tau_4 + \tau_{10},\ d_m^{\kappa}(1b) = \tau_8 + \tau_{10}; \\ d_m^{\kappa}(3c) &= \tau_1 + \tau_3 + \tau_5 + \tau_8 + \tau_7 + \tau_9 + \tau_{10}. \end{aligned}\right\} \qquad (4.45)$$

The basis functions for the first arm of the star were calculated with Eq. (4.22), whereas the basis functions for the other arms of the star were calculated with Eq. (4.24) (see Table 3).

The line ΓM. The wave vector $\mathbf{k} = \mu(\mathbf{b}_1 + \mathbf{b}_2)$ belongs to the 12-arm star $\{\mathbf{k}_1\}$. The group of the wave vector has the four one-dimensional irreducible representations $\tau_1, \ldots, \tau_4$. The composition of the mechanical representation of the group $G_\mathbf{k}$ is

TABLE 4. The Basis Functions of the Mechanical Representations for $\mathbf{k} = \mu(\mathbf{b}_1 + \mathbf{b}_2)$

Representation	Atoms 1 (a)	1 (b)	3 (c) 1	2	3
τ_1			100	010	
τ_1'			010	100	
τ_1''	110	110			110
τ_2			001	$00\bar{1}$	
τ_3			100	$0\bar{1}0$	
τ_3'			010	$\bar{1}00$	
τ_3''	$1\bar{1}0$	$1\bar{1}0$			$1\bar{1}0$
τ_4			001	001	
τ_4'	001	001			001

TABLE 5. The Basis Functions of the Mechanical Representation for $\mathbf{k} = \mu\ \mathbf{b}_1 + \mathbf{b}_2 + \mathbf{b}_3)$ $(\varepsilon = e^{i(2\pi/3)})$

Representation	Atoms 1 (a)	1 (b)	3 (c) 1	2	3
τ_1	111	111	100	010	001
τ_1'			011	101	110
τ_2			$01\bar{1}$	$\bar{1}01$	$1\bar{1}0$
τ_3	$1\varepsilon\varepsilon^2$	$1\varepsilon\varepsilon^2$	100	$0\varepsilon 0$	$00\varepsilon^2$
	$\varepsilon 1\varepsilon^2$	$\varepsilon 1\varepsilon^2$	$\varepsilon 00$	010	$00\varepsilon^2$
τ_3'			010	00ε	$\varepsilon^2 00$
			00ε	100	$0\varepsilon^2 0$
τ_3''			001	$\varepsilon 00$	$0\varepsilon^2 0$
			$0\varepsilon 0$	001	$\varepsilon^2 00$

$$\left.\begin{array}{c} d_m^{\mathbf{k}}(1a) = \tau_1 + \tau_3 + \tau_4,\ d_m^{\mathbf{k}}(1b) = \tau_1 + \tau_3 + \tau_4; \\ d_m^{\mathbf{k}}(3c) = 3\tau_1 + \tau_2 + 3\tau_3 + 2\tau_4. \end{array}\right\} \quad (4.46)$$

The basis functions for the first arm of the star are listed in Table 4.

The line Γ*R*. The wave vector $\mathbf{k} = \mu(\mathbf{b}_1 + \mathbf{b}_2 + \mathbf{b}_3)$ belongs to the eight-arm star $\{\mathbf{k}_9\}$. The group of the wave vector has three irreducible representations: two are one-dimensional (τ_1 and τ_2) and one is two-dimensional (τ_3). The composition of the mechanical representation of the group $G_{\mathbf{k}}$ is

$$\left.\begin{array}{c} d_m^{\mathbf{k}}(1a) = \tau_1 + \tau_3,\ d_m^{\mathbf{k}}(1b) = \tau_1 + \tau_3; \\ d_m^{\mathbf{k}}(3c) = 2\tau_1 + \tau_2 + 3\tau_3. \end{array}\right\} \quad (4.47)$$

TABLE 6. Connectivity Relations in the Group O_h^1 for the Points Γ and **R** and Their Neighborhoods

Point Γ $\kappa=0$	$\{\kappa_8\}$ $\kappa=\mu b_3$	$\{\kappa_4\}$ $\kappa=\mu(b_1+b_2)$	$\{\kappa_9\}$ $\kappa=\mu\ (b_1+b_2+b_3)$
Point R $\kappa=\frac{1}{2}(b_1+b_2+ +b_3)$	$\{\kappa_7\}$ $\kappa=\frac{1}{2}(b_1+b_2)+ +\mu b_3$	$\{\kappa_5\}$ $\kappa=\mu\ (b_1+b_2)+ +\frac{1}{2}\ b_3$	—
τ_1 (1)	τ_1 (1)	τ_1 (1)	τ_1 (1)
τ_2 (1)	τ_2 (1)	τ_2 (1)	τ_2 (1)
τ_3 (1)	τ_3 (1)	τ_3 (1)	τ_2 (1)
τ_4 (1)	τ_4 (1)	τ_4 (1)	τ_1 (1)
τ_5 (2)	τ_1 (1)+τ_3 (1)	τ_1 (1)+τ_3 (1)	τ_3 (2)
τ_6 (2)	τ_2 (1)+τ_4 (1)	τ_2 (1)+τ_4 (1)	τ_3 (2)
τ_7 (3)	τ_4 (1)+τ_5 (2)	τ_1 (1)+τ_2 (1)+τ_4 (1)	τ_1 (1)+τ_3 (2)
τ_8 (3)	τ_3 (1)+τ_5 (2)	τ_1 (1)+τ_2 (1)+τ_3 (1)	τ_2 (1)+τ_3 (2)
τ_9 (3)	τ_2 (1)+τ_5 (2)	τ_2 (1)+τ_3 (1)+τ_4 (1)	τ_2 (1)+τ_3 (2)
τ_{10} (3)	τ_1 (1)+τ_5 (2)	τ_1 (1)+τ_3 (1)+τ_4 (1)	τ_1 (1)+τ_3 (2)

The basis functions for the first arm of the star are listed in Table 5.

The basis functions for all other points of a Brillouin zone can be calculated in a similar fashion.

In a deviation from a point of symmetry, the intrinsic symmetry of the wave vector is reduced so that the group of the wave vector becomes a subgroup of the group $G_{\mathbf{k}}$ of the point of symmetry. The irreducible representations of the group $G_{\mathbf{k}}$ on its subgroup are split. This means that the phonon frequencies split in a deviation of the wave vector from the point of symmetry. Naturally, the degree of splitting depends upon the direction in which the wave vector is inclined.

The so-called connectivity relations help to determine the form in which the branches of the phonon spectrum are split and, vice versa, how the branches merge at points of symmetry of a Brillouin zone. Tables 6

TABLE 7. Connectivity Relations in the Group $O_h{}^1$ for the Point **M** and Its Neighborhood

Point M $\kappa = \frac{1}{2}(b_1 + b_2)$	$\{\kappa_7\}$ $\kappa = \frac{1}{2}(b_1 + b_2) + \mu b_3$	$\{\kappa_6\}$ $\kappa = \frac{1}{2} b_2 + \mu b_1$	$\{\kappa_4\}$ $\kappa = \mu (b_1 + b_2)$
τ_1 (1)	τ_1 (1)	τ_1 (1)	τ_1 (1)
τ_2 (1)	τ_2 (1)	τ_2 (1)	τ_2 (1)
τ_3 (1)	τ_2 (1)	τ_4 (1)	τ_3 (1)
τ_4 (1)	τ_1 (1)	τ_3 (1)	τ_4 (1)
τ_5 (1)	τ_3 (1)	τ_1 (1)	τ_3 (1)
τ_6 (1)	τ_4 (1)	τ_2 (1)	τ_4 (1)
τ_7 (1)	τ_4 (1)	τ_4 (1)	τ_1 (1)
τ_8 (1)	τ_3 (1)	τ_3 (1)	τ_2 (1)
τ_9 (2)	τ_5 (2)	τ_2 (1) + τ_3 (1)	τ_2 (1) + τ_4 (1)
τ_{10} (2)	τ_5 (2)	τ_1 (1) + τ_4 (1)	τ_1 (1) + τ_3 (1)

and 7 list the relations of the irreducible representations of the group of the wave vector of the points Γ, **R**, and **M** of symmetry of a Brillouin zone and their neighborhoods in the case of the space group $O_h{}^1$. These relations are also the connectivity relations. The Kovalev notation [50] for the stars and the irreducible representations was used. Proceeding in the table from the left to the right, it is realized how the irreducible representation of a point of symmetry (Γ, **R**, **M**) is split into irreducible representations of the corresponding group $G_{\mathbf{k}}$. To the right of the symbol τ of an irreducible representation, its dimensionality is indicated.

When we apply this table to the analysis of the phonon spectrum of a crystal, we must use only those irreducible representations of any point of symmetry, which appear in the mechanical representation. The calculation of the basis of these irreducible representations for points of symmetry and the calculations of the connectivity relations are the basis of the symmetry analysis of the vibration spectrum of a crystal.

5. Selection Rules in Coherent Single-Phonon Scattering

The intensity of single-phonon scattering. Having established the relation between the number of the phonon branch and the subscripts of the irreducible representations of the space group of a crystal, we rewrite the expressions for the cross section of single-phonon scattering at a crystal (which were stated in Section 2) in terms of group theory. We are particularly interested in the cross section of coherent scattering as described by Eq. (2.23). As has been shown in the preceding section, the subscript J is a composite subscript consisting of the number ν of the irreducible representation and the number λ of the basis function of this representation (strictly speaking, this is valid only for the case in which the given irreducible representation is contained only once in the mechanical representation; the general case of repeated representations will be considered below). Since the phonon frequency $\omega_J(\mathbf{q})$ depends only upon ν but is independent of the number of the degenerate phonon mode denoted by the subscript λ, we restate Eq. (2.23) in the form

$$\frac{1}{N}\frac{d^2\sigma^{\pm}_{\mathrm{coh}}}{d\Omega d\varepsilon}\bigg|_{\varkappa=\mp\mathbf{q}+\mathbf{b}}=\frac{\hbar k}{2k_0}\sum_{\nu}\left\{\sum_{\lambda}|\varkappa\xi_{\nu\lambda}(\varkappa)|^2\right\}\frac{1}{\omega_\nu(\varkappa)}\times$$
$$\times\left(N_{\varkappa}+\frac{1}{2}\pm\frac{1}{2}\right)\delta(\varepsilon\pm\hbar\omega_\nu(\varkappa)). \tag{5.1}$$

This shows that the intensity $I_\nu^{\pm}(\varkappa)$ of a single phonon peak (intensity corresponding to scattering at all phonon modes) with the frequency ω_ν is defined by the factors

$$I^{\pm}_{\nu}(\varkappa)\sim F_\nu(\varkappa)\frac{1}{\omega_\nu(\varkappa)}\left(N_{\varkappa}+\frac{1}{2}\pm\frac{1}{2}\right), \tag{5.2}$$

from which

$$F_\nu(\varkappa)=\sum_{\lambda}|\varkappa\xi_{\nu\lambda}(\varkappa)|^2 \tag{5.3}$$

is characterized by the polarization vectors of the modes belonging to the particular irreducible representation and plays the role of some form factor of these phonon modes.

In terms of group theory, the quantity $\xi_{\nu\lambda}(\varkappa)$ defined by Eq. (2.22) can be stated in the form

$$\xi_{\nu\lambda}(\varkappa) = \sum_i a_i e^{i\varkappa x_i} \varepsilon \begin{pmatrix} \varkappa\nu \\ \lambda \end{pmatrix} i \Bigg), \tag{5.4}$$

where the notation $a_i = b_i e^{-W_i}(1/\sqrt{M_i})$ was introduced for the sake of brevity. When we stated Eq. (5.4), we took into account the fact that in the case of nonrepeated irreducible representations, the polarization vectors coincide with the atomic components of the basis functions [see Eq. (4.35)].

Since the quantity $F_\nu(\varkappa)$ has all the information on the phonon modes and this information is contained in the intensity of the neutron resonance, our attention must focus on the properties of this quantity.

The form factor $F_\nu(\varkappa)$ depends upon the polarization vectors of the phonons and hence, in the case of an irreducible repeated (in the mechanical presentation) representation ν of the group $G_{\mathbf{k}}$, the form factor must depend on the force constants of the crystal. But as will be shown below [173], the polarization vectors actually cancel in the sum of Eq. (5.3) defining $F_\nu(\varkappa)$, so that $F_\nu(\varkappa)$ is defined only by the symmetry properties of the crystal and of the particular phonon branch and is a convenient quantity for determining the selection rules in neutron scattering. Below we will derive relations which make it possible to determine all irreducible representations for which (at a given scattering vector $\varkappa$) the form factor $F_\nu(\varkappa)$ vanishes. These extinction rules of single-phonon peaks help to identify the peaks.

Selection rules. The derivation of selection rules [173] is based on the transformation of Eq. (5.4) for $\xi_{\nu\lambda}(\varkappa)$ with the aid of the projection operator

$$P_\nu(\varkappa) = \frac{l_\nu}{\| G_\varkappa \|} \sum_{g \in G_\varkappa} \overset{*}{\chi}{}^{\varkappa\nu}(g)\, T(g) \tag{5.5}$$

($\| G_\varkappa \|$ denotes the number of elements in the group $G_\varkappa$). This operator is characterized by the property that in acting upon the basis function of the νth irreducible representation of the group $\psi_\lambda{}^{\varkappa\nu}$, there is no change, whereas when the operator acts upon the basis function ψ_λ of another irreducible representation, the result is zero, i.e.,

$$P_\nu(\varkappa)\, \psi_\lambda^{\varkappa\nu'} = \delta_{\nu\nu'}\, \psi_\lambda^{\varkappa\nu'}. \tag{5.6}$$

With the notation of the preceding section, we represent the function $\psi_\lambda^{\varkappa\nu}$ as a column of atomic components

$$\psi_\lambda^{\varkappa\nu'} = \sum_{j\beta} e^\beta \begin{pmatrix} \varkappa\nu' \\ \lambda \end{pmatrix} j \Big) \psi_\varkappa^{j\beta}, \tag{5.7}$$

where the quantities $\psi_\varkappa^{j\beta}$ are transformed with the matrices $D_m^{\varkappa}(g)$ of the mechanical representation in accordance with Eq. (4.14). Substituting Eqs. (5.5) and (5.7) in Eq. (5.6), we obtain relations for the atomic components

$$\delta_{\nu\nu'} e^\alpha \left(\begin{matrix} \varkappa\nu' \\ \lambda' \end{matrix} \middle| i \right) = \sum_{j\beta} \frac{l_\nu}{\| G_\varkappa \|} \sum_{g \in G_\varkappa} \overset{*}{\chi}{}^{\varkappa\nu}(g) \{ d_m^\varkappa(g) \}_{i\alpha,\, j\beta}\, e^\beta \left(\begin{matrix} \varkappa\nu' \\ \lambda' \end{matrix} \middle| j \right) \tag{5.8}$$

by virtue of the symbolic equation (5.6).

We set in this equation $\nu = \nu'$ and substitute for $\xi_{\nu\lambda}{}^\alpha(\varkappa)$ the expression obtained for $e^\alpha(\begin{smallmatrix}\varkappa\nu\\ \lambda\end{smallmatrix} \mid i)$ in Eq. (5.4). When the explicit form of Eq. (4.15) for the matrix of the mechanical representation is taken into account, the summation can be extended over the atomic subscript i with the aid of the $\delta_{i,gj}$– symbol in Eq. (4.15). The resulting expression for $\xi_{\nu\lambda}{}^\alpha(\varkappa)$ is then conveniently written in the form

$$\xi_{\nu\lambda}^\alpha(\varkappa) = \sum_j a_j \sum_\beta A_j^{\alpha\beta}(\nu, \varkappa)\, e^\beta \left(\begin{matrix} \varkappa\nu \\ \lambda \end{matrix} \middle| j \right), \tag{5.9}$$

where

$$A_j^{\alpha\beta}(\nu, \varkappa) = \frac{l_\nu}{\| G_\varkappa^\circ \|} \sum_{g \in G_\varkappa^\circ} \overset{*}{\chi}{}^{\varkappa\nu}(g)\, e^{i\varkappa g x_j} R_h^{\alpha\beta} \tag{5.10}$$

[In Eq. (5.10) the summation is extended over the translations of the group $G_\varkappa$, with the summation restricted to the rotational elements of this group.]

Now we will make use of another consequence of Eq. (5.8). Let us set $\nu \neq \nu'$ in it. We multiply the resulting relation with $a_i e^{i\varkappa x_i}$ and perform the summation over the subscript i. This results in the expression

$$\sum_j a_j \sum_\beta A_j^{\alpha\beta}(\nu, \varkappa)\, e^\beta \left(\begin{matrix} \varkappa\nu' \\ \lambda' \end{matrix} \middle| j \right) = 0, \tag{5.11}$$

establishing a relation between the quantities referring to the various ν and ν'. Equations (5.9) and (5.11) make it possible to transform the expression for the form factor $F_\nu(\varkappa)$. After substituting in it Eq. (5.9), Eq. (5.3) assumes the form

$$F_\nu(\varkappa) = \sum_\lambda \left| \sum_\alpha \varkappa^\alpha \sum_j a_j \sum_\beta A_j^{\alpha\beta}(\nu, \varkappa)\, e^\beta \begin{pmatrix} \varkappa\nu \\ \lambda \end{pmatrix} j \right|^2.$$

Because of Eq. (5.11), in the latter equation one can include the sum over all irreducible representations ν' ($\neq \nu$) and rewrite it in the form

$$F_\nu(\varkappa) = \sum_{\nu'\lambda'} \left| \sum_{\alpha\beta} \varkappa^\alpha \sum_j a_j A_j^{\alpha\beta}(\nu, \varkappa)\, e^\beta \begin{pmatrix} \varkappa\nu' \\ \lambda' \end{pmatrix} j \right|^2. \quad (5.12)$$

The sum over the subscripts $\nu'\lambda'$ is a sum over the basis functions of all irreducible representations of the group $G_\varkappa$. We denote it as sum $\Sigma_{\mu\lambda_\mu}$ and rewrite the initial expression in the form

$$F_\nu(\varkappa) = \sum_{\alpha\gamma} \varkappa^\alpha \varkappa^\gamma \sum_{ij} a_i a_j \sum_{\beta\delta} A_j^{\alpha\beta}(\nu, \varkappa)\, \overset{*}{A}{}_i^{\gamma\delta}(\nu, \varkappa) \times$$
$$\times \sum_{\mu\lambda_\mu} e^\beta \begin{pmatrix} \varkappa\mu \\ \lambda_\mu \end{pmatrix} j \, \overset{*}{e}{}^\delta \begin{pmatrix} \varkappa\mu \\ \lambda_\mu \end{pmatrix} i. \quad (5.13)$$

Making use of the completeness relations Eq. (4.27) for the basis functions, we can state the final expression for the form factor

$$F_\nu(\varkappa) = \sum_{\alpha\beta} \varkappa^\alpha B^{\alpha\beta}(\nu, \varkappa)\, \varkappa^\beta, \quad (5.14)$$

where the Elliott–Thorpe tensor [173]

$$B^{\alpha\beta}(\nu,\varkappa) = \sum_j \left(A_j(\nu, \varkappa)\, A_j^+(\nu, \varkappa)\right)^{\alpha\beta} a_j^2 \quad (5.15)$$

is expressed through the matrix $A_j(\nu, \varkappa)$ defined by Eq. (5.10).

We realize that the final expression for $F_\nu(\varkappa)$ does not depend upon the basis functions of the νth irreducible representation of the group $G_\varkappa$ but is expressed only through its characters, the matrices of the rotations of the group $G_\varkappa$, and the coordinates of the atoms of the primitive cell of the crystal.

In practice, Eq. (5.14) is valuable because the tensor $B^{\alpha\beta}(\nu,\varkappa)$ for all irreducible representations can be calculated at each scattering vector. If it turns out that this tensor vanishes for certain representations, it is obvious that the observed peak of coherent single-phonon scattering at the $\varkappa$ assumed is not related with the phonons which are described by these representations. After sorting through the available experimental observations of the vector $\varkappa$ (at a particular wave vector of the phonon), the neutron peak can be identified; i.e., the classification symbol, the number of the irreducible representation, can be assigned to the observed phonon frequency (when the wave vector is known).

We will now generalize this result to the case in which a particular νth irreducible representation appears r times ($r > 1$) in the composition of the mechanical representation. In this case, the subscript J of the phonon branch contains the number ν of the irreducible representation and the number n of the phonon branch. The number of phonon branches is equal to rl_ν, and r different phonon frequencies $\omega_{\nu p}(\varkappa)$ correspond to them ($p = 1,\ldots, r$), with each of the phonon frequencies having a l_ν-fold degeneracy. The intensity of a single-phonon resonance with a given frequency $\omega_{\nu p}$ is defined, as above, by Eq. (5.2) (with the replacements $\omega_\nu \to \omega_{\nu p}$, $N_\varkappa \to N_{\varkappa p}$), where the form factor

$$F_{\nu p}(\varkappa) = \sum_\lambda |\varkappa \xi_{\nu p\lambda}(\varkappa)|^2 \tag{5.16}$$

depends not only upon ν but also upon the number of the phonon frequency belonging to the repeated representation.

In the case of a repeated representation, the polarization vectors are the superposition of all sets of basis functions of this representation and are defined with Eq. (4.36). The elements of the corresponding unitary matrix depend upon the force constants, and when Eq. (5.16) is calculated, one cannot eliminate them. This means that the single-phonon scattering cross section, which is proportional to the form factor of Eq. (5.16), depends also upon the force constants and Eq. (5.16) is not very suitable for the identification of the neutron peaks by way of extinctions. However, one can introduce some expression $F_\nu(\varkappa)$ which represents the sum of the quantities of Eq. (5.16) over all numbers p of the phonon peaks belonging to the νth representation. Since this sum is composed of non-negative quantities, it can become zero only when all its terms vanish simultaneously. Thus, if this quantity vanishes for some scattering vector $\varkappa$, this means that peaks resulting from any phonons characterized by the given repeated representations do not appear in the scattering.

Therefore, let us introduce the quantity

$$F_\nu^{(r)}(\varkappa) = \sum_n |\varkappa \xi_{\nu n}(\varkappa)|^2, \tag{5.17}$$

where n denotes the composite subscript $p\lambda$ which numbers all phonon branches belonging to a particular representation ν. According to the definition by Eq. (2.22), we have

$$\xi_{\nu n}(\varkappa) = \sum_i a_i \, e^{i\varkappa x_i} \mathbf{e}_{\nu n}(\varkappa \mid i), \tag{5.18}$$

where the polarization vector $\mathbf{e}_{\nu n}(\varkappa \mid i)$ is expressed through the basis functions of the νth representation with the aid of Eq. (4.36). By substituting Eq. (5.18) in Eq. (5.17) and recalling the relations expressed by Eq. (4.37) for the orthogonality in the case of the unitary matrix, we obtain the expression

$$F_\nu^{(r)}(\varkappa) = \sum_{s\lambda} |\varkappa \xi_{\nu s\lambda}(\varkappa)|^2, \tag{5.19}$$

where

$$\xi_{\nu s\lambda}^{\alpha} = \sum_i a_i \, e^{i\varkappa x_i} \varepsilon_s^{\alpha} \begin{pmatrix} \varkappa\nu \\ \lambda \end{pmatrix} \Bigg| i \Bigg) \tag{5.20}$$

is expressed through the basis functions of the repeated representation as in the case of a nonrepeated irreducible representation. It is now easy to perform an analogous calculation for Eq. (5.19). The result is

$$F_\nu^{(r)}(\varkappa) = \sum_{\alpha\beta} \varkappa^{\alpha} B^{\alpha\beta}(\nu, \varkappa) \varkappa^{\beta}, \tag{5.21}$$

where the tensor $B^{\alpha\beta}(\nu, \varkappa)$ is given by the previous Eqs. (5.15) and (5.10) deduced for a nonrepeated representation.

Thus, in the general case, the zeros of the tensor $B^{\alpha\beta}(\nu, \varkappa)$ define the extinction conditions of the neutron resonances associated with the irreducible representation ν. It turns out that the expression for the tensor does not contain polarization vectors which can depend upon the force constants, so that this expression is suitable for practical calculations.

Example. Let us continue the symmetry analysis which we began in the preceding section for the case of the oscillatory spectrum of a perovskite-type crystal. We calculate the Elliott–Thorpe tensor for a number of phonon modes of different symmetry and determine the extinctions of the neutron resonances for them. All calculations are based on Eqs. (5.10), (5.15), and (5.14).

The point Γ. The representation τ_8 is realized only at the 3(*c*) positions. In the case of atom 1, the calculation performed with Eq. (5.10) yilds the matrix

$$A_1 = \begin{pmatrix} 0 & 0 & 0 \\ 0 & \frac{1}{2}\left(e^{-i\varkappa x_1} - e^{-i\varkappa x_3}\right) & 0 \\ 0 & 0 & \frac{1}{2}\left(e^{-i\varkappa x_1} - e^{-i\varkappa x_2}\right) \end{pmatrix}.$$

The matrices A_2 and A_3 have a similar structure with corresponding positions of the zero on the main diagonal. After calculating the matrix $\Sigma j A_j A_j^+$, we find the Elliott–Thorpe tensor:

$$B(\tau_8, \varkappa) = \frac{a_3^2(c)}{2} \begin{pmatrix} |e^{i\varkappa x_3} - e^{i\varkappa x_2}|^2 & 0 & 0 \\ 0 & |e^{i\varkappa x_3} - e^{i\varkappa x_1}|^2 & 0 \\ 0 & 0 & |e^{i\varkappa x_1} - e^{i\varkappa x_2}|^2 \end{pmatrix}.$$

We represent the scattering vector $\varkappa$ in the form of a set of Miller indices: $\varkappa = 2\pi/a(h, k, l)$. Then the factor $F_{\tau 8}(\varkappa)$ [see Eq. (5.3)], which defines the intensity of the resonance, is written in the form

$$F_{\tau_8}(\varkappa) \sim \{h^2 |e^{i\varkappa x_3} - e^{i\varkappa x_2}|^2 + k^2 |e^{i\varkappa x_3} - e^{i\varkappa x_1}|^2 + l^2 |e^{i\varkappa x_1} - e^{i\varkappa x_2}|^2\}.$$

Now it is easy to verify that the resonances are extinguished at the following points of the reciprocal lattice:

$$(h00) \text{ and } (hhh). \tag{5.22}$$

In the case of the representation τ_{10}, all positions give a contribution to the tensor $B^{\alpha\beta}$ which turns out to be equal to

$$B(\tau_{10}, \varkappa) = a_{1a}^2 + a_{1b}^2 + a_{3c}^2 \times$$

$$\times \begin{pmatrix} 1 + \frac{|e^{i\varkappa x_2} + e^{i\varkappa x_3}|^2}{2} & 0 & 0 \\ 0 & 1 + \frac{|e^{i\varkappa x_1} + e^{i\varkappa x_3}|^2}{2} & 0 \\ 0 & 0 & 1 + \frac{|e^{i\varkappa x_1} + e^{i\varkappa x_2}|^2}{2} \end{pmatrix}.$$

Obviously, there are no extinctions in the case of the representation τ_{10}. Thus, neutron peaks appear near the $(h00)$ and (hhh) points of the reciprocal lattice because of phonons with the symmetry of τ_{10}; peaks for the phonons of τ_8 and τ_{10} must be observed near all the other lattice points.

The point R. $\varkappa = 2\pi/a(h + 1/2, k + 1/2, l + 1/2)$. We obtain for the representation τ_4 at the first atom of the 3(*c*) position:

$$A_1 = \frac{1}{3}\begin{pmatrix} e^{-i\varkappa x_1} & 0 & 0 \\ e^{-i\varkappa x_2} & 0 & 0 \\ e^{-i\varkappa x_3} & 0 & 0 \end{pmatrix}.$$

The first column of this matrix is shifted accordingly to the second and third place in the case of atoms 2 and 3. The tensor $B^{\alpha\beta}$ is

$$B(\tau_4, \varkappa) = \frac{a_{3c}^2}{3}\begin{pmatrix} 1 & e^{-i\varkappa(x_1-x_2)} & e^{-i\varkappa(x_1-x_3)} \\ e^{i\varkappa(x_1-x_2)} & 1 & e^{-i\varkappa(x_2-x_3)} \\ e^{i\varkappa(x_1-x_3)} & e^{i\varkappa(x_2-x_3)} & 1 \end{pmatrix},$$

so that

$$F_{\tau_4}(\varkappa) \sim \left|\left(h+\frac{1}{2}\right)e^{i\varkappa x_1} + \left(k+\frac{1}{2}\right)e^{i\varkappa x_2} + \left(l+\frac{1}{2}\right)e^{i\varkappa x_3}\right|^2.$$

Obviously, this expression does not become zero for any values of *hkl*. There are no extinctions.

We easily find in the case of the representation τ_6:

$$F_{\tau_6}(\varkappa) \sim \left\{\left(h+\frac{1}{2}\right)^2 + \left(k+\frac{1}{2}\right)^2 + \left(l+\frac{1}{2}\right)^2 - \right.$$
$$-\left(h+\frac{1}{2}\right)\left(k+\frac{1}{2}\right)\cos\varkappa(x_1-x_2) - \left(h+\frac{1}{2}\right)\left(l+\frac{1}{2}\right)\cos\varkappa(x_1-x_3) -$$
$$\left. -\left(k+\frac{1}{2}\right)\left(l+\frac{1}{2}\right)\cos\varkappa(x_2-x_3)\right\}.$$

and from it we obtain the extinction conditions

$$h = k = l,$$
$$h = k = -l - 1, h = l = -k - 1, k = l = -h - 1. \qquad (5.23)$$

There are no extinctions in the case of the representation τ_7 because

$$F_{\tau_7}(\varkappa) \sim \left\{\left(h+\frac{1}{2}\right)^2 + \left(k+\frac{1}{2}\right)^2 + \left(l+\frac{1}{2}\right)^2\right\}.$$

TABLE 8. Extinction of the Neutron Peaks in the Case of the Phonons at the Point M

Representation	Extinctions		
	$\varkappa=\frac{2\pi}{a}\left(h+\frac{1}{2},\ k+\frac{1}{2},\ l\right)$	$\varkappa=\frac{2\pi}{a}\left(h+\frac{1}{2},\ k,\ l+\frac{1}{2}\right)$	$\varkappa=\frac{2\pi}{a}\left(h,\ k+\frac{1}{2},\ l+\frac{1}{2}\right)$
τ_1	none	none	none
τ_3	$h=k$ or $h+k+1=0$	$l=h$ or $h+l+1=0$	$k=l$ or $l+k+1=0$
τ_4	$l=0$	$k=0$	$h=0$
τ_5	$h=k$ or $h+k+1=0$	$l=h$ or $h+l+1=0$	$k=l$ or $l+k+1=0$
τ_6	$l=0$	$k=0$	$h=0$
τ_7	none	none	none
τ_9	$l=0$	$k=0$	$h=0$
τ_{10}	none	none	none

In the case of the representation τ_8, the conditions of extinction are the same as in the case of τ_6. In the case of the representation τ_{10}, we have

$$F_{\tau_{10}}(\varkappa) \sim a_{1a}^2\left\{\left(h+\frac{1}{2}\right)^2+\left(k+\frac{1}{2}\right)^2+\left(l+\frac{1}{2}\right)^2\right\}+a_{3c}^2\left\{\left(h+\frac{1}{2}\right)^2+\right.$$
$$+\left(k+\frac{1}{2}\right)^2+\left(l+\frac{1}{2}\right)^2+\left(h+\frac{1}{2}\right)\left(k+\frac{1}{2}\right)\cos\varkappa(\mathbf{x}_1-\mathbf{x}_2)+$$
$$\left.+\left(h+\frac{1}{2}\right)\left(l+\frac{1}{2}\right)\cos\varkappa(\mathbf{x}_1-\mathbf{x}_3)+\left(k+\frac{1}{2}\right)\left(l+\frac{1}{2}\right)\cos\varkappa(\mathbf{x}_2-\mathbf{x}_3)\right\},$$

which proves that there are no extinctions.

Thus, there are no extinctions at the point **R** in the case of the representations τ_4, τ_7, and τ_{10}, but there are extinctions in the case of the representations τ_6 and τ_8.

The point M. The extinction for the reflections $\varkappa = \mathbf{b} + \mathbf{k}_L$, which correspond to three arms of Eq. (4.44) of the star $\{\mathbf{k}_{11}\}$, are listed in Table 8.

We also state the results of calculating the extinction conditions for the Brillouin zone points situated on the lines **ΓM** and **ΓR**.

The line ΓM. The wave vector $\mathbf{k}_L$ which is situated on that line belongs to the 12-arm star $\{\mathbf{k}_4\}$. Its first arm $\mathbf{k}_1 = \mu(\mathbf{b}_1 + \mathbf{b}_2)$ corresponds to the scattering vector $\varkappa = 2\pi/a(h + \mu,\ k + \mu,\ l)$. The extinction conditions of the neutron peaks are

$$\left.\begin{array}{l} \tau_1 \text{ — no extinctions;} \\ \tau_2 \text{ — } h \text{ and } k \text{ of the same parity, or } l = 0; \\ \tau_3 \text{ — } h = k; \\ \tau_4 \text{ — } l = 0. \end{array}\right\} \quad (5.24)$$

The line ΓL. The wave vector $\mathbf{k}_L$ which is situated on this line belongs to the 8-arm star $\{\mathbf{k}_9\}$. Its first arm $\mathbf{k}_1 = \mu(\mathbf{b}_1 + \mathbf{b}_2 + \mathbf{b}_3)$ corresponds to the scattering vector $\varkappa = 2\pi/a(h + \mu,\ k + \mu,\ l + \mu)$. The extinction conditions of the neutron peaks are

$$\left.\begin{array}{l} \tau_1 \text{ — no extinctions;} \\ \tau_2 \text{ — } h, k, l \text{ of the same parity} \\ \text{or} \quad h = k, h = l, k = l; \\ \tau_3 \text{ — } h = k = l. \end{array}\right\} \quad (5.25)$$

6. Phonons and Structural Phase Transitions

The symmetry aspects of the theory of harmonic vibrations of a crystal are closely related to the problem of structural phase transitions involving the mechanism of a soft mode. We will start our considerations of this problem with an exposition of the general symmetry classification of structure transformations in crystals.

The soft mode and the parameters of order. Structure transformations of crystals are usually divided into two types. In transformations of the first type, a phase develops, which is obtained from the initial crystal by a continuous static shifting of the atoms from their equilibrium positions. Obviously, such a transition, which is termed distortion (or displacement-type transition), leads to a crystal the space group of which is a subgroup of the symmetry group of the initial phase. The symmetry of the distortion phase is independent of the magnitude of the atomic displacements at least as long as the shifted atoms do not occupy any singular symmetry positions which in the crystal might appear in displacements of the order of interatomic distances. In order to describe distortion transformations, we use the phenomenological Landau theory of phase transitions; taking into account changes in the symmetry which arise in the transition is the main feature of this theory.

In phase transitions of the second type, the symmetry of the final phase is unrelated to the symmetry of the initial phase by subgroup relations, and the atoms perform diffusion movements over distances of the order of the interatomic distances in such a phase transition. Such structural transitions are usually termed reconstructive transitions. They are well-pronounced phase transitions of the first type. In analyzing such transitions, one usually assumes that the energy concepts are much more important than the symmetry concepts.

The microscopic mechanisms of these two types of phase transitions are different. It is generally accepted that many distortion-type phase transitions pass through a soft-mode mechanism and are related to an instability of the phonon spectrum of the initial (highly symmetrical) phase. Owing to the electron–lattice interaction or an anharmonicity, the phonon frequencies of the crystal become dependent upon the temperature, and the eigenfrequency of some of the phonons tends to zero when a certain temperature T_c is approached. Such phonons are termed soft phonons [143]. No energy is required for exciting the soft phonon mode at $T = T_c$, and therefore these modes arise in a macroscopic amount, corresponding to the appearance of statistical displacements of the atoms in the initial crystal. One also says that the soft mode "freezes" in the initial crystal (or condenses in it) and, hence, leads to a new critical phase. In this way the distortion phase is determined by the symmetry of the soft-phonon mode. This soft mode can interact with the acoustic long-wave mode and leads to the appearance of spontaneous macroscopic deformations which are accompanied by relative displacements of the atoms.

As has been described in Section 4, any phonon mode is characterized by an irreducible representation of the space group of the crystal; therefore the symmetry of the distortion-type phase can be defined by the irreducible representation of the initial crystal's group, which describes the soft mode. Naturally, these concepts lead to the conclusion that for describing a structural phase transition of the distortion type, one must consider the expression for the free energy of the crystal in the form of an expansion in amplitudes of the normal vibrations of the initial crystal after separating, first of all, the normal vibration corresponding to the soft mode.

The problem of stating the free energy of the crystal near the point of a phase transition can be approached in the following more constructive way. Assume that the space group changes from G_0 to G_1 as a consequence of a phase transition. We express the coordinates $\mathbf{R}_{ni}$ of the atoms in the distortion phase through the coordinates $\mathbf{R}_{ni}°$ in the symmetric phase and by the displacements $\mathbf{u}_{ni}$:

$$\mathbf{R}_{ni} = \mathbf{R}_{ni}^{\circ} + \mathbf{u}_{ni}, \tag{6.1}$$

and expand the vector $\mathbf{u}_{ni}$ in series in the basis functions of the irreducible representations of the group G_0. As has been described in Section 4, the irreducible representation of the space group is characterized by the star of the wave vector **k** and the number ν of the irreducible representation. We assume that the resulting displacements are characterized by a certain star of the wave vector {**k**} (this is usually the case in the known structural transitions). In the expansion of the vectors $\mathbf{u}_{ni}$ one must include only those irreducible representations of the group G_0 which belong to this star. Therefore, the expansion of the displacements must have the following form:

$$\mathbf{u}_{ni} = \sum_{\nu} \sum_{L\lambda} C_{L\lambda}^{\nu} \, \boldsymbol{\varepsilon}\begin{pmatrix} \mathbf{k}_L \, \nu \\ \lambda \end{pmatrix} i \Big) \, e^{i\mathbf{k}_L \mathbf{t}_n}. \tag{6.2}$$

where $\varepsilon(\mathbf{k}\,{}_{\lambda}^{L\nu} \mid i)$ denotes the atomic component of the basis function of the νth irreducible representation at the arm $\mathbf{k}_L$. A calculation for a particular crystal is possible with Eqs. (4.22) and (4.24). Obviously, only those irreducible representations of the group G_0 which appear in the composition of the mechanical representation must be included in the expansion of Eq. (6.2) (see Section 4).

Equation (6.2) is a precise equation and expresses any atomic displacements in a crystal through the mixing coefficients $C_{L\lambda}{}^{\nu}$ of the basis functions. In accordance with the concept of a soft mode, it is obvious that only one term corresponding to the irreducible representation describing the soft mode must remain in the sum over the irreducible representations ν in Eq. (6.2). As a result, the expression for the atomic displacements substantially simplifies and encompasses only the sum over the basis functions of the space group, which are numbered by the composite subscript $L\lambda$:

$$u_{nt} = \sum_{L\lambda} C_{L\lambda}^{[\nu]} \varepsilon \left(\begin{matrix} \kappa_L & \nu \\ \lambda & \end{matrix} \middle| i \right) e^{i\kappa_L t_n}. \tag{6.3}$$

The irreducible representation corresponding to the soft mode is termed the relevant representation and, obviously, defines the symmetry of the distortion phase.

The expansion of the physical quantities characterizing the system (quantities resulting from a continuous phase transition) in basis functions of irreducible representations for the initial phase was suggested by Landau on the basis of the general theory of phase transitions of the second kind or of those of the first kind which are close to the second kind. The mixing coefficients of the basis functions form a multi-component parameter of the order $\{C_\lambda{}^{L\nu}\}$. This parameter is equal to zero for $T > T_c$ and differs from zero at $T < T_c$. At the transition point proper, the parameter is equal to zero for phase transitions of the second kind and finite in the case of phase transitions of the first kind. (Let us recall that all considerations are valid for phase transitions when a subgroup relation exists between the groups G_0 and G_1.)

Expansion of the free energy. It follows from a comparison of Eq. (6.3) with the expansion of the atomic displacements in normal vibration modes that the amplitudes of the normal vibrations of the soft phonon modes are the order parameters in a structural transition. The order parameters are the system's macroscopic characteristics through which, according to Landau, the free energy of the crystal can be expressed. Near a phase transition, where all components of the order parameter are small, the free energy F can be expanded in series in powers of these components. This expansion can be stated in the following symbolic form [59]:

$$F = F_0 + \frac{1}{2} r \sum_\lambda C_\lambda^2 + \frac{1}{3} \sum_p v_p f_p^{(3)} \{C_\lambda\} + \frac{1}{4} \sum_p u_p f_p^{(4)} \{C_\lambda\} + \dots \tag{6.4}$$

where $f_f^{(3)}$, $f_p^{(4)}$, ... denote third-degree, fourth-degree, etc. polynomials of the components of the order parameter. Each polynomial with a fixed subscript p is invariant relative to all elements of the space group G_0, and the polynomial expansion proper is obtained from the condition that the free energy F is invariant relative to the group G_0.

According to the definition of the order parameter, all C_λ are equal to zero for the initial phase and therefore F_0 represents the energy of this phase. A phase transition in the system will take place when the coefficient at the quadratic term changes its sign at some temperature T_c. The temperature dependence r can be assumed to be linear in $T - T_c$:

$$r \sim (T - T_c), \tag{6.5}$$

whereas the coefficients at the higher terms v_p, u_p, ... can be assumed as weakly temperature-dependent.

Depending upon the specific form of the expansion of the free energy for a given crystal and upon the coefficients v_p, u_p, ..., a phase transition in the system can be of the second kind or of the first kind. In a transition of the first kind, the equilibrium value of the order parameter is at the transition point obtained from the condition that the free energy of the two phases must be equal and from the condition that the expansion of Eq. (6.4) must have an extremum:

$$\partial F / \partial C_\lambda = 0. \tag{6.6}$$

The extremum point must be a minimum for a phase transition of the second kind. It is generally accepted that a function of several variables has a minimum, provided that the determinants of the principal minors of the matrix of the second derivatives are positive, i.e.,

$$F_{11} > 0, \quad \begin{vmatrix} F_{11} & F_{12} \\ F_{21} & F_{22} \end{vmatrix} > 0, \quad \begin{vmatrix} F_{11} & F_{12} & F_{13} \\ F_{21} & F_{22} & F_{23} \\ F_{31} & F_{32} & F_{33} \end{vmatrix} > 0, \tag{6.7}$$

where $F_{\lambda\lambda'} = \partial^2 F / \partial C_\lambda \partial C_{\lambda'}$ at C_λ values satisfying Eq. (6.6).

Let us illustrate the general scheme of analyzing a phase transition with the Landau theory by the example of a structural transition in a crystal of the perovskite type, $SrTiO_3$. Perovskites were selected to illustrate the symmetry techniques of the theory of lattice vibrations and phase transitions in Sections 4-6. Structural phase transitions in them will be considered fully in Chapter 3.

A phase transition from the cubic phase $O_h{}^1$ to the tetragonal phase D_{4h} occurs at $T_c = 105$ K in $SrTiO_3$. It was established in experiments that basically the oxygen atoms forming octahedra are shifted in such a structural transition, and the displacements are in a first approximation described by a rotation of the octahedra over a certain angle around the

⟨001⟩ directions. The octahedra are rotated to opposite sides in neighboring cells in directions perpendicular to the axis of rotation and parallel to it (see Figs. 41 and 42). It is readily recognized that such a shifting in the crystal is described by a single-arm star $\{\mathbf{k}_{13}\}$ with $\mathbf{k} = 1/2(\mathbf{b}_1 + \mathbf{b}_2 + \mathbf{b}_3)$ which corresponds to the point **R** of the Brillouin zone of the initial crystal. The observed displacement must be described by the basis functions of the irreducible representations of the group $G_{\mathbf{k}}$, which are compiled in Table 2. It follows from this table that the third basis function of the three-dimensional irreducible representation τ_8 also describes the displacement observed in $SrTiO_3$. Consequently, the relevant irreducible representation accounting for the particular structural transition is τ_8 of the star $\{\mathbf{k}_{13}\}$. The symmetry of the soft mode leading to the structural transition also corresponds to this representation.

We use the Landau method to describe the thermodynamics of a structural transition. The free energy for a three-component order parameter $\{C_1, C_2, C_3\}$ transforming according to the representation τ_8 has the form

$$F = F_0 + \frac{1}{2} r (C_1^2 + C_2^2 + C_3^2) + \frac{1}{4} u_1 (C_1^4 + C_2^4 + C_3^4) + \\ + \frac{1}{2} u_2 (C_1^2 C_2^2 + C_1^2 C_3^2 + C_2^2 C_3^2). \tag{6.8}$$

The equation $\partial F/\partial C_\lambda = 0$ has the following three solutions for which the conditions of the minimum of the free energy are stated in parentheses:

$$\{00\,C\}\colon C^2 = -r/u_1 \\ (u_1 > 0,\ u_2 - u_1 > 0); \tag{6.9}$$

$$\{CC0\}\colon C^2 = -r/(u_1 + u_2) \\ (u_1 > 0,\ u_1^2 - u_2^2 > 0,\ u_2 - u_1 > 0); \tag{6.10}$$

$$\{CCC\}\colon C^2 = -r/(u_1 + 2u_2) \\ (u_1 > 0,\ u_1 - u_2 > 0,\ u_1 + u_2 > 0,\ u_1 + 2u_2 > 0). \tag{6.11}$$

It is easy to verify that the system of inequalities (6.10) is inconsistent, which means that a phase of the type $\{CC0\}$ is unstable. The stability regions of the two phases $\{00C\}$ and $\{CCC\}$ on the planes of the parameters u_1 and u_2 are indicated in Fig. 4. Thus, depending upon the parameters at the quaternary terms in the expansion of the free energy, one or the other of the two phases exists. As indicated above, the $\{00C\}$ phase is observed in crystals of the type of $SrTiO_3$. The $\{CCC\}$ phase, which corresponds to rhombohedral distortion of the crystal, is observed

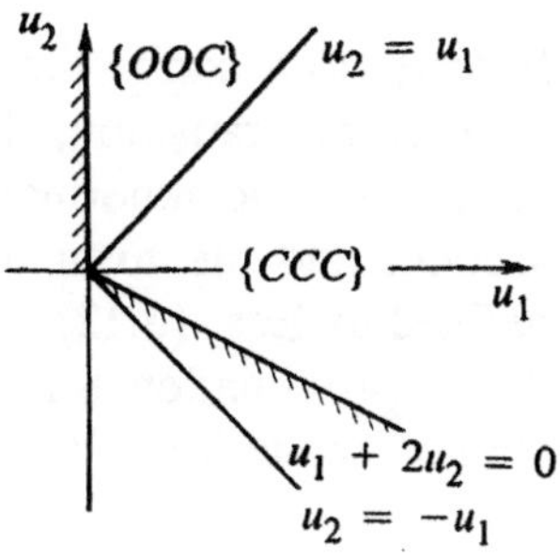

Fig. 4. Stability regions of the phases corresponding to the free energy of Eq. (6.8) in the approximation of an average field.

in crystals of the type of $PrAlO_3$. In both cases, the observed structural transition is a phase transition of the second kind. This agrees with the theoretical temperature dependences of the order parameter as expressed by Eqs. (6.9) and (6.11): C vanishes at the point of the phase transition $T = T_c$.

Neutron scattering at a soft mode. We leave now the phenomenological description of the phase transition for some time and consider the problem of a soft mode on the microscopic level. Information on the phonon states in a crystal is contained in the phonon Green's function defined by Eq. (3.1). This definition is suitable for the harmonic approximation in the description of the crystal vibrations, as well as for introducing anharmonic effects or the interaction of vibrations with other degrees of freedom, e.g., electrons. In the harmonic approximation, the Green's function of Eq. (3.1) is given by Eq. (3.16), which can be restated in the form

$$G^{\circ\alpha\beta}_{ni,n'i'}(\omega)=\frac{1}{\sqrt{M_iM_{i'}}}\,\frac{1}{N}\sum_{\mathbf{q}J}\overset{*}{e}{}^{\alpha}_{J}(\mathbf{q}\,|\,i)e^{\beta}_{J}(\mathbf{q}|i')\,e^{-i\mathbf{q}(\mathbf{R}^{\circ}_{n}-\mathbf{R}^{\circ}_{n'})}G^{\circ}_{J}(\mathbf{q},\,\omega), \quad (6.12)$$

where

$$G^{\circ}_{J}(\mathbf{q},\ \omega) = 1/[\omega^2 - \omega^2_J(\mathbf{q})]. \quad (6.13)$$

When anharmonicity is brought into account, the expression for $G_J{}^{\circ}(\mathbf{q}, \omega)$ is replaced by

$$G^{\circ}_{J}(\mathbf{q},\ \omega) = 1/[\omega^2 - \omega^2_J(\mathbf{q}) - \Pi_J(\mathbf{q},\ \omega)], \quad (6.14)$$

where $\Pi_J(\mathbf{q},\omega)$ denotes the polarization operator leading to a shift of the harmonic frequencies and to their attenuation as a function of temperature.

In order to describe the crystal vibrations at a high temperature and near a structural transition, the pseudoharmonic approximation is often used; in it, only the renormalization of the phonon frequencies is taken into account and the form of the Green's function coincides with the form defined by Eqs. (6.12) and (6.13) of the harmonic approximation. We use this approximation for calculating the cross section of coherent single-phonon scattering. We write on the basis of Eq. (3.61),

$$\frac{d^2\sigma^{\text{kor}}}{d\Omega d\varepsilon}=\frac{k}{k_0}\sum_{ll'} b_l b_{l'}\, e^{-W_l-W_{l'}}\, e^{i\varkappa(\mathbf{R}_l^0-\mathbf{R}_{l'}^0)}\varkappa^\alpha\varkappa^\beta\times$$
$$\times\frac{1}{e^{\beta\varepsilon}-1}\,\frac{1}{\pi}\,\mathrm{Im}\,G_{ll'}^{\alpha\beta}(-\varepsilon/\hbar). \tag{6.15}$$

After substituting an expression of the form of Eq. (6.12) instead of $G_{ll'}{}^{\alpha\beta}$, we rewrite the equation for the cross section in the following form:

$$\frac{d^2\sigma}{d\Omega d\varepsilon}=\frac{k}{k_0}\sum_{\mathbf{q}J}\sum_{\mathbf{b}}\delta_{\varkappa-\mathbf{q},\,\mathbf{b}}\,|\varkappa\xi_J(\varkappa)|^2\,S_J(\mathbf{q},\,-\varepsilon/\hbar), \tag{6.16}$$

where

$$\xi_J(\varkappa)=\sum_i b_i\,\frac{e^{-W_i}}{\sqrt{M_i}}\,e^{i\varkappa x_i}\,\mathbf{e}_J(\mathbf{q}\,|\,i); \tag{6.17}$$

$$S_J(\mathbf{q},\,\omega)=\frac{1}{e^{-\beta\hbar\omega}-1}\,\frac{1}{\pi}\,\mathrm{Im}\,G_J(\mathbf{q},\,\omega). \tag{6.18}$$

The vector quantity $\xi_J(\varkappa)$ is the structure amplitude of the inelastic single-phonon scattering, and $S_J(\mathbf{q},\ \omega)$ denotes a dynamic factor which is proportional to the generalized dynamic susceptibility of the system.

Equation (6.16) is additive in the phonon modes which are numbered by the subscript J. We separate the contribution describing scattering at a soft mode in a symmetric crystal phase at $T > T_c$ (the subscript of this mode will be omitted in what follows). Let us consider the dynamic factor $S(\mathbf{q},\ \omega)$ corresponding to this mode. The Green's function for the soft mode is stated in the form

$$G(\mathbf{q},\ \omega)=1/[\omega^2-\omega^2(\mathbf{q})+i\Gamma\omega] \tag{6.19}$$

and we assume that the frequency $\omega^2(\mathbf{q})$ at small $\mathbf{q}$ has the following dependence on both the wave vector and the temperature:

$$\omega^2(\mathbf{q}) = \omega^2(0) + \sum_{\alpha\beta} T^{\alpha\beta} q_\alpha q_\beta; \tag{6.20}$$

$$\omega^2(0) \sim (T - T_c) \tag{6.21}$$

(**q** is calculated from the wave vector **k** of the superlattice appearing below T_c). The attenuation of the soft mode is accounted for by **Γ**. This form of the imaginary part of $G(\mathbf{q}, \omega)$ is a generalization of the Green's function $G^\circ(\mathbf{q}, \omega)$ given by Eq. (6.13) after introducing an infinitely small imaginary addition to the frequency ($\omega \rightarrow \omega + i\delta$) and subsequent replacement of δ by the finite quantity **Γ**. Here we are not interested in the physical nature of the attenuation of the soft mode and that assume **Γ** is a phenomenological parameter that does not depend on the temperature or on the wave vector.

In the neighborhood of the phase transition ($T - T_c << T_c$), we have $\hbar\omega(0) << kT_c$, and therefore the exponential in Eq. (6.18) can be expanded in a series of the small quantity $\beta\hbar\omega$ for transition energies $\omega \sim \omega(0)$; the dynamic factor of the soft phonon assumes the form

$$S(\mathbf{q}, \omega) = \frac{kT}{\hbar} \cdot \frac{1}{\pi} \cdot \frac{\Gamma}{[\omega^2 - \omega^2(\mathbf{q})]^2 + \Gamma^2\omega^2}. \tag{6.22}$$

By virtue of Eq. (6.16), the dependence of the cross section of scattering at the soft mode on the transferred energy ε is described by Eq. (6.22). When we consider the conditions for the minimum of the denominator of the expression, we observe that two different situations can develop depending on the relation between the attenuation **Γ** and the frequency $\omega(\mathbf{q})$ of the soft mode. For $\omega^2(\mathbf{q}) > (1/2)\gamma^2$, Eq. (6.22) has two symmetric peaks at the frequencies $\omega = \pm\,\omega(\mathbf{q})$, and at $\omega^2(\mathbf{q}) < (1/2)\Gamma^2$ there is one peak at the frequency $\omega = 0$. The width of the peaks depends upon the relations between the parameters $\omega^2(\mathbf{q})$ and Γ^2. It is easy to state the analytic expression of Eq. (6.22) in the limit $(1/2)\Gamma^2 >> \omega^2(\mathbf{q})$:

$$S(\mathbf{q}, \omega) = \frac{kT}{\hbar\,\omega^2(\mathbf{q})} \cdot \frac{1}{\pi} \cdot \frac{\Gamma'}{\omega^2 + \Gamma'^2}, \tag{6.23}$$

where

$$\Gamma' = \omega^2(\mathbf{q})/\Gamma. \tag{6.24}$$

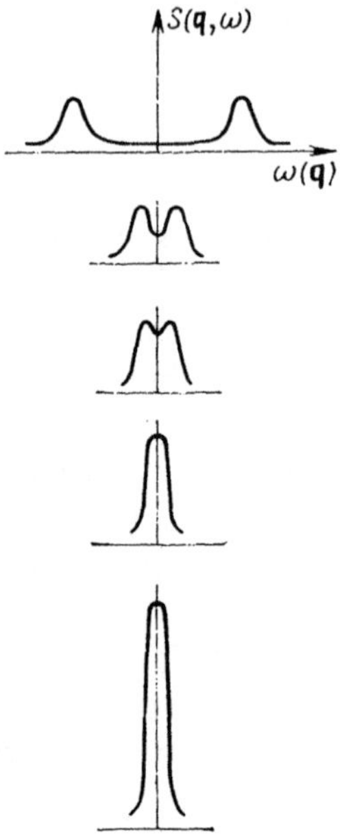

Fig. 5. Evolution of the scattering cross section at an attenuating soft mode on approaching I_c.

In the case of arbitrary relations between $\omega^2(\mathbf{q})$ and Γ^2, the evolution of the function of Eq. (6.22) with temperature (which modifies the relation between $\omega^2(\mathbf{q})$ and Γ^2) is schematically indicated in Fig. 5. Far from T_c, there are two peaks corresponding to scattering with absorption or with emission of a soft phonon. The peaks merge in proportion to the approach to T_c and become a single peak at $\omega^2(\mathbf{q}) = (1/2)\Gamma^2$.

Farther on, the width of this peak decreases in proportion to the decease in the frequency $\omega^2(\mathbf{q})$ while $\mathbf{q}$ is fixed, but the height of the peak grows correspondingly like $\sim [1/\omega^4(\mathbf{q})]$. It will be shown below (Chapter 3) that the result expressed by Eq. (6.22) can be phenomenologically interpreted in a simple way: the dynamic factor $S(\mathbf{q}, \omega)$ corresponds to the response function of a damped oscillator. When the two peaks of the response function merge into a single peak, one speaks of an over-attenuated oscillator.

The experimental results obtained at crystals experiencing a phase transition with distortions confirm the pattern illustrated in Fig. 5, but the situation is more complicated in a number of cases (see Chapter 3): When there are two separated peaks corresponding to scattering at a soft phonon, a central peak is observed and its intensity increases in proportion to the approximation to T_c. The central peak was explained in a large number of papers and, as a result, it was shown that this effect can be described by introducing frequency-dependent attenuation of the soft mode. For this purpose, it suffices to select the polarization operator in the Green's function of a soft phonon [113] in the form

$$\Pi(\mathbf{q}, \omega) = \Delta(q) - i\omega\Gamma - \frac{\gamma\delta^2}{\gamma - i\omega}, \tag{6.25}$$

where γ and δ denote some parameters with the dimension of frequency. This leads to the following form of the dynamic factor:

$$S(\mathbf{q}, \omega) = \frac{kT}{\hbar}\frac{1}{\pi}\frac{\Gamma + \dfrac{\delta^2\gamma}{\omega^2+\gamma^2}}{\left[\omega^2 - \omega_\infty^2(\mathbf{q}) + \dfrac{\delta^2\gamma^2}{\omega^2+\gamma^2}\right]^2 + \omega^2\left[\Gamma + \dfrac{\delta^2\gamma}{\omega^2+\gamma^2}\right]^2}, \tag{6.26}$$

where $\omega_\infty{}^2(\mathbf{q}) \equiv \omega^2(\mathbf{q}) + \Delta(\mathbf{q})$.

As will be shown in Chapter 3, Eq. (6.26) corresponds to the response function of a damped oscillator interacting with some degree of freedom and having the relaxation time $\tau = 1/\gamma$; the quantity δ is a measure of the coupling.

Equation (6.26) defines two lateral peaks and one central peak. This can be analytically recognized when the damping coefficients $\mathbf{\Gamma}$ and γ are constant:

$$\Gamma \ll \delta^2/\gamma, \quad \omega_\infty^2 \gg \gamma. \tag{6.27}$$

Then $S(\mathbf{q}, \omega)$ can be represented as a sum of two terms:

$$S(\mathbf{q}, \omega) = \frac{kT}{\hbar\pi}\left\{\frac{\Gamma}{[\omega^2 - \omega_\infty^2(\mathbf{q})]^2 + \Gamma^2\omega^2} + \frac{\delta^2}{\omega_0^2(\mathbf{q})\,\omega_\infty^2(\mathbf{q})}\cdot\frac{\gamma'}{\omega^2+\gamma'^2}\right\}, \tag{6.28}$$

where

$$\omega_0^2(\mathbf{q}) = \omega_\infty^2(\mathbf{q}) - \delta^2; \tag{6.29}$$

$$\gamma' = \gamma\,(\omega_0^2(\mathbf{q})/\omega_\infty^2(\mathbf{q})). \tag{6.30}$$

The first term coincides with the previously examined expression (6.22), whereas the second term describes the central peak whose width decreases (while its height increases) in proportion to the approximation to T_c when

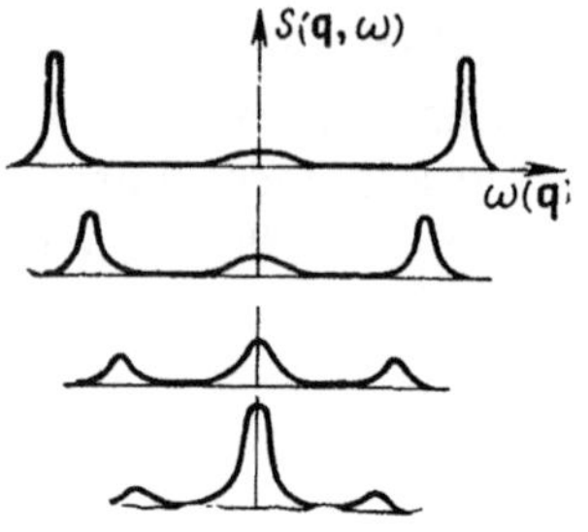

Fig. 6. Evolution of the cross section of scattering at an attenuating soft mode when the interaction of the soft mode with a degree of freedom having relaxation is taken into account. The approach to T_c at decreasing temperatures corresponds to a transition from the upper curves to the lower curves.

$\omega_0^2(\mathbf{q}) \to 0$ at $\mathbf{q} = 0$. The ratio of the intensities I_c of the central peak and I_{ph} of the two lateral peaks is given by the expression

$$I_c/2I_{ph} = \delta^2/\omega_0^2(\mathbf{q}). \tag{6.31}$$

The evolution of the function with the temperature upon approaching T_c is in the general case described by curves of the type illustrated in Fig. 6. The physical mechanism of coupling a soft mode with a degree of freedom having relaxation will be discussed in Chapter 3.

Now let us return to the general Eq. (6.16) for the cross section of single-phonon scattering. This expression gives the angular and energy dependencies of the cross section, including the neighborhood of the phase transition. When we restrict our considerations to a neighborhood close to T_c and small momenta $\mathbf{q}$, Eq. (6.16) can be integrated over the energy ε while the ε dependence of $\mathbf{q}$ and $\varkappa$ remains disregarded. We separate the contribution of the soft mode from Eq. (6.16). We obtain an approximation of the angular distribution of the scattering cross section at the soft mode

$$\frac{d\sigma}{d\Omega} = \sum_J |\varkappa\xi(\varkappa)|^2 \frac{kT}{\hbar} \int_{-\infty}^{\infty} \frac{d\omega}{\omega} \left\{\frac{1}{\pi} \operatorname{Im} G(\mathbf{q}, \omega)\right\}. \tag{6.32}$$

In the case of the Green's function $G^\circ(\mathbf{q}, \omega)$ of the harmonic approximation, the above integral is equal to $1/\omega^2(\mathbf{q})$. Assuming that for the Green's function of Eq. (6.19), in which the attenuation of the soft mode was taken into account, the normalization

$$\int_{-\infty}^{\infty} \frac{d\omega}{\omega} \frac{1}{\pi} \operatorname{Im} G(\mathbf{q}, \omega) = \frac{1}{\omega^2(\mathbf{q})}, \tag{6.33}$$

is approximately satisfied, we obtain in place of Eq. (6.32) the following expression:

$$\frac{d\sigma}{d\Omega} = \sum_J |\varkappa \xi_J(\varkappa)|^2 \frac{kT}{\omega^2(\mathbf{q})}, \tag{6.34}$$

where $\omega^2(\mathbf{q})$ is defined by Eqs. (6.20) and (6.21).

It follows from the result expressed by Eq. (6.24) that for $T \to T_c$ a peak develops in the angular dependence of the cross section near the point of the Brillouin zone (where a soft phonon exists). Besides that, when the wave vector $\mathbf{q}$ is fixed, a sharp increase in the cross section occurs as the temperature approaches T_c. Thus, Eq. (6.34) reflects all the features of the well-known critical scattering resulting from fluctuations of the order parameter.

Starting from the microscopic pattern of the vibrations of the crystal lattice, we obtained Eq. (6.34) for the angular distribution of the scattering cross section near T_c. First the double differential cross section giving also the energy distribution was calculated. Based on the macroscopic fluctuation pattern, we obtain in the following section an expression analogous to Eq. (6.34) for the cross section of critical scattering. We will not be interested in the energy distribution but will consider interaction effects of the fluctuations and their appearance in the angular distribution of the critical scattering.

7. The Critical Scattering of Neutrons

Fluctuations of the order parameter. Let us consider the cross section of neutron scattering at a crystal near the point of a structural phase transition. It is generally accepted that, when in the general case the phase-transition point of the second kind is approached, fluctuations of the order parameter develop in the system and the extent of the fluctuations is given by the correlation length ξ. The correlation length diverges at the transition point proper, and this means that long-range order develops in the system or, as one occasionally says, the symmetry of the system is spontaneously disturbed. The fluctuations of the order parameter lead to the so-called critical scattering of the neutrons, and its cross section increases sharply as the temperature T_c is approached. The fluctuations have macroscopic dimensions near T_c and macroscopic lifetimes, and therefore the critical scattering they cause is concentrated under small angles and appears to be quasi-elastic.

Assume that a structural phase transition involves the irreducible representation $\{\mathbf{k}\}\nu$ of the space group G of the crystal, where $\{\mathbf{k}\}$ denotes the star of the wave vector with the arms $\mathbf{k}_L$ ($L = 1, 2, \ldots, l_{\mathbf{k}}$). At $T < T_c$, the atomic displacements can be represented as a superposition of the basis functions of this irreducible representation:

$$\mathbf{u}_{ni} = \frac{1}{\sqrt{NM_i}} \sum_{L\nu} C_{L\lambda}\, \varepsilon\left(\begin{matrix}\mathbf{k}_L \nu \\ \lambda\end{matrix}\Big| i\right) e^{i\mathbf{k}_L \mathbf{t}_n} \tag{7.1}$$

[by contrast to Eq. (6.3), we separated the normalizing factor $(NM_i)^{-1/2}$]. The mixing coefficients $C_{L\lambda}$ are also the order parameters of the system. They are equal to zero above T_c but their instantaneous values may be nonvanishing. These instantaneous values of $C_{L\lambda}$ also describe the fluctuations of the order parameter. Their values averaged over time are equal to zero for $T > T_c$ so that, according to Eq. (7.1), the average values of the atomic displacements tend to zero.

Let us consider a wave packet of fluctuations of the order parameter, with wave vectors near the arms of the star $\{\mathbf{k}\}$ which characterizes the structure resulting from the transition; we write the instantaneous value of the atomic displacement at $T > T_c$ in the form

$$\mathbf{u}_{ni} = \frac{1}{\sqrt{NM_i}} \sum_{L\lambda} \sum_{\mathbf{q}} C_{L\lambda}(\mathbf{q})\, \varepsilon\left(\begin{matrix}\mathbf{k}_L \nu \\ \lambda\end{matrix}\Big| i\right) e^{i(\mathbf{k}_L+\mathbf{q})\mathbf{t}_n}. \tag{7.2}$$

where the summation over $\mathbf{q}$ is extended over small neighborhoods of each arm $\mathbf{k}_L$ of the star because it is generally accepted that only the long-wave fluctuations in the system are important.

Equation (7.2) is an approximation firstly because it contains Fourier components only for small sections of the Brillouin zone and, secondly, the polarization vectors $\varepsilon(\begin{smallmatrix}\mathbf{k}_L+\mathbf{q}\nu \\ \lambda\end{smallmatrix} \mid i)$ for the wave vector $\mathbf{k}_L + \mathbf{q}$ were replaced by the polarization vector $\varepsilon(\begin{smallmatrix}\mathbf{k}_L\nu \\ \lambda\end{smallmatrix} \mid i)$ of the center of gravity of the wave packet. This makes it possible to describe the entire packet of fluctuations by the relevant irreducible representation of the group G which is involved in the phase transition.

The probability $\rho\{C\}$ of a certain configuration of fluctuations to appear is given by the Gibbs distribution

$$\rho\{C\}=e^{-F\{C\}/kT}/\mathrm{Sp}\, e^{-F\{C\}/kT}, \tag{7.3}$$

where $F\{C\}$ denotes the Ginzburg–Landau functional which is the expansion of the free energy in powers of the (fluctuating) order parameter:

$$F\{C\}=\sum_{\mathbf{q}} A_\nu(\mathbf{q})\sum_{L\lambda} C^*_{L\lambda}(\mathbf{q})\, C_{L\lambda}(\mathbf{q})+\mathcal{H}_{int}\{C\}. \tag{7.4}$$

where $\mathcal{H}_{int}\{C\}$ denotes terms of third, fourth, etc. order in $C_{L\lambda}(\mathbf{q})$, with the coefficients v, u, ..., etc. weakly depending upon the temperature, and

$$A_\nu(\mathbf{q}) = a\,(T - T_c) + \sum_{\alpha\beta} q_\alpha q_\beta T^\nu_{\alpha\beta} + \dots \tag{7.5}$$

The term with $\mathbf{q} = 0$ in the sum of Eq. (7.4) corresponds to the Landau expansion of Eq. (6.4) in powers of the condensed order parameter. Terms depending upon q_α appear in $A_\nu(\mathbf{q})$ because the inhomogeneous three-dimensional distribution of the order parameter in the Ginzburg–Landau functional was taken into account.

Let us note that, since the atomic displacements $\mathbf{u}_{ni}$ are real, the quantities $C_{L\lambda}(\mathbf{q})$ satisfy the condition

$$C^*_{L\lambda}(\mathbf{q}) = C_{L\lambda}(-\mathbf{q}), \tag{7.6}$$

whereby the momentum conservation law is satisfied in the quadratic term of the expansion of Eq. (7.4). One must bear in mind that the total momentum must also be conserved in all terms of $\mathcal{H}_{\mathrm{int}}(C\}$.

Obviously the average values $C_{L\lambda}(\mathbf{q})$ of the order parameter with the distribution of Eq. (7.3) are equal to zero but the correlation functions differ from zero, particularly the single-particle function, which is defined as follows:

$$\langle C^*_{L\lambda}(\mathbf{q})\, C_{L'\lambda'}(\mathbf{q}')\rangle=\int \delta C \rho\{C\}\, C^*_{L\lambda}(\mathbf{q})\, C_{L'\lambda'}(\mathbf{q}'), \tag{7.7}$$

where $\delta C = \prod_{L\lambda}\prod_{\mathbf{q}} dC_{L\lambda}(\mathbf{q})$, and the integral is understood to be continuous. In the Gaussian approximation, in which interaction terms of the fluctuations in the function of Eq. (7.4) are disregarded, the continuous integral of Eq. (7.7) is reduced to the usual Gauss integrals and is equal to

$$\langle C^{*}_{L\lambda}(\mathbf{q})\, C_{L'\lambda'}(\mathbf{q}')\rangle_0 = \delta_{LL'}\,\delta_{\lambda\lambda'}\,\delta_{\mathbf{q}\mathbf{q}'}\,(kT/A_\nu(\mathbf{q})) \tag{7.8}$$

(the subscript zero denotes Gaussian approximation of the correlation function).

The quantity $1/A_\nu(\mathbf{q})$ is the generalized susceptibility and is in the isotropic case reduced to the well-known Ornstein–Zernicke function:

$$1/[q^2 + (1/\xi)^2],\quad \xi^2 = 1/[a\,(T - T_c)], \tag{7.9}$$

where ξ denotes the correlation length. As we will see below, the result given by Eq. (7.8) defines the cross section of scattering at fluctuations.

The cross section of critical scattering. We use the general equation (3.60) which expresses the cross section of neutron scattering through the correlator of the atomic displacements and substitute the expansion for the fluctuating displacements of the atoms as per Eq. (7.23). The resulting expression for the cross section is stated in the form

$$\frac{d^2\sigma}{d\Omega d\varepsilon} = \frac{k}{k_0}\sum_{ij} b_i b_j \frac{e^{-W_i - W_j}}{\sqrt{M_i M_j}}\, e^{i\varkappa(\mathbf{x}_i - \mathbf{x}_j)}\,\frac{1}{N}\sum_{\mathbf{q}\mathbf{q}'}\;\sum_{L\lambda L'\lambda'}\;\sum_{\alpha\beta}\varkappa^\alpha \varkappa^\beta \times$$
$$\times\, \overset{*}{\varepsilon}{}^{\alpha}\left(\begin{matrix}\kappa_L\,\nu\\ \lambda\end{matrix}\middle| i\right)\varepsilon^{\beta}\left(\begin{matrix}\kappa_{L'}\,\nu\\ \lambda'\end{matrix}\middle| j\right) N\sum_{\mathbf{b}}\delta_{\varkappa-\kappa_L-\mathbf{q},\,\mathbf{b}}\; N\sum_{\mathbf{b}'}\delta_{\varkappa-\kappa_{L'}-\mathbf{q}',\,\mathbf{b}'}\times$$
$$\times\frac{1}{2\pi\hbar}\int_{-\infty}^{\infty} e^{-\frac{1}{\hbar}\varepsilon t}\,\langle C^{*}_{L\lambda}(\mathbf{q}, t)\, C_{L'\lambda'}(\mathbf{q}', 0)\rangle\, dt. \tag{7.10}$$

The product of the two δ functions is nonvanishing only for $L = L'$ because two arms $\mathbf{k}_L$ and $\mathbf{k}_{L'}$ of a star cannot differ on the vector of the reciprocal lattice and the wave vectors $|\mathbf{q}|$ and $|\mathbf{q}'|$ are small compared with $|\mathbf{k}_L|$. Let us consider the static limit in the correlation of the fluctuations when the dependence of the order parameter upon the time t can be disregarded. In this case, the integral over the time is equal to the $\delta(\varepsilon)$ function. Further, taking into account that the correlation function of Eq. (7.8) is diagonal in the subscripts L, λ, and $\mathbf{q}$, we can rewrite Eq. (7.10) in the following form:

$$\frac{d^2\sigma}{d\Omega d\varepsilon} = N \sum_{L\lambda} |\varkappa \xi_{\nu\lambda}(\varkappa)|^2 \sum_{\mathbf{q}} \langle C^*_{L\lambda}(\mathbf{q}) C_{L\lambda}(\mathbf{q})\rangle \times \sum_{\mathbf{b}} \delta_{\varkappa - \kappa_L - \mathbf{q},\, \mathbf{b}} \times \delta(\varepsilon), \quad (7.11)$$

where we introduced the notation

$$\xi_{\nu\lambda}(\varkappa) = \sum_i b_i \frac{e^{-W_i}}{\sqrt{M_i}} e^{i\varkappa \mathbf{x}_i} \varepsilon\left(\begin{matrix}\varkappa\nu\\ \lambda\end{matrix}\middle| i\right). \quad (7.12)$$

By integrating over the energy ε, we obtain the cross section $d\sigma_\varkappa/d\Omega$ for the critical scattering of neutrons over a Bragg superlattice reflection $\varkappa_0 \equiv \mathbf{b} + \mathbf{k}_L$:

$$\frac{1}{N} \cdot \frac{d\sigma_\varkappa}{d\Omega} = \sum_\lambda |\varkappa_0 \xi_{\nu\lambda}(\varkappa_0)|^2 \frac{kT}{a(T - T_c) + \sum_{\alpha\beta} q^\alpha q^\beta T_\nu^{\alpha\beta}}, \quad (7.13)$$

where $\mathbf{q} = \varkappa - \varkappa_0$.

Let us note the great similarity between Eqs. (7.13) and (6.34) which were obtained in the macroscopic and microscopic descriptions of the neighborhood of the phase transition. This similarity is converted into an identity when Eq. (7.5) for $A_\nu(\mathbf{q})$, the self-energy of the fluctuations, and Eq. (6.2) for $\omega^2(\mathbf{q})$, the square of the frequency of the soft mode, are compared. The two approaches to the description of the structural phase transition coincide, provided that we assume $\omega^2(\mathbf{q}) \propto A(\mathbf{q})$.

The following should be noted in regard to the derivation of Eq. (7.13). The expression was obtained under the assumption of a static approximation when the fluctuations were taken into account, i.e., when the spectral density of the correlator of the fluctuations was reduced to the $\delta(\varepsilon)$ function. As a matter of fact, Eq. (7.13) holds even in the case of a more general form of the spectral density of the fluctuation correlator, even when there are two peaks corresponding to soft phonons. It is important that the Fourier integral appearing in Eq. (7.10) was represented as a product of a static correlator $\langle C_{L\lambda}{}^*(\mathbf{q})\ C_{L\lambda}(\mathbf{q})\rangle$ with some frequency function normalized to unity. This follows from the hypothesis of dynamic scaling and will be discussed on the example of dynamic magnetic fluctuations (see Section 17) and also when the experimental results on the critical scattering in structural phase transitions are expounded (see Chapter 3).

Equation (7.13) describes a critical effect of the Ornstein–Zernicke type. The intensity of the critical scattering depends upon the quantity

$$F_\nu(\varkappa_0) = \sum_\lambda |\varkappa_0 \xi_{\nu\lambda}(\varkappa_0)|^2, \tag{7.14}$$

which was defined as the cross section of coherent single-phonon scattering corresponding to the irreducible representation ν. A representation through the Elliott–Thorpe tensor $B^{\alpha\beta}(\nu, \varkappa)$ of Eq. (5.14) [tensor defined by Eq. (5.15)] was obtained in Section 5 for this quantity.

The factor $F_\nu(\varkappa_0)$ determines the extinction in the case of single-phonon neutron scattering. We infer from Eq. (7.13) that this factor defines the extinctions of critical scattering at fluctuations which are characterized by the given irreducible representation of the group G. By studying the extinctions of the critical scattering near $\varkappa_0 = \mathbf{b} + \mathbf{k}_L$ for various $\mathbf{b}$, we can identify the critical fluctuations (i.e., we can find the number of the irreducible representation) corresponding to the phase transition.

Interaction effects of the fluctuations. When the single-particle correlation function of Eq. (7.8) was calculated, the interaction of the fluctuations was disregarded by restricting the considerations to the Gaussian approximation. As a result, according to Eq. (7.9), the correlation length near the point of the phase transition follows the law $\xi \propto (T - T_c)^{-1/2}$. At the same times, the behavior of the order parameter in the condensed phase is described in the Landau theory (see section 6) by the expression $C \propto (T - T_c)^{-1/2}$, with the fluctuations disregarded. As has been explained (see [59]), such a behavior of ξ and C is suitable only for a neighborhood sufficiently far from the point T_c. In the immediate vicinity of T_c, the behavior of these quantities changes considerably because of the strong interaction of the fluctuations whose density increases in proportion to the approach to the point of the phase transition.

It is assumed that the correlation length ξ, the order parameter C, the susceptibility χ, the heat capacity C_V, and the other thermodynamic quantities of the system near T_c follow a power law which is characterized by the critical indices ν, β, γ, α, etc.:

$$\xi \sim (T - T_c)^{-\nu},\ C \sim (T_c - T)^{\beta},\ \chi \sim (T - T_c)^{-\gamma},\ C_V \sim (T - T_c)^{-\alpha}. \tag{7.15}$$

An index η, which characterizes the three-dimensional behavior of the correlation function at the transition point proper, is of importance in addition to the critical indices introduced. Whereas for $T > T_c$ the correlation function decreases exponentially with the distance,

$$\langle C(R)\, C(0)\rangle \sim e^{-R/\xi} \tag{7.16}$$

(this expression also serves as the definition of the correlation length ξ), a power law is assumed at $T = T_c$,

$$\langle C(R)\, C(0)\rangle \sim \frac{1}{R^{d-2+\eta}}. \tag{7.17}$$

where d denotes the dimensionality of the space in which the system undergoing the phase conversion is located, and η denotes the critical correlation index.

The theory which disregards the interaction of the fluctuations [i.e., the term $\mathcal{H}_{int}\{C\}$ in the Ginzburg–Landau Hamiltonian of Eq. (7.4)] leads to the following values of the critical indices:

$$\nu = \frac{1}{2},\ \eta = 0,\ \beta = 1/2,\ \gamma = 1,\ \ \alpha = 0. \tag{7.18}$$

It was noted some time ago that very close to T_c, the critical indices differ substantially from the classical values of Eq. (7.18). Before a consistent theory of phase transitions accounting for the interaction of fluctuations had been developed, relations had been established for the critical indices on the basis of the similarity hypothesis. This hypothesis states that near T_c, where the correlation length is the greatest of all the system-characterizing parameters with the dimension of length, the free energy is a homogeneous function of the thermodynamic parameters. We state a few of these relations

$$\alpha + 2\beta + \gamma = 2,\ \ (2 - \eta)\,\nu = \gamma,\ \ 2 - \alpha = d\nu. \tag{7.19}$$

They make it possible to express all critical indices through a pair of indices which can be assumed to be independent. Usually the indices ν and η referring to the correlation length are considered such fundamental indices.

It was shown in numerous physics and computer experiments that the critical indices do not depend upon the specific interaction in the system and upon the specific form of the phase transition in it but rather depend only upon the dimension of the space and the number of components of the order parameter. The determination of the universal behavior of the critical indices is one of the most important theoretical problems.

At the present time there exist two methods of consecutively introducing the interactions of fluctuations in a phase transition: one method is based on the summation of so-called parquet diagrams (see [86]) and the other method is based on the group renormalization technique (see [15]). Both methods make use of the fact [60] that in the theory of phase transitions, the dimension $d = 4$ of the space is singled out insofar as for it the critical indices of the system coincide with the classical indices of Eq. (7.18).

Wilson has suggested that, when the dimensionality differs only slightly from 4, the critical indices will be changed only slightly. Based on this assumption, Wilson [36] has developed a mathematical calculation scheme in the form of an expansion in the parameter $\varepsilon = 4 - d$, which is assumed to be small. By switching in the final expressions to $\varepsilon = 1$, the values of the critical indices can be obtained for the real (three-dimensional) space.

Thus, the following expressions were obtained for the so-called vector model (n denotes the component parameter of order, anisotropy being disregarded):

$$\nu = \frac{1}{2} + \frac{n+2}{4(n+8)}\varepsilon, \quad \eta = \frac{n+2}{2(n+8)^2}\varepsilon^2. \tag{7.20}$$

It follows from the expressions cited that the theoretical results confirm that the critical behavior is universal: the critical indices depend only upon the number of the components of the order parameter and upon the dimension of the space.

Anisotropic interactions are another factor affecting the critical indices; there the symmetry of the anisotropic interactions rather than the values of the parameters in the Ginzburg–Landau Hamiltonian is important.

We consider as an example the critical indices for the structural phase transition in perovskites; the transition is characterized by the representation τ_8 of the star of the wave vector $\{\mathbf{k}_{13}\}$. The three-component order parameter determines the Ginzburg–Landau Hamiltonian of Eq. (6.8). Equation (7.20) of the vector model cannot be used in this case because anisotropic interactions are represented in the Hamiltonian.

The equilibrium phases are determined in Section 6 in the average-field approximation, and the stability regions of these phases on the planes

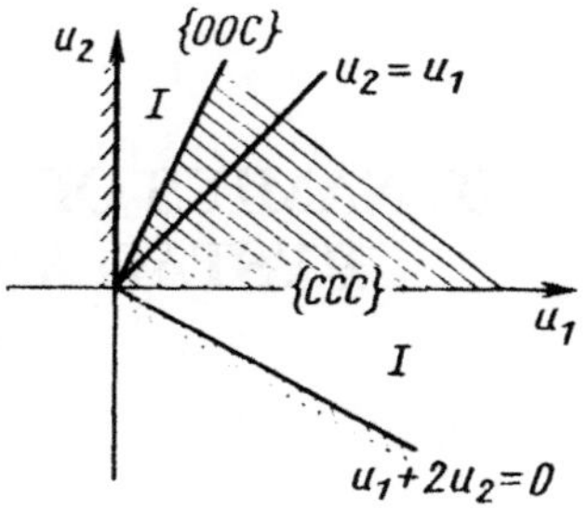

Fig. 7. Stability region of the phases corresponding to the free energy of Eq. (6.8) and form of the phase transition when the when the interaction of fluctuations is taken into account.

of the parameters u_1 and u_2 of the functional of Eq. (6.8) are indicated (see Fig. 4). The phase transition is a transition of the second kind for any values of the parameters u_1 and u_2. The form of the phase transitions defined by the Hamiltonian of Eq. (6.8) changes substantially when the fluctuations are brought into account (Fig. 7). Only the shaded region of Fig. 7 corresponds to phase transitions of the second kind. The regions which are adjacent to the boundaries of the stability interval of the functional for the free energy (denoted in Fig. 7 by the Roman number I) correspond to phase transitions of the first kind.

In the region of phase transitions of the second kind, the critical index for the correlation length turns out to be [15]

$$\nu=\frac{1}{2}+\frac{5}{44}\varepsilon. \tag{7.21}$$

This critical index coincides with the critical index of the vector model [Eq. (7.2)] when the number n of components of the order parameter is $n = 3$. Thus, at the value $n = 3$, an anisotropic interaction of cubic symmetry does not affect the critical behavior of the system: the critical index is defined only by the number n and the dimension of the space. An analysis shows that in the case of an n-component vector model with cubic anisotropy, the behavior at the critical point depends upon the number of components in the following way: at $n < 4$, there is coincidence with the behavior of the isotropic vector model and the critical index ν is given by Eq. (7.20); at $n > 4$, the critical behavior of the system is different and the index ν [15] is equal to

$$\nu=\frac{1}{2}+(n-1)\varepsilon/6n. \tag{7.22}$$

Chapter 2

DYNAMICS OF THE MAGNETIC SYSTEM OF A CRYSTAL AND INELASTIC SCATTERING OF NEUTRONS

In this chapter, various aspects of inelastic scattering of neutrons at crystals with a magnetic order are considered. The theory of spin waves in magnetic structures of any type is expounded in detail. General formulas for the cross section of single-quantum scattering of both polarized and non-polarized neutrons at spin waves are derived for magnetic materials which can be characterized through Shubnikov symmetry. Also the general formulas for the polarization vector of an inelastically scattered neutron beam are derived. Great attention is paid to the problems of the symmetry of crystals with magnetic order. A method of describing and deriving the spin-wave modes with the theory of symmetry is presented.

A review is given of the experimental results of research on the dispersion curves of spin waves in various classes of magnetic materials, namely dielectric materials, rare-earth metals, transition metals, quasi-one-dimensional systems, etc. The theoretical and experimental research made with the aid of neutron spectroscopy on nonlinear magnetic excitations, i.e., the solitons, is treated in a separate section.

The principles of the theory of critical scattering at the fluctuations of a magnetic order parameter are explained and the neutron-diffraction research of the critical parameters of physical quantities changing in a magnetic phase transition is reviewed.

The theoretical and experimental research on the critical dynamics of magnetic materials is reviewed. Results of this research are described with the hypothesis of dynamic scaling. This chapter is essentially an exhaustive exposition of the present state of neutron spectroscopy of magnetic materials.

8. The General Expressions for the Cross Section of Magnetic Scattering

The matrix element of the magnetic interaction. The magnetic interaction of a neutron with an atom must be understood as the interaction of the magnetic field generated by the neutron with the currents of the electrons constituting the unfilled shells of the atoms. The matrix element of this interaction between the states of the neutron with the moments $\mathbf{k}_0$ and $\mathbf{k}$ was calculated by Halpern and Johnson [189] as early as in 1939; their result is the basis of the entire theory of neutron diffraction at crystals with a magnetic order. A similar derivation of the matrix element for the entire crystal can be found in [43]; we state only the final expression:

$$V_{\mathbf{k}\mathbf{k}_0} = -\frac{4\pi\hbar^2}{m} r_0 \gamma \sum_l F_l(\varkappa) e^{-i\varkappa \mathbf{R}_l} (\mathbf{S}_l, \mathbf{S}_n - (\mathbf{e}\mathbf{S}_n)\mathbf{e}). \quad (8.1)$$

The notation is interpreted as follows: $\mathbf{S}_l$ denotes the spin of the lth atom; $\mathbf{S}_n$ denotes the spin of the neutron; r_0 denotes the electromagnetic radius of the electron, with $r_0 = e^2/mc^2$; $\gamma = -1.91$ denotes the magnetic moment of the neutron expressed in nuclear magnetons; $\varkappa$ denotes the scattering vector; and $\mathbf{e} = \varkappa/\varkappa$ denotes the unit scattering vector. The summation over l is extended over all magnetic atoms of the crystal. The quantity $F_l(\varkappa)$ is the magnetic form factor of the atom, i.e., the Fourier component of the spin density of the atom.

Equation (8.1) corresponds to the interaction of a neutron with a system of isolated atoms having localized magnetic moments. The case of metals in which a nonlocalized spin density exists will be discussed in Section 14. When an orbital magnetic moment of an atom in a crystal exists, the matrix element can also be calculated [247], but it does not have the simple form of Eq. (8.1) because it depends upon the wave functions of the electrons in the crystal. But in the case of rare-earth elements, in which the uncompensated f shells of the atoms are highly shielded by external electrons, the orbital contribution can be taken into account by the transition to the total moment $\mathbf{J}$ of the atom with the aid of the de Gennes formula:

$$\mathbf{S} \rightarrow (g-1)\,\mathbf{J}. \tag{8.2}$$

The expressions for the cross section. Elastic scattering. Let us substitute Eq. (8.1) in Eq. (1.10). We obtain a general expression for the cross section of magnetic scattering at a crystal:

$$\frac{d^2\sigma}{d\Omega d\varepsilon} = (r_0\gamma)^2 \frac{k}{k_0} \sum_{ll'} F_l(\varkappa) F_{l'}(\varkappa)\, e^{i\varkappa(\mathbf{R}_l^\circ - \mathbf{R}_{l'}^\circ)} \sum_{\alpha\beta} (\delta_{\alpha\beta} - e_\alpha e_\beta) \times$$

$$\times \frac{1}{2\pi\hbar} \int_{-\infty}^{\infty} dt e^{-i\frac{\varepsilon}{\hbar}t} \langle S_l^\alpha(t)\, e^{i\varkappa u_l(t)}\, S_{l'}^\beta(0)\, e^{-i\varkappa u_{l'}(0)} \rangle. \tag{8.3}$$

In deriving this expression, we averaged over the polarization states of the neutrons in the incident beam and therefore the factor $\delta_{\alpha\beta} - e_\alpha e_\beta$ appeared.

The quantities $u_l^\alpha(t)$ and $S_l^\alpha(t)$ are the Heisenberg representation of the displacement operator u_l^α and the spin operator S_l^α of the lth atom at the time t, i.e.,

$$u_l^\alpha(t) = e^{i\frac{\mathscr{H}}{\hbar}t} u_l^\alpha e^{-i\frac{\mathscr{H}}{\hbar}t}, \quad S_l^\alpha(t) = e^{i\frac{\mathscr{H}}{\hbar}t} S_l^\alpha e^{-i\frac{\mathscr{H}}{\hbar}t}. \tag{8.4}$$

When the interaction between the atomic displacements and the spins is ignored in the Hamiltonian of the scatterer, statistical averaging over the spins and the atomic positions can be independently performed and Eq. (8.3) can be rewritten in the following form:

$$\frac{d^2\sigma}{d\Omega d\varepsilon} = (r_0\gamma)^2 \frac{k}{k_0} \sum_{ll'} F_l(\varkappa) F_{l'}(\varkappa)\, e^{-W_l - W_{l'}}\, e^{i\varkappa(\mathbf{R}_l^\circ - \mathbf{R}_{l'}^\circ)} \times$$

$$\times \sum_{\alpha\beta} (\delta_{\alpha\beta} - e_\alpha e_\beta) \frac{1}{2\pi\hbar} \int_{-\infty}^{\infty} dt e^{-i\frac{\varepsilon}{\hbar}t} \langle S_l^\alpha(t)\, S_{l'}^\beta(0) \rangle. \tag{8.5}$$

In this approximation, the cross section of magnetic scattering is expressed through the Fourier component of the spin correlator.

Disregarding the spin correlations at the various instants of time corresponds to elastic scattering. Indeed, if

$$\langle S_l^{\alpha}(t)\, S_{l'}^{\beta}(0)\rangle = \langle S_l^{\alpha}\rangle \langle S_{l'}^{\beta}\rangle, \tag{8.6}$$

we obtain from Eq. (8.6) an expression which is proportional to the δ function:

$$\frac{d^2\sigma}{d\Omega d\varepsilon} = (r_0\gamma)^2 \sum_{ll'} F_l(\varkappa)\, F_{l'}(\varkappa)\, e^{-W_l - W_{l'}}\, e^{i\varkappa(\mathbf{R}_l^{\circ} - \mathbf{R}_{l'}^{\circ})} \times$$
$$\times \sum_{\alpha\beta} (\delta_{\alpha\beta} - e_\alpha e_\beta) \langle S_l^{\alpha}\rangle \langle S_{l'}^{\beta}\rangle\, \delta(\varepsilon).$$

By integrating this expression over the energy ε, we obtain the angular distribution of elastic magnetic scattering; this distribution is conveniently stated in the form

$$\frac{d\sigma}{d\Omega} = (r_0\gamma)^2 |\mathcal{F}(\varkappa)|^2, \tag{8.7}$$

where

$$\mathcal{F}(\varkappa) = \sum_l F_l(\varkappa)\, e^{-W_l}\, e^{-i\varkappa\mathbf{R}_l^{\circ}} \{\mathbf{S}_l - (\mathbf{e}\mathbf{S}_l)\,\mathbf{e}\} \tag{8.8}$$

[$\mathcal{F}(\varkappa)$ denotes the amplitude of elastic magnetic scattering]. In the latter expression, we expressed through $\mathbf{S}_l$ the classical vector with the components $\langle S_l^{\alpha}\rangle$ which are the average values of the spin operator at the particular temperature and at the atom l. Equations (8.7) and (8.8) are the initial equations in magnetic neutron diffraction. One must add to these expressions only the relation expressing the spin $\mathbf{S}_{ni}$ of the ith atom in the nth primitive cell of a crystal with a magnetic structure through the set of vectors $\mathbf{S}_{0i}^{L}$ given in the zeroth cell ($l \equiv ni$) [42]:

$$\mathbf{S}_{ni} = \sum_L e^{i\mathbf{k}_L \mathbf{t}_n}\, \mathbf{S}_{0i}^{L}. \tag{8.9}$$

The summation over L is extended over all arms $\mathbf{k}_L$ of the star of the wave vector, with the star characterizing the magnetic structure. The vectors $\mathbf{S}_{0i}^{L}$ are termed the arm contributions of the magnetic structure.

When we substitute Eq. (8.9) in Eq. (8.7) and extend the summation over the translations $\mathbf{t}_n$, we obtain the fundamental formula of magnetic neutron diffraction:

$$\frac{1}{N}\frac{d\sigma}{d\Omega} = (r_0\gamma)^2 \sum_L (\overset{*}{\mathbf{M}}{}^L_{\varkappa} \mathbf{M}^L_{\varkappa})\, N \sum_{\mathbf{b}} \delta_{\varkappa,\, \mathbf{b}+\kappa_L}, \tag{8.10}$$

where $\delta_{\varkappa,\mathbf{b}+\mathbf{k}_L}$ denotes the Kronecker symbol indicating the conditions for the development of magnetic Bragg reflections:

$$\varkappa = \mathbf{b} + \kappa_L, \tag{8.11}$$

whereas the vector $\mathbf{M}_{\varkappa}{}^L$ determines the intensity of each of the reflections and is equal to

$$\mathbf{M}^L_{\varkappa} = \mathbf{f}^L_{\varkappa} - (\mathbf{e}\mathbf{f}^L_{\varkappa})\,\mathbf{e}; \tag{8.12}$$

$$\mathbf{f}^L_{\varkappa} = \sum_i F_i(\varkappa)\, e^{-W_i} e^{-i\varkappa x_i} \mathbf{S}^L_{0i}. \tag{8.13}$$

Since elastic scattering at magnetic structures was described in detail in the second volume [42], we now focus out attention on inelastic magnetic scattering of neutrons.

9. The Cross Section of Scattering at Spin Waves

Ferromagnets. First let us consider inelastic scattering of neutrons under emission and absorption of a spin wave in a very simple ferromagnetic material consisting of magnetic atoms of a single type. We start from the general expression (8.5) for the cross section, in which we state the sum over α and β in explicit form:

$$\frac{d^2\sigma}{d\Omega d\varepsilon} = (r_0\gamma)^2 F^2(\varkappa) \frac{k}{k_0} e^{-2W} \sum_{ll'} e^{i\varkappa(\mathbf{R}^\circ_l - \mathbf{R}^\circ_{l'})} \frac{1}{2\pi\hbar} \int_{-\infty}^{\infty} dt e^{-i\frac{\varepsilon}{\hbar}t} \times$$

$$\times \Big\{(1-(\mathbf{em})^2)\langle S^z_l(t) S^z_{l'}(0)\rangle + (1+(\mathbf{em})^2)\frac{1}{4}\langle S^-_l(t) S^+_{l'}(0) + S^+_l(t) S^-_{l'}(0)\rangle\Big\}. \tag{9.1}$$

We recall that the nondiagonal correlators of the types $\langle S_l^z(t) S_{l'}{}^{\pm}(0)\rangle$ and $\langle S_l{}^{\pm}(t) S_{l'}{}^{\pm}(0)\rangle$ are equal to zero in a ferromagnetic material (the z axis was chosen parallel to the spontaneous moment of the crystal, and $S_l{}^{\pm} = S_{l'}{}^x \pm iS_l{}^y$).

The longitudinal correlator includes the time-independent part $\langle S_l^z\rangle$ $\langle S_{l'}^z\rangle$ which leads to elastic scattering. Inelastic scattering at the spin waves is related to the correlation of the transverse components. This part of the cross section is given by the expression

$$\frac{d^2\sigma_{SW}}{d\Omega d\varepsilon}=\frac{1}{4}(r_0\gamma)^2F^2(\varkappa)\,e^{-2W}\frac{k}{k_0}[1+(\mathbf{em})^2]\sum_{ll'}e^{i\varkappa(\mathbf{R}_l^0-\mathbf{R}_{l'}^0)}\times$$

$$\times\frac{1}{2\pi\hbar}\int_{-\infty}^{\infty}dt\,e^{-i\frac{\varepsilon}{\hbar}t}\langle S_l^-(t)\,S_{l'}^+(0)+S_l^+(t)\,S_{l'}^-(0)\rangle, \tag{9.2}$$

in which the transverse correlators still must be calculated.

Let us perform the calculation for the very simple case of a Heisenberg ferromagnet which is described by the exchange Hamiltonian

$$\mathcal{H}=-\sum_{l\neq l'}J(\mathbf{R}_l-\mathbf{R}_{l'})\,\mathbf{S}_l\mathbf{S}_{l'}\,. \tag{9.3}$$

At low temperatures, much lower than the Curie temperature T_c, one conveniently uses the so-called spin-wave approximation in which the spin operators $\mathbf{S}_l$ are approximated by the Bose creation operator b_l^+ and the Bose annihilation operator b_l of the spin deviations. The relation between them is given by the following approximating expressions [197]:

$$S_l^+=\sqrt{2S_l}\,b_l,\; S_l^-=\sqrt{2S_l}\,b_l^+,\; S_l^z=S_l-b_l^+\,b_l, \tag{9.4}$$

which are valid only at low temperatures where the average value $\langle b_l^+b_l\rangle$ of the spin deviation at a lattice point is much smaller than the atomic spin S. Under these conditions, the exchange Hamiltonian is approximated by the quadratic form of the Bose operators and this quadratic form is diagonalized by expansion of the operators b_l and b_l^+ in Fourier series:

$$\mathcal{H}=\sum_{\mathbf{q}}\varepsilon_{\mathbf{q}}\,b_{\mathbf{q}}^+\,b_{\mathbf{q}}, \tag{9.5}$$

where $\varepsilon_{\mathbf{q}}$ denotes the energy of the spin wave with the dispersion law

$$\varepsilon_{\mathbf{q}}=2S\sum_{\mathbf{t}}J(\mathbf{t})\,(1-e^{i\mathbf{qt}})\,. \tag{9.6}$$

The spin correlators can be easily calculated with the aid of Eq. (9.4):

$$\langle S_l^-(t)\, S_{l'}^+(0)\rangle = \frac{2S}{N}\sum_{\mathbf{q}} e^{-i\mathbf{q}(\mathbf{R}_l-\mathbf{R}_{l'})}\, e^{i\frac{\varepsilon_{\mathbf{q}}}{\hbar}t} N_{\mathbf{q}};$$

$$\langle S_l^+(t)\, S_{l'}^-(0)\rangle = \frac{2S}{N}\sum_{\mathbf{q}} e^{i\mathbf{q}(\mathbf{R}_l-\mathbf{R}_{l'})}\, e^{-i\frac{\varepsilon_{\mathbf{q}}}{\hbar}t} (1+N_{\mathbf{q}}), \tag{9.7}$$

where

$$N_{\mathbf{q}} = \frac{1}{e^{\varepsilon_{\mathbf{q}}/kT}-1} \tag{9.8}$$

(N_q denotes the Bose distribution function of the spin waves). Equations (9.2) and (9.7) lead to the following expression for the cross section of inelastic scattering:

$$\frac{d^2\sigma_{SW}}{d\Omega d\varepsilon} = \frac{S}{2}(r_0\gamma)^2 F^2(\varkappa)\, e^{-2W}\frac{k}{k_0}[1+(\mathbf{em})^2]\times$$
$$\times\sum_{\mathbf{b}} N\sum_{\mathbf{q}}\{N_{\mathbf{q}}\,\delta_{\varkappa-\mathbf{q},\,\mathbf{b}}\,\delta(\varepsilon-\varepsilon_{\mathbf{q}}) + (1+N_{\mathbf{q}})\,\delta_{\varkappa+\mathbf{q},\,\mathbf{b}}\,\delta(\varepsilon+\varepsilon_{\mathbf{q}})\}. \tag{9.9}$$

The first term in the braces describes scattering under absorption of a spin wave, and the second term, scattering under emission of a spin wave. It follows from Eq. (9.9) that in the magnetic scattering cross section there appear sharp peaks whose positions in regard to angle and energy are given by the momentum and energy conservation laws:

$$\left.\begin{aligned}\varkappa &= \mp\mathbf{q}+\mathbf{b};\\ \varepsilon &= \mp\varepsilon_{\mathbf{q}}.\end{aligned}\right\} \tag{9.10}$$

The intensity of an inelastic magnetic peak is given by the equation

$$\frac{1}{N}\frac{d^2\sigma_{SW}}{d\Omega d\varepsilon}\bigg|_{\varkappa=\mp\mathbf{q}+\mathbf{b}} = \frac{1}{2}S(r_0\gamma)^2F^2(\varkappa)\, e^{-2W}\frac{k}{k_0}[1+(\mathbf{em})^2]\times$$
$$\times\left(N_{\varkappa}+\frac{1}{2}\pm\frac{1}{2}\right)\delta(\varepsilon\pm\varepsilon_{\mathbf{q}}), \tag{9.11}$$

where the upper sign refers to the process with emission of a spin wave and the lower sign to the process with absorption of a spin wave. The height of the inelastic magnetic peak depends through the factor 1 +

$(\mathbf{em})^2$ upon the orientation of the scattering vector with respect to the direction of the spontaneous moment; in this way, the observed inelastic peaks can be identified as magnetic peaks in an experiment.

Strictly speaking, Eq. (9.11) is valid at low temperatures $T << T_c$ because when we derived this equation, we used the approximation of Eq. (9.4) which establishes a relation between the spin operators and the Bose operators. The exact expressions are series of Bose operators such as, for instance, the well-known Goldstein–Primakoff relations [197]. Their use in the calculation of the spin correlators would lead to additional terms in the expression for the cross section of inelastic scattering, with the terms describing processes involving the emission of two or more magnons. But the cross sections of these multimagnon processes do not have such clearly pronounced singularities as the single-magnon process [68]. Besides that, their contribution to the total cross section of inelastic scattering is small at low temperatures. As far as the single-magnon cross section is concerned, its calculation at higher temperatures with the spin-wave approximation disregarded shows that the form of Eq. (9.11) for the cross section is preserved in certain approximations (the random-phase approximation). The parameters S, i.e., the amplitude of the atomic spin, must be replaced by its average value $\langle S_l^z \rangle$ at the particular temperature. Then the energy of the spin wave must be renormalized and depends upon the temperature through the quantity $\langle S_l^z \rangle$ which is proportional to the spontaneous magnetization. This renormalization is as follows:

$$\varepsilon_{\mathbf{q}} \to \widetilde{\varepsilon}_{\mathbf{q}} = 2 \langle S^z \rangle \sum_{\mathbf{t}} J(\mathbf{t}) (1 - e^{i\mathbf{k}\mathbf{t}}). \qquad (9.12)$$

Thus, the singular form of single-magnon scattering is preserved at any temperature up to T_c but the intensity of the inelastic peak decreases with the temperature like $\langle S^z \rangle$, whereas the other (diffuse) part of the inelastic scattering increases in proportion to the approach to T_c.

Arbitrary magnetic structures. In order to determine the spin waves in a crystal with an arbitrary magnetic structure, one must separate the so-called quantization axes, i.e., directions in which the spins are aligned in the case of total magnetic order, i.e., at the temperature zero. The spin oscillations leading to the spin waves must take place relative to the corresponding quantization axes.

In the general case, for each lattice point occupied by a magnetic atom one must introduce a separate quantization axis parallel to the average position of the spin and one must associate with each atom l a local coordinate system such that the z axis of the system coincides with the equilibrium position of the spin at the lattice point. The transition from the spin operator stated in the general coordinate system associated with the crystal axes to the local coordinate system is made with the aid of the transformation matrix d_l so that

$$S_l^{\alpha} = \sum_{\beta} d_l^{\alpha\beta} S_l'^{\beta}, \tag{9.13}$$

where $S_l'^{\beta}$ denotes the components of the spin vector of atom l in the local coordinate system.

In the local system, the exchange Hamiltonian of Eq. (9.3) is a quadratic function of the operators $S_l'^{\beta}$ of the very general form

$$\mathcal{H} = -\sum_{ll'} J(\mathbf{R}_l - \mathbf{R}_{l'}) \{ D_{ll'}^{++} S_l'^{+} S_{l'}'^{+} + D_{ll'}^{--} S_l'^{-} S_{l'}'^{-} + \\ + D_{ll'}^{+-} S_l'^{+} S_{l'}'^{-} + D_{ll'}^{-+} S_l'^{-} S_{l'}'^{+} + D_{ll'}^{zz} S_l'^{z} S_{l'}'^{z} \}, \tag{9.14}$$

where $D_{ll'}{}^{\alpha\beta}$ is expressed through the matrices of the unitary transformation of Eq. (9.13). Terms of the type $S_l'^{\pm} S_{l'}'^{z}$ in the Hamiltonian of Eq. (9.14) cancel as a consequence of the condition that the magnetic structure given at $T = 0$ corresponds to the minimum of the exchange energy. The subscript l must be split into subscripts numbering the sublattice and the atoms in it.

In order to separate the spin-wave excitations of a system, which can be described by the Hamiltonian of Eq. (9.14), one can now switch from the spin operators given in the local coordinate system to the Bose operators as this was done in the case of the ferromagnetic material. For the set of atoms related to a single magnetic sublattice, we obtain in analogy to the transformation of Eq. (9.4):

$$S_l'^{+} = \sqrt{2S}\, b_l, \quad S_l'^{-} = \sqrt{2S}\, b_l^{+}, \quad S_l'^{z} = S - b_l^{+} b_l. \tag{9.15}$$

These relations must be stated for each sublattice by introducing the corresponding Bose operators of the spin deviations. The spin-wave spectrum of the magnetic material is obtained by diagonalizing the resulting quadratic form of the Bose operators.

In order to calculate the cross sections of scattering at the spin waves, the spin operators in the general equation (8.5) must be expressed in the local coordinate system. Then the double differential cross section of inelastic scattering at the spin waves is described in terms containing the correlators of the transverse spin components:

$$\frac{d^2\sigma_{SW}}{d\Omega d\varepsilon}=\frac{1}{4}(r_0\gamma)^2\frac{k}{k_0}\sum_{ll'}F_l(\varkappa)F_{l'}(\varkappa)e^{-W_l-W_{l'}}e^{i\varkappa(\mathbf{R}_l-\mathbf{R}_{l'})}\times$$

$$\times\sum_{\alpha\beta}(\delta_{\alpha\beta}-e_\alpha e_\beta)\frac{1}{2\pi\hbar}\int_{-\infty}^{\infty}dt\, e^{-i\frac{\varepsilon}{\hbar}t}\{d_l^{\alpha-}d_{l'}^{\beta-}\langle S_l'^{+}(t)S_{l'}'^{+}(0)\rangle+$$

$$+d_l^{\alpha+}d_{l'}^{\beta+}\langle S_l'^{-}(t)S_{l'}'^{-}(0)\rangle+d_l^{\alpha-}d_{l'}^{\beta+}\langle S_l'^{+}(t)S_{l'}'^{-}(0)\rangle+$$

$$+d_l^{\alpha+}d_{l'}^{\beta-}\langle S_l'^{-}(t)S_{l'}'^{+}(0)\rangle\},\qquad(9.16)$$

where the notation $d_l^{\alpha\pm}=d_l^{\alpha x}\pm id_l^{\alpha y}$ was introduced for the sake of brevity. The transition from the spin operators to the Bose amplitudes of the spin deviations in accordance with Eq. (9.15) makes it possible to state the cross section of scattering at spin waves but this can be completed only in specific cases. We consider below two typical examples: a collinear two-sublattice antiferromagnetic material and a very simple spiral structure.

A two-sublattice antiferromagnetic material. The features of inelastic scattering at an antiferromagnetic structure can be illustrated by the example of a two-sublattice antiferromagnetic material. Assume that in the primitive cell of the crystal there are two magnetic atoms each of which is associated with its proper magnetic sublattice. The primitive magnetic cell of such a structure coincides with the crystal cell so that the wave vector of the structure is $\mathbf{k}=0$.

Let us use the above-developed method for analyzing this structure. In the first sublattice, all the local systems are equivalent to the basic system, and therefore we can set $d_1{}^{\alpha\beta}=\delta_{\alpha\beta}$. The local system of the second sublattice is obtained from the first by transformation of the axes: $x_2=x_1$, $y_2=-y_1$, $z_2=-z_1$. The matrices d_1 and d_2 of the unitary transformation to the two local coordinate systems are of the form

$$d_1=\begin{pmatrix}1&0&0\\0&1&0\\0&0&1\end{pmatrix},\quad d_2=\begin{pmatrix}1&0&0\\0&-1&0\\0&0&-1\end{pmatrix}.\qquad(9.17)$$

We will assume that, as it happens ever so often in real cases, the strongest exchange force exists between those atoms of the various sublattices that are closest neighbors. In this case, the exchange Hamiltonian in the local systems is:

$$\mathcal{H} = -\sum_{nn'} J(S_{1n}'^{+} S_{2n'}'^{+} + S_{1n}'^{-} S_{2n'}'^{-} + 2S_{1n}'^{z} S_{2n'}'^{z}), \tag{9.18}$$

where nn' denotes summation over all pairs of crystal cells which are nearest neighbors.

Let us now switch from the spin operators $S_{1n}{}^{\alpha}$ and $S_{2n}{}^{\alpha}$ to the Bose amplitudes b_n and c_n with formulas (9.15):

$$\begin{aligned} S_{1n}'^{+} &= \sqrt{2S}\, b_n, \quad S_{1n}'^{z} = S - b_n^{+} b_n, \\ S_{2n}'^{+} &= \sqrt{2S}\, c_n, \quad S_{2n}'^{z} = S - c_n^{+} c_n \end{aligned} \tag{9.19}$$

and expand the latter in Fourier series:

$$b_n = \frac{1}{\sqrt{N}} \sum_{\mathbf{q}} e^{i\mathbf{q}\mathbf{R}_n} b_{\mathbf{q}}, \quad c_n = \frac{1}{\sqrt{N}} \sum_{\mathbf{q}} e^{-i\mathbf{q}\mathbf{R}_n} c_{\mathbf{q}}.$$

We obtain the exchange Hamiltonian of Eq. (9.18) in the spin-wave approximation:

$$\mathcal{H}_{\mathrm{SW}} = 2SJz \sum_{\mathbf{q}} \{b_{\mathbf{q}}^{+} b_{\mathbf{q}} + c_{\mathbf{q}}^{+} c_{\mathbf{q}} - \gamma_{\mathbf{q}} (b_{\mathbf{q}} c_{\mathbf{q}} + b_{\mathbf{q}}^{+} c_{\mathbf{q}}^{+})\}; \tag{9.20}$$

$$\gamma_{\mathbf{q}} = \frac{1}{z} \sum_{\delta} e^{i\mathbf{q}\delta}, \tag{9.21}$$

where δ denotes the radius vector of the nearest atoms; and z denotes the number of the nearest neighbors.

Diagonalization of the quadratic form of Eq. (9.20) is obtained with the well-known u–v transformation:

$$\left.\begin{aligned} b_{\mathbf{q}} &= u_{\mathbf{q}} \xi_{\mathbf{q}} + v_{\mathbf{q}} \eta_{\mathbf{q}}^{+}; \\ c_{\mathbf{q}} &= v_{\mathbf{q}} \xi_{\mathbf{q}}^{+} + u_{\mathbf{q}} \eta_{\mathbf{q}} \end{aligned}\right\} \tag{9.22}$$

having coefficients which satisfy the equations

$$\left.\begin{aligned} \gamma_{\mathbf{q}} (u_{\mathbf{q}}^2 + v_{\mathbf{q}}^2) &= 2u_{\mathbf{q}} v_{\mathbf{q}}; \\ u_{\mathbf{q}}^2 - v_{\mathbf{q}}^2 &= 1. \end{aligned}\right\} \tag{9.23}$$

It follows from these formulas that:

$$u_{\mathbf{q}}^2=\frac{1}{2}\left[1+\frac{1}{\sqrt{1-\gamma_{\mathbf{q}}^2}}\right],\quad v_{\mathbf{q}}^2=\frac{1}{2}\left[-1+\frac{1}{\sqrt{1-\gamma_{\mathbf{q}}^2}}\right]; \tag{9.24}$$

$$\mathcal{H}_{\mathrm{SW}}=\sum_{\mathbf{q}}\varepsilon_{\mathbf{q}}(\xi_{\mathbf{q}}^+\xi_{\mathbf{q}}+\eta_{\mathbf{q}}^+\eta_{\mathbf{q}}),\quad \varepsilon_{\mathbf{q}}=2SzJ\sqrt{1-\gamma_{\mathbf{q}}^2}. \tag{9.25}$$

Thus, in a collinear two-sublattice antiferromagnetic material there exist two degenerate branches of spin waves with a dispersion law $\varepsilon_{\mathbf{q}}$ which is linear at small momenta.

The cross section of scattering at the spin waves was calculated with Eq. (9.16) in which Eqs. (9.17) and (9.19) must be brought into account. Then we obtain

$$\frac{d^2\sigma_{\mathrm{SW}}}{d\Omega d\varepsilon}=N\frac{1}{2}S(r_0\gamma)^2F^2(\varkappa)\,\mathrm{e}^{-2W}\frac{k}{k_0}[1+(\mathbf{em})^2]\frac{1}{2\pi\hbar}\int_{-\infty}^{\infty}dt\mathrm{e}^{-i\frac{\varepsilon}{\hbar}t}\times$$

$$\times\sum_{\mathbf{b}}\sum_{\mathbf{q}}\{\delta_{\varkappa+\mathbf{q},\,\mathbf{b}}[\mathrm{e}^{+i\varkappa(\mathbf{r}_1-\mathbf{r}_2)}\langle b_{\mathbf{q}}(t)\,c_{\mathbf{q}}(0)\rangle+e^{-i\varkappa(\mathbf{r}_1-\mathbf{r}_2)}\langle c_{\mathbf{q}}^+(t)b_{\mathbf{q}}^+(0)\rangle]+$$

$$+\delta_{\varkappa-\mathbf{q},\,\mathbf{b}}[\mathrm{e}^{i\varkappa(\mathbf{r}_1-\mathbf{r}_2)}\langle b_{\mathbf{q}}^+(t)\,c_{\mathbf{q}}^+(0)\rangle+\mathrm{e}^{-i\varkappa(\mathbf{r}_1-\mathbf{r}_2)}\langle c_{\mathbf{q}}(t)\,b_{\mathbf{q}}(0)\rangle]\}. \tag{9.26}$$

After calculating the correlators with the aid of the transformation equation (9.22), we can represent the inelastic cross section in its final form [171]:

$$\frac{d^2\sigma_{\mathrm{SW}}}{d\Omega d\varepsilon}=NS(r_0\gamma)^2F^2(\varkappa)\,\mathrm{e}^{-2W}\frac{k}{k_0}[1+(\mathbf{em})^2]\times$$

$$\times\cos\varkappa(\mathbf{r}_1-\mathbf{r}_2)\sum_{\mathbf{b}}\sum_{\mathbf{q}}\frac{\gamma_{\mathbf{q}}}{\sqrt{1-\gamma_{\mathbf{q}}^2}}\{N_{\mathbf{q}}\delta_{\varkappa-\mathbf{q},\,\mathbf{b}}\delta(\varepsilon-\varepsilon_{\mathbf{q}})+$$

$$+(N_{\mathbf{q}}+1)\,\delta_{\varkappa+\mathbf{q},\,\mathbf{b}}\delta(\varepsilon+\varepsilon_{\mathbf{q}})\}. \tag{9.27}$$

where $N_{\mathbf{q}}$ denotes the Bose distribution function of the spin waves with the energy $\varepsilon_{\mathbf{q}}$; and **m** denotes the unit vector of the spontaneous magnetization of the sublattice.

In formal terms, the result of Eq. (9.27) is very similar to the result expressed by Eq. (9.11) for a ferromagnetic material. In both cases, the angular dependence of the cross section upon the orientation of the scattering vector relative to the vector **m** (magnetization of the crystal or magnetization of the sublattice) is given by the factor $1 + (\mathbf{em})^2$. In the case of the antiferromagnetic material, there exists an additional angular dependence of the cross section given by the factor $\gamma_{\mathbf{q}}/\sqrt{1 - \gamma_{\mathbf{q}}^2} \cos \varkappa \times (\mathbf{r}_1 - \mathbf{r}_2)$.

The simple spiral. In the case of spiral structures and other incommensurate structures to which the concept of a sublattice is not applicable, a different situation is encountered. We consider as an example a simple spiral with the wave vector $\mathbf{k}_0$ parallel to the z axis and with spins on a perpendicular plane.

When the basic coordinate system is chosen so that the z axis is parallel to the axis of the helicoid and that the x and y axes are on the base plane, the local system at the lattice point l must be chosen so that the z_l' axis is parallel to the spin vector of the lattice point l and that the y_l' axis is on the base plane; then the x_l' axis will be parallel to the z axis. The spin operators in the local and basic coordinate systems are related by the transformations:

$$\left.\begin{aligned} S_l^x &= S_l^{'z} \cos(\varkappa_0 \mathbf{R}_l) - S_l^{'y} \sin(\varkappa_0 \mathbf{R}_l); \\ S_l^y &= S_l^{'z} \sin(\varkappa_0 \mathbf{R}_l) + S_l^{'y} \cos(\varkappa_0 \mathbf{R}_l); \\ S_l^z &= -S_l^{'x}, \end{aligned}\right\} \tag{9.28}$$

with which one can state the exchange Hamiltonian of the structure:

$$\begin{aligned} \mathcal{H} = -\sum_{ll'} J(\mathbf{R}_l - \mathbf{R}_{l'}) \{S_l^{'x} S_{l'}^{'x} + (S_l^{'y} S_{l'}^{'y} + S_l^{'z} S_{l'}^{'z}) \times \\ \times \cos \varkappa_0 (\mathbf{R}_l - \mathbf{R}_{l'})\}. \end{aligned} \tag{9.29}$$

In the spin-wave approximation of Eq. (9.15), a reduction to the quadratic form of the Bose operators takes place:

$$\mathcal{H}_{\mathrm{SW}} = \sum_{\mathbf{q}} \left\{ B_{\mathbf{q}} b_{\mathbf{q}}^+ b_{\mathbf{q}} + \frac{1}{2} A_{\mathbf{q}} (b_{\mathbf{q}} b_{-\mathbf{q}} + b_{\mathbf{q}}^+ b_{-\mathbf{q}}^+) \right\}, \tag{9.30}$$

where $A_{\mathbf{q}}$ and $B_{\mathbf{q}}$ denote combinations of the Fourier components $J(\mathbf{q})$ of the exchange integral $\mathbf{J}(\mathbf{R}_l - \mathbf{R}_l')$:

$$B(\mathbf{q}) = S\left\{2J(\varkappa_0) - J(\mathbf{q}) - \frac{1}{2}[J(\varkappa_0 + \mathbf{q}) + J(\varkappa_0 - \mathbf{q})]\right\};$$

$$A(\mathbf{q}) = S\left\{\frac{1}{2}[J(\varkappa_0 + \mathbf{q}) + J(\varkappa_0 - \mathbf{q})] - J(\mathbf{q})\right\}. \tag{9.31}$$

Diagonalization is obtained with the aid of the canonical transformation

$$b_{\mathbf{q}} = u_{\mathbf{q}} \xi_{\mathbf{q}} - v_{\mathbf{q}} \xi^{+}_{-\mathbf{q}}, \quad b^{+}_{\mathbf{q}} = u_{\mathbf{q}} \xi^{+}_{\mathbf{q}} - v_{\mathbf{q}} \xi_{-\mathbf{q}}, \tag{9.32}$$

where $u_{\mathbf{q}}$ and $v_{\mathbf{q}}$ are obtained from the system of equations

$$\left.\begin{aligned} 2B_{\mathbf{q}} u_{\mathbf{q}} v_{\mathbf{q}} &= A_{\mathbf{q}} (u^2_{\mathbf{q}} + v^2_{\mathbf{q}}); \\ u^2_{\mathbf{q}} - v^2_{\mathbf{q}} &= 1, \end{aligned}\right\} \tag{9.33}$$

which has the solution

$$u^2_{\mathbf{q}} = \frac{1}{2}\left(1 + \frac{B_{\mathbf{q}}}{\sqrt{B^2_{\mathbf{q}} - A^2_{\mathbf{q}}}}\right), \quad v^2_{\mathbf{q}} = \frac{1}{2}\left(\frac{B_{\mathbf{q}}}{\sqrt{B^2_{\mathbf{q}} - A^2_{\mathbf{q}}}} - 1\right). \tag{9.34}$$

The diagonalized quadratic form is as follows:

$$\mathcal{H}_{SW} = \sum_{\mathbf{q}} \varepsilon_{\mathbf{q}} b^{+}_{\mathbf{q}} b_{\mathbf{q}}, \quad \varepsilon_{\mathbf{q}} = \sqrt{B^2_{\mathbf{q}} - A^2_{\mathbf{q}}}. \tag{9.35}$$

Thus, in the magnetic structure considered, there is one branch of spin waves with the dispersion [367]:

$$\varepsilon_{\mathbf{q}} = 2S \sqrt{[J(\boldsymbol{\kappa}_0) - J(\mathbf{q})]\left[J(\boldsymbol{\kappa}_0) - \frac{1}{2}(J(\boldsymbol{\kappa}_0 + \mathbf{q}) + J(\boldsymbol{\kappa}_0 - \mathbf{q}))\right]}. \tag{9.36}$$

This is the acoustic branch but it has one singularity: at $\mathbf{q} = \mathbf{k}_0$, its energy is equal to zero. This fact is fully understandable because the spiral structure by itself is a "frozen" spin wave with the wave vector $\mathbf{k}_0$ and no energy is required for its excitation.

The scattering cross section can be calculated with the general equation (9.16) when one substitutes in it the matrix elements defined by the transformation of Eq. (9.28). One easily obtains the final result:

$$\begin{aligned} \frac{d^2\sigma_{SW}}{d\Omega d\varepsilon} &= N \frac{S}{2} (r_0 \gamma)^2 F^2(\boldsymbol{\varkappa}) e^{-2W} \frac{k}{k_0} \times \\ &\times \sum_{\mathbf{q}} \sum_{\mathbf{b}} \Big\{ \frac{1}{4} [1 + (\mathbf{em})^2] (\delta_{\boldsymbol{\varkappa}+\mathbf{q}+\mathbf{k}_0, \mathbf{b}} + \delta_{\boldsymbol{\varkappa}+\mathbf{q}-\mathbf{k}_0, \mathbf{b}}) \times \\ &\times (u_{\mathbf{q}} + v_{\mathbf{q}})^2 [(N_{\mathbf{q}} + 1)\delta(\varepsilon + \varepsilon_{\mathbf{q}}) + N_{\mathbf{q}} \delta(\varepsilon - \varepsilon_{\mathbf{q}})] + [1 - (\mathbf{em})^2] \times \\ &\times \delta_{\boldsymbol{\varkappa}+\mathbf{q}, \mathbf{b}} (u_{\mathbf{q}} - v_{\mathbf{q}})^2 [(N_{\mathbf{q}} + 1)\delta(\varepsilon + \varepsilon_{\mathbf{q}}) + (N_{\mathbf{q}} \delta(\varepsilon - \varepsilon_{\mathbf{q}})] \Big\}. \end{aligned} \tag{9.37}$$

In this expression, **m** is the unit vector of the spiral, i.e., a pseudovector parallel to the plane on which the spin rotation takes place.

At a high temperature, when the number of spin waves is $N_{\mathbf{q}} >> 1$, the intensity of the various components in the cross section is given by the factors $[1 \pm (\mathbf{em})^2]\,(u_{\mathbf{q}} \pm v_{\mathbf{q}})^2$ and Eq. (9.37) assumes the form of the corresponding expression of [5].

10. Single-Magnon Scattering of Polarized Neutron at an Arbitrary Magnetic Structure

Spin waves in a crystal with a finite magnetic cell. We obtained in the preceding section the spectra of the spin waves of very simple magnetic structures and devised a scheme for calculating the magnon spectra and the cross section of single-magnon scattering in some structure with a magnetic order. The method is based on stating the atomic components of the spin vector in a local coordinate system whose z axis is chosen parallel to the quantization axis corresponding to this atom. The transition to the new coordinate system is effected with the aid of the transformation matrix d_l given for each magnetic atom as per Eq. (9.13). The magnetic structure defines the form of the transformation matrices. When we have found them, the separation of the spin-wave excitations is reduced to diagonalizing the Hamiltonian of Eq. (9.14) (to which anisotropic interactions can be added).

The diagonalization of the quadratic form of Eq. (9.14) in the spin-wave approximation of Eq. (9.15) depends greatly upon whether the magnetic structure has a finite (commensurate) magnetic cell. In the case of a commensurate magnetic cell, the magnetic atoms can be numbered by the composite subscript $l = \{ni\}$, where n denotes the number of the magnetic cell and i denotes the number of the atom in the cell. Since the magnetic order is repeated in each of the cells, Eq. (9.13) is reduced to

$$S_{ni}^{\alpha} = \sum_{\beta} d_i^{\alpha\beta} S_{ni}^{\prime\beta}. \qquad (10.1)$$

The matrix $d_i^{\alpha\beta}$ of the local transformation does not depend upon the number of the cell, and this detail makes it possible to state the Hamiltonian of the spin-wave approximation of some structure of this type in the following general form:

$$\mathcal{H}_{SW} = \sum_{\mathbf{q}} \sum_{ij} \Big\{ B_{ij}(\mathbf{q})\, b_i^+(\mathbf{q})\, b_j(\mathbf{q}) +$$

$$+ \frac{1}{2} A_{ij}(\mathbf{q})\, b_i(-\mathbf{q})\, b_j(\mathbf{q}) + \frac{1}{2} A_{ij}^*(\mathbf{q})\, b_i^+(-\mathbf{q})\, b_j^+(\mathbf{q}) \Big\}. \tag{10.2}$$

The Bose operators $b_i^+(\mathbf{q})$ and $b_i(\mathbf{q})$ are the Fourier components of the spin deviations introduced by the expressions (9.15):

$$b_{in} = \frac{1}{\sqrt{N}} \sum_{\mathbf{q}} e^{i\mathbf{q}\mathbf{t}_n} b_i(\mathbf{q}), \quad b_{in}^+ = \frac{1}{\sqrt{N}} \sum_{\mathbf{q}} e^{-i\mathbf{q}\mathbf{t}_n} b_i^+(\mathbf{q}). \tag{10.3}$$

It is essential that $\mathbf{t}_n$ denotes the translation vectors of the magnetic lattice, and therefore $\mathbf{q}$ belongs to the first Brillouin zone of the reciprocal magnetic lattice.

Equation (10.2) is the most general form of the spin-wave Hamiltonian corresponding to the exchange Hamiltonian of Eq. (9.14) and of some anisotropic energy represented by the quadratic form in spin operators. The coefficients $A_{ij}(\mathbf{q})$ and $B_{ij}(\mathbf{q})$ can be easily stated for a given magnetic structure and are expressed through the parameters of the initial Hamiltonian and the matrix of the local transformations. Let us note the general properties of these coefficients:

$$A_{ij}(\mathbf{q}) = A_{ji}(-\mathbf{q}), \quad B_{ij}(\mathbf{q}) = B_{ji}^*(\mathbf{q}), \tag{10.4}$$

which follow from the hermiticity of the Hamiltonian $\mathcal{H}_{SW}$.

The diagonalization of the quadratic form of Eq. (10.2) can be obtained with the u–v transformation:

$$b_i(\mathbf{q}) = \sum_{\Lambda} [u_{i\Lambda}(\mathbf{q})\, \xi_\Lambda(\mathbf{q}) + v_{i\Lambda}^*(-\mathbf{q})\, \xi_\Lambda^+(-\mathbf{q})], \tag{10.5}$$

where $\xi_\Lambda(\mathbf{q})$ and $\xi_\Lambda^+(\mathbf{q})$ also denote Bose operators. It is convenient to use below the matrix notation of the equations by introducing 2σ-component columns and lines (σ denotes the number of the magnetic atoms in the magnetic cell of the crystal):

$$\Psi(\mathbf{q}) = \begin{pmatrix} b(\mathbf{q}) \\ b^+(-\mathbf{q}) \end{pmatrix}, \quad \Psi^+(\mathbf{q}) = (b^+(\mathbf{q}),\ b(-\mathbf{q})). \tag{10.6}$$

where $b(\mathbf{q})$ denotes the σ-component column of the operators $b_i(\mathbf{q})$.

The equation of motion of the operators b and b^+ is stated in matrix form as follows:

$$i\hbar \frac{d\Psi(\mathbf{q})}{dt} = \mathrm{H}(\mathbf{q})\,\Psi(\mathbf{q}), \qquad (10.7)$$

where $H(\mathbf{q})$ is a $2\sigma \times 2\sigma$ matrix:

$$\mathrm{H}(\mathbf{q}) = \begin{pmatrix} B(\mathbf{q}) & A^*(-\mathbf{q}) \\ -A(\mathbf{q}) & -B^*(-\mathbf{q}) \end{pmatrix}, \qquad (10.8)$$

composed of the $\sigma \times \sigma$ matrices $A_{ij}(\mathbf{q})$ and $B_{ij}(\mathbf{q})$.

We choose the components of the eigenvector of the matrix $H(\mathbf{q})$ as coefficients $u_{i\Lambda}$ and $v_{i\Lambda}$ of the canonical transformation of Eq. (10.5); they can be obtained from an equation of [37]:

$$\begin{pmatrix} B(\mathbf{q}) & A^*(\mathbf{q}) \\ -A(\mathbf{q}) & -B^*(-\mathbf{q}) \end{pmatrix} \begin{pmatrix} u_\Lambda(\mathbf{q}) \\ v_\Lambda(\mathbf{q}) \end{pmatrix} = \hbar\omega_\Lambda(\mathbf{q}) \begin{pmatrix} u_\Lambda(\mathbf{q}) \\ v_\Lambda(\mathbf{q}) \end{pmatrix}. \qquad (10.9)$$

where $u_\Lambda(\mathbf{q})$ and $v_\Lambda(\mathbf{q})$ denote the σ-component columns of the quantities $u_{i\Lambda}(\mathbf{q})$ and $v_{i\Lambda}(\mathbf{q})$. The subscript Λ numbers the eigenvalues $\hbar\omega_\Lambda(\mathbf{q})$ and the eigenvectors of the matrix $\mathrm{H}(\mathbf{q})$. Then the Hamiltonian of Eq. (10.2) is reduced to diagonal form:

$$\mathcal{H}_{\mathrm{SW}} = \sum_{\mathbf{q}}\sum_{\Lambda} \hbar\omega_\Lambda(\mathbf{q})\,\xi_\Lambda^+(\mathbf{q})\,\xi_\Lambda(\mathbf{q}). \qquad (10.10)$$

The eigenenergies $\hbar\omega_\Lambda(\mathbf{q})$ are determined from the equation

$$\det \begin{vmatrix} B(\mathbf{q}) - \hbar\omega_\Lambda(\mathbf{q}) & A^*(-\mathbf{q}) \\ -A(\mathbf{q}) & -B^*(-\mathbf{q}) - \hbar\omega_\Lambda(\mathbf{q}) \end{vmatrix} = 0, \qquad (10.11)$$

which has the property that in addition to the solution $\hbar\omega_\Lambda(\mathbf{q})$, there holds also the other solution $-\hbar\omega_\Lambda(\mathbf{q})$. Thus, the $2\sigma \times 2\sigma$ matrix (10.8) for $\mathrm{H}(\mathbf{q})$ has σ positive eigenvalues $\hbar\omega_\Lambda(\mathbf{q})$ which define the spectrum of the spin waves. Only positive values $\hbar\omega_\Lambda(\mathbf{q})$ must be considered in the summation over Λ in Eqs. (10.10) and (10.5).

Let us note that Eqs. (10.9) are invariant with respect to the transformation

$$u_{i\Lambda}(\mathbf{q}) \to v^{*}_{i\Lambda}(-\mathbf{q}),\quad v_{i\Lambda}(\mathbf{q}) \to u^{*}_{i\Lambda}(-\mathbf{q}),\quad \omega_{\Lambda}(\mathbf{q}) \to -\omega_{\Lambda}(-\mathbf{q}), \tag{10.12}$$

which provides a relation between the solution with positive and negative eigenvalues. Let us also note the normalization condition for the coefficients of the $u-v$ transformation; the coefficients are most conveniently described in matrix form:

$$U^{-1}U = 1,\ UU^{-1} = 1, \tag{10.13}$$

where U and U^{-1} are given by the expressions

$$\Psi = U\xi,\ \xi = U^{-1}\Psi, \tag{10.14}$$

from which we find, taking into account the explicit form of the transformation expressed by Eq. (10.5), that:

$$U = \begin{pmatrix} u(\mathbf{q}) & v^{*}(-\mathbf{q}) \\ v(\mathbf{q}) & u^{*}(-\mathbf{q}) \end{pmatrix},\quad U^{-1} = \begin{pmatrix} u^{*}(\mathbf{q}) & -v^{*}(\mathbf{q}) \\ -v(-\mathbf{q}) & u(-\mathbf{q}) \end{pmatrix}. \tag{10.15}$$

The cross section of single-magnon scattering. The cross section of single-magnon scattering can be calculated when one switches in the general equation (9.16) from the spin operators to the Bose amplitudes with the aid of Eqs. (9.15), (10.3), and (10.5). However, Eq. (9.16) holds only for non-polarized neutrons. Since we want to obtain the cross section of scattering of polarized neutrons, we will start from the general equation (8.5) for the cross section and substitute in it the expression for the amplitude of magnetic scattering of Eq. (8.1). The averaging over the neutron spins in the incident beam must be performed with proper regard for the matrix of the spin density $\rho = 1/2(1 + \mathbf{p}_0\sigma)$, where $\mathbf{p}_0$ denotes the polarization vector of the beam and σ denotes a vector composed of the Pauli matrices (neutron spin $\mathbf{S}_n = 1/2\sigma$).

The general expression for the cross section of magnetic scattering can be restated in the following form:

$$\frac{d^2\sigma}{d\Omega d\varepsilon} = (r_0\gamma)^2 \frac{k}{k_0} \sum_{ll'} F_l(\varkappa) F_{l'}(\varkappa)\, e^{-W_l - W_{l'}}\, e^{i\varkappa(\mathbf{R}_l^{\circ} - \mathbf{R}_{l'}^{\circ})} \times$$

$$\times \sum_{\alpha\beta}\sum_{\gamma\gamma'} K^{\gamma\gamma'} (\delta_{\alpha\gamma} - e_{\alpha} e_{\gamma})(\delta_{\beta\gamma'} - e_{\beta} e_{\gamma'}) \times$$

$$\times \frac{1}{2\pi\hbar} \int_{-\infty}^{\infty} dt e^{-i\frac{\varepsilon}{\hbar}t} \langle S_i^{\alpha}(t) S_{i'}^{\beta}(0) \rangle, \tag{10.16}$$

where

$$K^{\nu\nu'} = \mathrm{Sp}(\rho\sigma^{\nu}\sigma^{\nu'}) = \delta_{\nu\nu'} + i\varepsilon_{\nu\nu'\mu} p_{\mu}^{0}. \tag{10.17}$$

When the trace in Eq. (10.17) was calculated, we made use of the following relations for the Pauli matrices [75, 125]:

$$\frac{1}{2}\mathrm{Sp}\,\sigma^{\nu}\sigma^{\nu'} = \delta_{\nu\nu'}, \quad \frac{1}{2}\mathrm{Sp}(\sigma^{\nu}\sigma^{\nu'}\sigma^{\nu''}) = i\varepsilon_{\nu\nu'\nu''}, \tag{10.18}$$

where $\varepsilon_{\gamma\gamma'\gamma''}$ denotes an antisymmetric unit tensor of rank 3. It is easy to verify that in the case $\mathbf{p}_0 = 0$, Eq. (10.16) is transformed into Eq. (8.5) for the cross section of nonpolarized neutrons.

After switching to the local coordinate systems with the transformation of Eq. (10.1), we separate the contribution which single-magnon scattering renders to the cross section:

$$\frac{d^2\sigma_{SW}}{d\Omega d\varepsilon} = \frac{1}{2}(r_0\gamma)^2 \frac{k}{k_0} \sum_{ii'} \sqrt{S_i S_{i'}}\, F_i(\varkappa) F_{i'}(\varkappa)\, e^{-W_i - W_{i'}} e^{i\varkappa(\mathbf{r}_i - \mathbf{r}_{i'})} \times$$

$$\times \sum_{\alpha\beta}\sum_{\nu\nu'} K^{\nu\nu'} (\delta_{\alpha\nu} - e_{\alpha} e_{\nu})(\delta_{\beta\nu'} - e_{\beta} e_{\nu'}) \sum_{nn'} e^{i\varkappa(\mathbf{t}_n - \mathbf{t}_{n'})} \times$$

$$\times \frac{1}{2\pi\hbar} \int_{-\infty}^{\infty} dt e^{-i\frac{\varepsilon}{\hbar}t} \{ d_i^{\alpha -} d_{i'}^{\beta -} \langle b_{in}(t) b_{i'n'}(0) \rangle +$$

$$+ d_i^{\alpha +} d_{i'}^{\beta +} \langle b_{in}^{+}(t) b_{i'n'}^{+}(0) \rangle + d_i^{\alpha -} d_{i'}^{\beta +} \langle b_{in}(t) b_{i'n'}^{+}(0) \rangle +$$

$$+ d_i^{\alpha +} d_{i'}^{\beta -} \langle b_{in}^{+}(t) b_{i'n'}(0) \rangle \}. \tag{10.19}$$

Further treatment of this expression requires the transition to the operators ξ_Λ and ξ_Λ^+ in accordance with Eqs. (10.3) and (10.5); after that, the correlators are determined through the frequencies $\omega_\Lambda(\mathbf{q})$ of the spin waves and the distribution function $N_\Lambda(\mathbf{q})$. After very simple but lengthy calculations, Eq. (10.19) is reduced to the following form [37]:

$$\frac{d^2\sigma_{SW}}{d\Omega d\varepsilon}=\frac{1}{4}(r_0\gamma)^2\frac{k}{k_0}\sum_{\mathbf{q}\Lambda}\sum_{\mathbf{b}}S_\Lambda(\varkappa)\{(1+N_\Lambda(\mathbf{q}))\,\delta_{\varkappa+\mathbf{q},\,\mathbf{b}}\,\delta(\varepsilon+\hbar\omega_\Lambda(\mathbf{q}))+$$
$$+N_\Lambda(\mathbf{q})\,\delta_{\varkappa-\mathbf{q},\,\mathbf{b}}\,\delta(\varepsilon-\hbar\omega_\Lambda(\mathbf{q}))\},\tag{10.20}$$

where

$$S_\Lambda(\varkappa)=\sum_{\alpha\beta}\sum_{\gamma\gamma'}K^{\gamma\gamma'}(\delta_{\alpha\gamma}-e_\alpha e_\gamma)(\delta_{\beta\gamma'}-e_\beta e_{\gamma'})\,\overset{*}{\varphi}{}^{\alpha}_{\varkappa\Lambda}\varphi^{\beta}_{\varkappa\Lambda};\tag{10.21}$$

$$\varphi^{\alpha}_{\varkappa\Lambda}=\sum_i\sqrt{S_i}\,F_i(\varkappa)\,\mathrm{e}^{-W_i}\,\mathrm{e}^{-\mathrm{i}\varkappa r_i}\,(d_i^{\alpha-}\,u_{i\Lambda}(\varkappa)+d_i^{\alpha+}\,v_{i\Lambda}(\varkappa)).\tag{10.22}$$

Using Eq. (10.17) for $K^{\gamma\gamma'}$, we state the expression for $S_\Lambda(\varkappa)$ in its final form

$$S_\Lambda(\varkappa)=(\mathfrak{M}^*_{\varkappa\Lambda}\mathfrak{M}_{\varkappa\Lambda})+\mathrm{i}\,[\mathfrak{M}^*_{\varkappa\Lambda}\times\mathfrak{M}_{\varkappa\Lambda}]\,\mathbf{p}_0,\tag{10.23}$$

where

$$\mathfrak{M}_{\varkappa\Lambda}=\varphi_{\varkappa\Lambda}-(\mathbf{e}\varphi_{\varkappa\Lambda})\,\mathbf{e}.\tag{10.24}$$

Thus, Eqs. (10.2) and (10.23) determine the cross section of single-magnon scattering of polarized neutrons in any complicated magnetic structure (which, however, has a finite magnetic cell!). The general expression for the cross section is surprisingly simple. The intensity of scattering under emission or absorption of a spin wave is given by the factor $S_\Lambda(\varkappa)$ which, as follows from its definition in Eq. (10.23), is real and positive. Owing to the relations

$$\varphi_{-\varkappa\Lambda}=\varphi^*_{\varkappa\Lambda},\quad \mathfrak{M}_{-\varkappa\Lambda}=\mathfrak{M}^*_{\varkappa\Lambda},\tag{10.25}$$

which follow from Eq. (10.22), the factor $S_\lambda(\varkappa)$ has the following important property:

$$S_\Lambda(-\varkappa)=(\mathfrak{M}^*_{\varkappa\Lambda}\mathfrak{M}_{\varkappa\Lambda})-\mathrm{i}\,[\mathfrak{M}^*_{\varkappa\Lambda}\times\mathfrak{M}_{\varkappa\Lambda}]\,\mathbf{p}_0,\tag{10.26}$$

which means that the sign changes in the term depending upon the polarization vector. In some particular cases, this leads to opposite contributions to the cross section under emission and absorption of magnons [40].

Such a situation arises in a very simple crystal with magnetic order, namely in a single-sublattice ferromagnetic material. In this case Eq. (10.22) for the structure amplitude of the spin-wave mode has the form

$$\varphi^{\alpha}_{\varkappa\Lambda} = \sqrt{S}\, F(\varkappa)\, e^{-W} (d^{\alpha -} u_{\Lambda}(\varkappa) + d^{\alpha +} v_{\Lambda}(\varkappa)), \tag{10.27}$$

where $\mathbf{d} = (1, -i, 0)$ (the vector $\mathbf{d}$ is composed of the components $d^{-\alpha}$ of the matrix $d^{\alpha\beta} = \delta_{\alpha\beta}$). The quantities u, v are in this case equal to $u_{\Lambda} = 1$, $v_{\Lambda} = 0$ or $u_{\Lambda} = 0$, $v_{\Lambda} = 1$. By calculating the expression for $S_{\lambda}(\varkappa)$ according to Eqs. (10.23) and (10.26), we obtain the cross section of scattering at a ferromagnetic material [69]:)

$$\frac{d^2\sigma_{SW}}{d\Omega d\varepsilon} = \frac{1}{2} S (r_0\gamma)^2 F^2(\varkappa) \frac{k}{k_0} e^{-2W} \sum_{\mathbf{qb}} \{[1 + (\mathbf{em})^2 - 2(\mathbf{em})(\mathbf{ep}_0)] \times$$
$$\times (1 + N(\mathbf{q}))\, \delta_{\varkappa+\mathbf{q},\mathbf{b}}\, \delta(\varepsilon + \hbar\omega(\mathbf{q})) +$$
$$+ [1 + (\mathbf{em})^2 + 2(\mathbf{em})(\mathbf{ep}_0)]\, N(\mathbf{q})\, \delta_{\varkappa-\mathbf{q},\mathbf{b}}\, \delta(\varepsilon - \hbar\omega(\mathbf{q}))\}. \tag{10.28}$$

Now let us switch to another polarization effect in inelastic magnetic scattering, namely to the change of the polarization vector in the scattered beam.

Polarization vector of the scattered beam. One can calculate $\mathbf{p}$ from the following general equation [43]:

$$\mathbf{p}\frac{d^2\sigma}{d\Omega d\varepsilon} = \frac{m^2}{(2\pi)^3\hbar^5} \frac{k}{k_0} \int_{-\infty}^{\infty} dt e^{-i\frac{\varepsilon}{\hbar}t} \langle V^{+}_{\mathbf{kk}_0}(t)\, \sigma V_{\mathbf{kk}_0}(0)\rangle. \tag{10.29}$$

After substituting the amplitude of magnetic scattering according to Eq. (8.1), we can express this relation through the correlators of the atomic spins:

$$p_{\mu}\frac{d^2\sigma}{d\Omega d\varepsilon} = (r_0\gamma)^2 \frac{k}{k_0} \sum_{ll'} F_l(\varkappa) F_{l'}(\varkappa)\, e^{-W_l - W_{l'}}\, e^{i\varkappa(\mathbf{R}^{\circ}_l - \mathbf{R}^{\circ}_{l'})} \times$$
$$\times \sum_{\alpha\beta}\sum_{\gamma\gamma'} \Lambda^{\gamma\mu\gamma'} (\delta_{\alpha\gamma} - e_{\alpha} e_{\gamma})(\delta_{\beta\gamma'} - e_{\beta} e_{\gamma'}) \times$$
$$\times \frac{1}{2\pi\hbar} \int_{-\infty}^{\infty} dt e^{-i\frac{\varepsilon}{\hbar}t} \langle S^{\alpha}_{l}(t)\, S^{\beta}_{l'}(0)\rangle, \tag{10.30}$$

where

$$\Lambda^{\nu\mu\nu'} = \mathrm{Sp}\,(\rho\sigma^{\nu}\sigma^{\mu}\sigma^{\nu'}) = i\varepsilon_{\nu\mu\nu'} + \sum_{\lambda} [\delta_{\nu\mu}\,\delta_{\nu'\lambda} + \delta_{\lambda\nu}\,\delta_{\nu'\mu} - \delta_{\nu\nu'}\,\delta_{\mu\lambda}], \quad (10.31)$$

with the cross section $d^2\sigma/d\Omega d\varepsilon$ defined by Eq. (10.16).

If we want to calculate the polarization of the beam scattered in a single-magnon process, we must separate in Eq. (10.30) terms with correlators of the transverse spin components. Thus, the expression for $d^2\sigma/d\Omega d\varepsilon$ of Eq. (10.19) must be written in on the left side of Eq. (10.30), and the right side must be written in a similar way. Afterwards, by equating the coordinates at the singular factors $\delta_{\varkappa\pm\mathbf{q},\mathbf{b}}\delta(\varepsilon \pm \hbar\omega_\Lambda(\mathbf{q}))$ on the left side and the right side of Eq. (10.29), we obtain an expression for the polarization vector $\mathbf{p}_{\varkappa\Lambda}$ of the beam scattered with absorption or emission of a single magnon of energy $\omega_\Lambda(\mathbf{q})$:

$$p^{\mu}_{\varkappa\Lambda} = \frac{\sum\limits_{\alpha\beta}\sum\limits_{\nu\nu'} \Lambda^{\nu\mu\nu'}\,(\delta_{\alpha\nu} - e_\alpha e_\nu)\,(\delta_{\beta\nu'} - e_\beta e_{\nu'})\,\overset{*}{\varphi}{}^{\alpha}_{\varkappa\Lambda}\,\varphi^{\beta}_{\varkappa\Lambda}}{\sum\limits_{\alpha\beta}\sum\limits_{\nu\nu'} K^{\nu\nu'}\,(\delta_{\alpha\nu} - e_\alpha e_\nu)\,(\delta_{\beta\nu'} - e_\beta e_{\nu'})\,\overset{*}{\varphi}{}^{\alpha}_{\varkappa\Lambda}\,\varphi^{\beta}_{\varkappa\Lambda}}.$$

After substituting Eqs. (10.31) and (10.17) for Λ and K, we finally obtain

$$\mathbf{p}_{\varkappa\Lambda} = \frac{-i\,[\mathfrak{M}^{*}_{\varkappa\Lambda} \times \mathfrak{M}_{\varkappa\Lambda}] + \mathfrak{M}^{*}_{\varkappa\Lambda}(\mathfrak{M}_{\varkappa\Lambda}\,\mathbf{p}_0) + \mathfrak{M}_{\varkappa\Lambda}(\mathfrak{M}^{*}_{\varkappa\Lambda}\,\mathbf{p}_0) - (\mathfrak{M}^{*}_{\varkappa\Lambda}\,\mathfrak{M}_{\varkappa\Lambda})\,\mathbf{p}_0}{(\mathfrak{M}^{*}_{\varkappa\Lambda}\,\mathfrak{M}_{\varkappa\Lambda}) + i\,[\mathfrak{M}^{*}_{\varkappa\Lambda} \times \mathfrak{M}_{\varkappa\Lambda}]\,\mathbf{p}_0}. \quad (10.32)$$

It follows from this expression that in the general case, spontaneous polarization of the scattered beam develops in single-magnon scattering:

$$\overset{\circ}{\mathbf{p}}_{\varkappa\Lambda} = \frac{-i\,[\mathfrak{M}^{*}_{\varkappa\Lambda} \times \mathfrak{M}_{\varkappa\Lambda}]}{(\mathfrak{M}^{*}_{\varkappa\Lambda}\,\mathfrak{M}_{\varkappa\Lambda})}. \quad (10.33)$$

More specifically, in the case of a single-sublattice ferromagnet we obtain the well-known expression

$$\overset{\circ}{\mathbf{p}}_{\varkappa\Lambda} = \mp \frac{2\,(\mathbf{em})\,\mathbf{e}}{1 + (\mathbf{em})^2} \quad (10.34)$$

with the aid of Eq. (10.27). The upper sign refers to scattering under emission, and the lower sign to scattering under absorption, of a spin wave [69]. An antiferromagnetic crystal also leads to spontaneous polarization, provided that the corresponding vector $\mathfrak{M}_{\varkappa\Lambda}$ is complex. The spontaneous polarization of the beam can be detected by recognizing the magnon and phonon peaks in the inelastic scattering of nonpolarized neutrons.

The general formulas for the cross section of single-magnon scattering [Eq. (10.20) together with Eq. (10.23)] and for the polarization Eq. (10.32) of the scattered beam are conveniently compared with the corresponding equations for the cross section of elastic scattering at a magnetic structure. They are [34, 209]:

$$\frac{d\sigma^{\text{elas}}}{d\Omega} = (r_0 \gamma)^2 \sum_L \{(\overset{*}{\mathbf{M}}{}_\varkappa^L \mathbf{M}_\varkappa^L) + \mathrm{i}\,[\overset{*}{\mathbf{M}}{}_\varkappa^L \times \mathbf{M}_\varkappa^L]\,\mathbf{p}_0\}\, N \sum_{\mathbf{b}} \delta_{\varkappa,\, \mathbf{b}+\boldsymbol{\kappa}_L}; \tag{10.35}$$

$$\mathbf{p}_\varkappa^{\text{elas}} = \frac{-\mathrm{i}\,[\overset{*}{\mathbf{M}}{}_\varkappa^L \times \mathbf{M}_\varkappa^L] + \overset{*}{\mathbf{M}}{}_\varkappa^L (\mathbf{M}_\varkappa^L \mathbf{p}_0) + \mathbf{M}_\varkappa^L (\overset{*}{\mathbf{M}}{}_\varkappa^L \mathbf{p}_0) - (\overset{*}{\mathbf{M}}{}_\varkappa^L \mathbf{M}_\varkappa^L)\,\mathbf{p}_0}{(\overset{*}{\mathbf{M}}{}_\varkappa^L \mathbf{M}_\varkappa^L) + \mathrm{i}\,[\overset{*}{\mathbf{M}}{}_\varkappa^L \times \mathbf{M}_\varkappa^L]\,\mathbf{p}_0}. \tag{10.36}$$

Obviously, the vector structure is the same in the expressions for the cross section and for the polarization of elastic and inelastic scattering. The specific feature is that the vectors $\mathbf{M}_\varkappa{}^L$ and $\mathfrak{M}_{\varkappa\Lambda}$ are different. The vector $\mathbf{M}_\varkappa{}^L$, which is given by Eq. (8.12), is expressed through the magnetic structure amplitude $\mathbf{f}_\varkappa{}^L$ of Eq. (8.13) and derived from the equilibrium vectors of the atomic spins. The vector $\mathfrak{M}_{\varkappa\Lambda}$ of Eq. (10.24) is expressed through the structure amplitude of the spin-wave mode of Eq. (10.22) and derived from the atomic components of this mode, which are associated with the projections of the instantaneous positions of the atomic spin vector onto a plane perpendicular to the vector's equilibrium position.

Spin waves in crystals with an incommensurate magnetic structure. For magnetic structures of the spiral type or of the longitudinal spin-wave type one cannot determine a finite magnetic cell whose dimensions would be commensurate with the crystal cell, and our analysis of the spin-wave excitations and of neutron scattering cannot be employed. Nevertheless, the basic concept of a transition to local coordinate systems remains valid, but in the fundamental equation (9.13), each atom with the number l must have its proper transformation matrix $d_l{}^{\alpha\beta}$. In this situation one conveniently expands the quantity $d_l{}^{\alpha\beta}$ as a function of the position of the atom in a Fourier series:

$$d_{ni}^{\alpha\beta} = \sum_\mu d_i^{\alpha\beta}(\mathbf{k}_\mu)\, e^{i\mathbf{k}_\mu \mathbf{t}_n}. \tag{10.37}$$

where n denotes the number of the cell in the crystal in its paramagnetic phase and $\mathbf{t}_n$ denotes the corresponding translation in this crystal [not to be confused with the symbols n and $\mathbf{t}_n$ denoting in Eq. (10.11) and below the number of the cell and the translation within the magnetic cell!].

This expansion must be substituted in the Hamiltonian of the spin-wave approximation in which the Bose operators of the spin deviations must also be expanded in Fourier series with the period of the crystal lattice:

$$b_{ni} = \frac{1}{\sqrt{N}} \sum_{\mathbf{q}} e^{i\mathbf{q}\mathbf{t}_n} b_i(\mathbf{q}), \tag{10.38}$$

assuming that $\mathbf{q}$ is in the first Brillouin zone of the reciprocal lattice of the paramagnetic crystal.

It is easy to understand that the simple form of the spin-wave Hamiltonian obtained from some quadratic form in the spin operators must be

$$\begin{aligned} \mathcal{H}_{SW} = \sum_{\mathbf{q}} \sum_{ij} \sum_{\mu\nu} \Big\{ & B_{ij}(\mathbf{\kappa}_\mu, \mathbf{\kappa}_\nu, \mathbf{q})\, b_i^+(\mathbf{q}+\mathbf{\kappa}_\mu)\, b_j(\mathbf{q}-\mathbf{\kappa}_\nu) + \\ & + \frac{1}{2} A_{ij}(\mathbf{\kappa}_\mu, \mathbf{\kappa}_\nu, \mathbf{q})\, b_i(-\mathbf{q}-\mathbf{\kappa}_\mu)\, b_j(\mathbf{q}-\mathbf{\kappa}_\nu) + \\ & + \frac{1}{2} A_{ij}^*(\mathbf{\kappa}_\mu, \mathbf{\kappa}_\nu, \mathbf{q})\, b_i^+(-\mathbf{q}-\mathbf{\kappa}_\mu)\, b_j^+(\mathbf{q}-\mathbf{\kappa}_\nu) \Big\}, \end{aligned} \tag{10.39}$$

where the quantities A and B satisfy the conditions

$$\left. \begin{aligned} A_{ij}(\mathbf{\kappa}_\mu, \mathbf{\kappa}_\nu, \mathbf{q}) &= A_{ji}(\mathbf{\kappa}_\nu, \mathbf{\kappa}_\mu, \mathbf{q}); \\ B_{ij}(\mathbf{\kappa}_\mu, \mathbf{\kappa}_\nu, \mathbf{q}) &= B_{ji}^*(-\mathbf{\kappa}_\nu, -\mathbf{\kappa}_\mu, \mathbf{q}). \end{aligned} \right\} \tag{10.40}$$

Owing to the additions $\mathbf{k}_\mu$ to the momenta $\mathbf{q}$, the terms in the Hamiltonian with various momenta become entangled and the diagonalization process of the Hamiltonian is reduced to the determination of the eigenvalues of a $2\sigma \times 2\sigma$ matrix as in the case of structures with a finite magnetic cell. The Hamiltonian of Eq. (10.40) describes the transmutation processes through which the quasi-momentum can be changed by some quantity composed of the vectors $\mathbf{q}_\mu$. This reflects the fact that the quasimomentum $\mathbf{q}$, which is given in the reciprocal lattice space of the paramagnetic crystal, is not conserved in a noncommensurate magnetic lattice.

The possible diagonalization of the Hamiltonian of Eq. (10.39) depends upon the number of terms with various $\mathbf{k}_\mu$ appearing in the sum of Eq. (10.37) and upon the form of the coefficients A and B as functions of the vectors $\mathbf{k}_\mu$ and $\mathbf{k}_\nu$. More specifically, in the case of a structure of the "simple-spiral" type considered in Section 9, a comparison of Eqs. (9.28) and (10.37) shows that only three $\mathbf{k}_\mu$ vectors, namely 0, $\mathbf{k}_0$, and $-\mathbf{k}_0$, enter the expansion of Eq. (10.37). It is easy to show that in this case the Hamiltonian of Eq. (10.39) is reduced to the form of Eq. (9.30) and is fully diagonalized. In the case of noncommensurate magnetic structures of greater complexity, the scheme described in the present section must be employed.

11. Classification of Spin Waves in Terms of Symmetry

Corepresentations. It was shown in Chapter 1 that the classification of harmonic oscillations of a crystal lattice (phonons) involves irreducible representations of the space group of the crystal. Similarly, the oscillations of the magnetic lattice (spin waves) of a crystal with magnetic order can be classified with irreducible representations of the magnetic group of the crystal. The theory of representations has been developed only for magnetic Shubnikov groups which describe magnetic structures with finite magnetic cells commensurate with the cell of the paramagnetic phase of the crystal. Noncommensurate magnetic structures of the type of a simple spiral, a longitudinal spin wave, etc., cannot be described by Shubnikov groups and require for their description the so-called color groups for which the theory of representations is still in the development stage. Thus, the symmetry classification scheme of spin waves as outlined below refers only to magnetic materials with Shubnikov symmetry which, fortunately, can describe the vase majority of known crystals with magnetic order.

First let us provide the required information from the theory of Shubnikov group representations (see Vol. 2 of this three-volume work). A Shubnikov group M is constructed from the symmetry elements of Fedorov groups and the so-called anti-unitary operation θ of sign reversal of the spin. The operation θ cannot enter the group M, or can enter it as an independent element, or can enter only in a product with a conventional crystallographic element. Depending upon that, there are three types of Shubnikov groups:

$$\left.\begin{aligned} \mathrm{I}: M &= G; \\ \mathrm{II}: M &= G + \theta G; \\ \mathrm{III}: M &= G + a_0 G. \end{aligned}\right\} \tag{11.1}$$

where G denotes some Fedorov group (not necessarily coinciding with the space group of the particular magnetic crystal); a_0 denotes some anti-unitary element with the structure

$$a_0 = \theta u, \tag{11.2}$$

where u denotes one of the crystallographic elements. The group G forms a so-called unitary subgroup of the group M, and the set of elements a_0G forms an anti-unitary class. Shubnikov groups II (there exists 230 of them) are symmetry groups of paramagnetic crystals, whereas group I and III (there exist 1421 of them) describe crystals with magnetic order [51].

Shubnikov groups are conveniently employed when on makes use of the concept of corepresentations. The latter are derived with the usual representations of space groups. Let us consider Shubnikov group III and the irreducible representation D of its unitary subgroup G. The representation is determined with the aid of a relation which shows the transformation of the basis functions ψ_λ under the influence of an element $u \in G$:

$$u\psi_\lambda = \sum_\mu D_{\mu\lambda}(u)\,\psi_\mu. \tag{11.3}$$

The basis Φ_Λ of the corepresentation is obtained by adding another set of functions to the set of functions ψ_Λ

$$\varphi_\lambda \equiv a_0\psi_\lambda, \tag{11.4}$$

so that Φ_Λ is one of the functions of the set $\{\psi_\lambda, \varphi^\lambda\}$ which is composed of 21 independent functions (l denotes the dimensionality of the representation D).

The system of $2l \times 2l$ matrices D^c defined with the aid of the equation

$$g\Phi_\Lambda = \sum_M D^c_{M\Lambda}(g)\,\Phi_M, \tag{11.5}$$

where g denotes some element of the magnetic group M, is termed the corepresentation of the group M. It is easy to show that the matrices of the corepresentations for the elements of the unitary and nonunitary parts of the group M are constructed from the matrices of the representations D:

$$D^c(u) = \begin{pmatrix} D(u) & 0 \\ 0 & \overset{*}{D}(a_0^{-1} u a_0) \end{pmatrix},$$

$$D^c(a) = \begin{pmatrix} 0 & D(a a_0) \\ \overset{*}{D}(a_0^{-1} a) & 0 \end{pmatrix}. \tag{11.6}$$

For the non-unitary element it holds that $a \in a_0 G$ but the product of two non-unitary elements, e.g., aa_0, renders an unitary element.

Multiplication rules, which differ for the unitary and nonunitary elements, are easily obtained when one uses the explicit form of the corepresentation matrices:

$$\left.\begin{aligned} &D^c(u_1)\, D^c(u_2) = D^c(u_1 u_2); \\ &D^c(u)\, D^c(a) = D^c(ua); \\ &D^c(a)\, \overset{*}{D^c}(u) = D^c(au); \\ &D^c(a_1)\, \overset{*}{D^c}(a_2) = D^c(a_1 a_2), \end{aligned}\right\} \tag{11.7}$$

and this indicates the difference of the corepresentations from the usual (unitary) representations of the space groups.

Thus, by virtue of Eq. (11.6), a corepresentation D^c is set into correspondence with each of the irreducible representations D of the unitary group G. The corepresentations derived in this way can be reducible in certain cases, and the reduction is effected with the aid of the unitary matrix U according to the scheme

$$\begin{aligned} D'^c(u) &= U^{-1} D^c(u)\, U; \\ D'^c(a) &= U^{-1} D^c(a)\, U^*. \end{aligned} \tag{11.8}$$

The matrices U effects the transition from the basis Φ_M of the corepresentation to the new basis $\Phi_{\Lambda'}$:

$$\Phi'_\Lambda = \sum_M U_{M\Lambda}\, \Phi_M. \tag{11.9}$$

Let us note that if the two representations D^c and D'^c are related by Eqs. (11.8), they are equivalent.

Two cases can be encountered in the analysis of the reducibility of corepresentations defined by Eqs. (11.6):

1) If the unitary representations $D(u)$ and $D^*(a_0{}^{-1}ua_0)$ are non-equivalent, the corepresentation $D^c(g)$ is irreducible;
2) if the representations $D(u)$ and $D^*(a_0{}^{-1}ua_0)$ are equivalent, the corepresentation $D^c(g)$ can be irreducible or reducible.

In practical terms, reducibility can be recognized on the basis of the relations [131]

$$\frac{1}{\|G\|}\sum_{a\in a_0G} \operatorname{Sp} D(a^2) = \begin{cases} 0, & \text{then } D^c \text{ is irreducible (case 1);} \\ -1, & \text{then } D^c \text{ is irreducible (case 2);} \\ 1, & \text{then } D^c \text{ is reducible (case 2).} \end{cases} \tag{11.10}$$

The spin-wave representation of the magnetic group. Corepresentations of magnetic groups are important in the physics of crystals with magnetic order in view of the Wigner theorem [14] which states that some corepresentation of a magnetic Shubnikov group M is associated with each of the eigenvalues of a Hamiltonian whose symmetry is described by a magnetic Shubnikov group M. Accordingly, each spin wave with a given wave vector $\mathbf{q}$ can be related to a corepresentation of the group $M_{\mathbf{q}}$ of the wave vector of the magnetic group M. This correspondence is the basis of the grouptheoretical classification of the spin waves [30, 300].

It has already been discussed in Section 9 that for determining spin waves in some magnetic structure, one must consider the spin vectors at each magnetic atom in the local coordinate system whose z axis is chosen parallel to the direction of the equilibrium position of the spin. The set of equilibrium vectors of the spin (or of the magnetic moment) of each of the atoms does form the magnetic structure of the crystal. The projections of the spin on a plane perpendicular to the direction of the average vector (quantization axes) of the spin of the atom participate in the spin-wave motion. Therefore, a spin-wave mode is given if two projections S'^x and S'^y of the spin onto the x and y axes chosen on that plane are given for each magnetic atom.

Consequently, in order to derive the spin-wave modes, one must choose the unit vectors of the spin-wave space, i.e., a $2\sigma N$-dimensional columns $\psi_{nj\gamma}$ with a single nonvanishing element (which is equal to 1) at the atom nj for the projection $\gamma = x, y$. In this case, σ denotes the number of the magnetic atoms in the primitive magnetic cell, and N denotes the number of such cells. When we consider a mode with a given wave vector $\mathbf{q}$, it suffices to concern ourselves with the 2σ-dimensional columns $\psi_{j\gamma}{}^{\mathbf{q}}$ corresponding to the atoms of the zeroth cell only. Any configuration $\{S_{ni}{}'^{\gamma}\}$ of spin-wave deviations of spins with the wave vector $\mathbf{q}$ is given by the $2\sigma N$-dimensional column which can be stated in the form [44]:

$$\{S_{ni}'^{\gamma}\}=\sum_{n}^{\oplus}\sum_{j\gamma}S_{j}'^{\gamma}\,\psi_{j\gamma}^{\mathbf{q}}\,e^{i\mathbf{q}\mathbf{t}_{n}}, \tag{11.11}$$

where $\mathbf{t}_n$ denotes the vector of translation from the zeroth magnetic cell to the nth cell.

Let us note that here, and everywhere in what follows, the wave vector $\mathbf{q}$ of the spin wave is taken from the first Brillouin zone of the magnetic lattice, which in this case can be one of the 36 Bravais lattices of Shubnikov symmetry.

Now let us have the element g of the group M act upon the column $\psi_{nj\gamma}$. Under the influence of the rotational part h of the element $g = \{h \mid \tau_n\}$, the spin of the jth atom in the zeroth cell is rotated, and the atom proper is transferred to another cell and its number changed to i. With the considerations resembling those used by us in Section 4 in the derivation of the mechanical representation, we restate the result of the effect of the element g upon the 2σ-dimensional column $\psi_{j\gamma}{}^{\mathbf{q}}$ in the following form

$$T(g)\,\psi_{j\gamma}^{\mathbf{q}}=e^{-i\mathbf{q}\mathbf{a}_{p}(g,\,j)}\,\delta_{h}\sum_{\gamma'}O_{ij}^{\gamma'\gamma}(h)\,\psi_{i\gamma'}^{\mathbf{q}}. \tag{11.12}$$

where $O_{ij}{}^{\gamma'\gamma}(h)$ denotes a part of the three-dimensional matrix $O_{ij}(h)$ formed by the matrix R_h of rotation of the element g and by the matrices d_j and d_i of the transition from the laboratory system to the local coordinate system:

$$O_{ij}(h)=d_{i}\,R_{h}\,d_{j}^{-1}. \tag{11.13}$$

The purpose of the operation $O_{ij}(h)$ is to transfer the spin vector of the jth atom from the local system, then to perform the rotation R_h on it, and to transfer the resulting vector into the local coordinate system for the atom i into which atom j is transformed by the group operation acting on it. The vector $\mathbf{a}_p(g, j)$ of the return translation is usually given by the relation

$$g\mathbf{x}_{j}=h\mathbf{x}_{j}+\tau_{h}=\mathbf{x}_{i}+\mathbf{a}_{p}(g,\,j). \tag{11.14}$$

Of course, one must bear in mind that $\mathbf{a}_p$ denotes the translation vector of the magnetic lattice.

Equation (11.12) can be restated in the form

$$T(g)\,\psi^{\mathbf{q}}_{j\gamma}=\sum_{i\delta}\{d^{\mathbf{q}}_{\mathrm{SW}}(g)\}_{i\delta,\,j\gamma}\,\psi^{\mathbf{q}}_{i\delta},\tag{11.15}$$

where

$$\{d^{\mathbf{q}}_{\mathrm{SW}}(g)\}_{i\delta,\,j\gamma}=e^{-i\mathbf{q}\mathbf{a}_p(g,\,j)}\,\delta_{i,\,gj}\,\delta_h\,O^{\delta\gamma}_{ij}(h),\tag{11.16}$$

i.e., there appears a $2\sigma\times 2\sigma$ matrix which effects some representation of the group $G_{\mathbf{q}}$. We term it the spin-wave representation. In the general case, it is reducible and can be expanded in the normal way in terms of irreducible representation $d^{\mathbf{q}\gamma}$ of the group $G_{\mathbf{q}}$:

$$d^{\mathbf{q}}_{\mathrm{SW}}(g)=\sum_{\nu}n^{\nu}_{\mathrm{SW}}\,d^{\mathbf{q}\nu}.\tag{11.17}$$

The character of the spin-wave representation can be calculated with the aid of the formula

$$\chi^{\mathbf{q}}_{\mathrm{SW}}(g)=\sum_{j}e^{-i\mathbf{q}\mathbf{a}_p(g,\,j)}\,\delta_{j,\,gj}\,\delta_h\,2O^{xx}_{jj}(h),\tag{11.18}$$

resulting from the definition by Eq. (11.16). We recalled that the matrix $O_{ij}{}^{\delta\gamma}(h)$ (a two-line matrix with the subscripts δ, $\gamma = x, y$) is an unitary matrix of rotation around the z axis of the local coordinate system attached to atom j; therefore, for it $O_{ij}{}^{yy} = O_{ij}{}^{xx}$ holds.

For the irreducible representations contained in $d_{SW}{}^{\mathbf{q}}$, we now set up linear combinations of the unit vectors $\psi_{j\gamma}{}^{\mathbf{q}}$ which are transformed according to the matrices of these representations. They are also the possible spin-wave modes of a given crystal with magnetic order. Working formulas can be derived with the usual technique of a projection operator as in the case of the phonon modes. In specific terms, one must insert in general equation (4.19) one of the unit vectors $\psi_{jx}{}^{\mathbf{q}} \pm i\psi_{jy}{}^{\mathbf{q}}$ of the spin-wave representation as a starting function ψ and thereafter use Eq. (11.12). This renders the following structure of the basis functions:

$$\left.\begin{aligned}\psi^{\mathbf{q}\nu}_{\lambda}&=\sum_{n}{}^{\oplus}\,\sigma\begin{pmatrix}\mathbf{q}\nu\\ \lambda\end{pmatrix}e^{i\mathbf{q}\mathbf{t}_n};\\ \sigma\begin{pmatrix}\mathbf{q}\nu\\ \lambda\end{pmatrix}&=\sum_{i}{}^{\oplus}\,\sigma\begin{pmatrix}\mathbf{q}\nu\\ \lambda\end{pmatrix|\,i\end{pmatrix},\end{aligned}\right\}\tag{11.19}$$

where $\sigma\begin{pmatrix} \mathbf{q}\nu & i \\ \lambda & \end{pmatrix} = \begin{pmatrix} \sigma^+ \\ \sigma^- \end{pmatrix}$ denotes a two-component atomic function:

$$\sigma^{\pm}\begin{pmatrix} \mathbf{q}\nu & i \\ \lambda & \end{pmatrix} = \sum_{g \in G_{\mathbf{q}}^{\circ}} \overset{*}{d}{}^{\mathbf{q}\nu}_{\lambda[\mu]}(g)\, e^{-i\mathbf{q}\mathbf{a}_p(g,\, j)}\, \delta_{i,\, gj}\, \delta_h \left(O^{xx}_{ij} \pm iO^{xy}_{ij}\right). \qquad (11.20)$$

In this equation, the subscripts j and μ are fixed and a set of linear independent basis functions which are transformed according to the νth irreducible representation of the group $G_{\mathbf{q}}$ corresponds to them. When a representation appears several times in the composition of $d_{SW}{}^{\mathbf{q}}$, we must obtain a corresponding number of linearly independent sets which we will number by the subscript s. Thus, the instructions for deriving the basis functions of the spin wave representation have much in common with the magnetic and mechanical representations. By contrast to the latter, now we have to calculate the matrix O_{ij} defined by Eq. (11.13). To this end, when the magnetic crystal structure is given, one must find the matrix d_i of the local transition from the laboratory coordinate system to the local coordinate system. This transition is most simply effected by rotating the z axis to the position of the spin vector of atom i. Consequently, the rotation is performed over some angle φ_i of the transition from the laboratory coordinate system to the local coordinate system. This transition is most simply effected by rotating the z axis to the position of the spin vector of atom i. Consequently, the rotation is performed over some angle φ_i on the (z, S_i) plane around the n_i axis situation on the (x, y) plane; the unit vector $\mathbf{n}_i$, which defines the axis of rotation, has the form

$$\mathbf{n}_i = \{S_i^y,\ -S_i^x,\ 0\}\, \frac{1}{\sqrt{S_i^{x^2} + S_i^{y^2}}}\,. \qquad (11.21)$$

The matrices d_i of the rotation can be stated with the geometrical identity $(\mathbf{r}' \equiv d\mathbf{r})$

$$\mathbf{r}' = \mathbf{n}(\mathbf{n}\mathbf{r}) + [\mathbf{n} \times [\mathbf{r} \times \mathbf{n}]] \cos\varphi + [\mathbf{n} \times \mathbf{r}] \sin\varphi \qquad (11.22)$$

and have the form

$$d = \begin{pmatrix} n_x^2 + n_y^2 \cos\varphi & n_x n_y (1 - \cos\varphi) & n_y \sin\varphi \\ n_x n_y (1 - \cos\varphi) & n_y^2 + n_x^2 \cos\varphi & -n_x \sin\varphi \\ -n_y \sin\varphi & n_x \sin\varphi & (n_x^2 + n_y^2)\cos\varphi \end{pmatrix}. \qquad (11.23)$$

Diagonalization of the spin-wave Hamiltonian. The basis functions of Eq. (11.20) can be used to diagonalize the Hamiltonian of Eq. (10.2) of a crystal with magnetic order in the harmonic (spin-wave) approximation. The problem is reduced to determining the corresponding unitary matrix and can be solved in three steps. In the first step, the relation between the matrix elements of the spin-wave matrix H(**q**) must be established. Since any element g of the $G_{\mathbf{q}}$ is a symmetry element of the Hamiltonian and therefore must commutate with it, the following relation holds:

$$\overset{+}{d}{}^{\mathbf{q}}_{SW}(g)\,\mathrm{H}(\mathbf{q})\,d^{\mathbf{q}}_{SW}(g) = \mathrm{H}(\mathbf{q}), \tag{11.24}$$

where $d_{SW}{}^{\mathbf{q}}(g)$ denotes the $2\sigma \times 2\sigma$ matrix of the spin-wave representation of the group $G_{\mathbf{q}}$, which matrix must be calculated with Eq. (11.16). The use of Eqs. (11.24) for the generators of the group $G_{\mathbf{q}}$ determines the form of the spin-wave matrix H(**q**), with the form resulting from the symmetry of the magnetic structure. This is a complete analog to studying the dynamic matrix of a crystal with the aid of the matrices of the mechanical representation of its space group.

The second step involves the derivation of a $2\sigma \times 2\sigma$ matrix composed of the basis functions of the spin-wave representation:

$$U'_{i\gamma,\,\Lambda}(\mathbf{q}) = \begin{pmatrix} \sigma^{+}\begin{pmatrix} \mathbf{q}\nu \\ \lambda \end{pmatrix} i \end{pmatrix} \\ \sigma^{-}\begin{pmatrix} \mathbf{q}\nu \\ \lambda \end{pmatrix} i \end{pmatrix} \end{pmatrix}. \tag{11.25}$$

where Λ denotes the composite subscript $\nu\lambda$ numbering these basis functions; and γ denotes a subscript having the value "plus" and "minus." When the functions σ^{+} and σ^{-} defined by Eq. (11.20) are normalized to unity, the matrix $U'(\mathbf{q})$ is unitary and must split the matrix H(**q**) into blocks corresponding to the irreducible representations ν entering into the spin-wave representation. Thus, we have

$$U'^{+}(\mathbf{q})\,\mathrm{H}(\mathbf{q})\,U'(\mathbf{q}) = \sum_{\nu}{}^{\oplus}\,\hbar\omega_{\nu}(\mathbf{q})\,E_{\nu} \oplus \sum_{\mu}{}^{\oplus}\,\mathrm{H}_{\mu}. \tag{11.26}$$

where E_{ν} denotes the unit matrix of the dimensionality of the νth irreducible representation, provided that the representation appears only once in the composition of the spin-wave representation; H_{μ} is the matrix block corresponding to the μth repeated representation. Its dimensionality

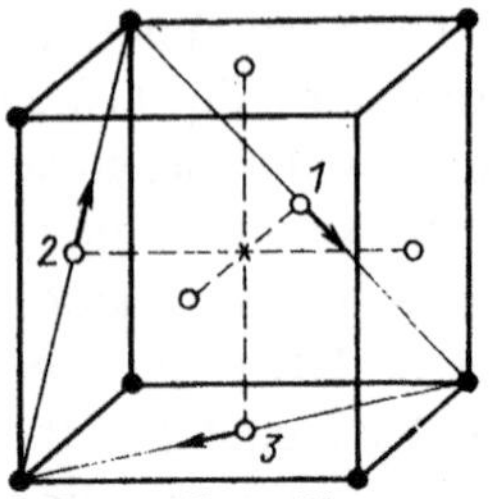

Fig. 8. Unit cell of the magnetic structure of the Mn_3NiN perovskite.

is $n_\mu l_\mu$, where n_ν denotes the multiplicity with which this representation appears. The blocks H_μ are nondiagonal, and their diagonalization cannot be obtained with symmetry methods.

The third step of the full diagonalization of the Hamiltonian involves the diagonalization of the blocks H_μ.

In the event that the matrix H(**q**) has been fully diagonalized, we obtain the spectrum of the spin waves with a given wave vector **q**, with the spectrum classified according to the irreducible representations of the group $G_{\mathbf{q}}$ of the unitary subgroup G of the magnetic symmetry group M. The presence of additional nonunitary symmetry elements in the group M can lead to additional degeneracy of the energy of the spin waves classified in accordance with the unitary subgroup. The intrinsic multiplicity of the spin-wave frequencies corresponds to the multiplicity of the corepresentations of the group of the wave vector of the full group of magnetic symmetry of the crystal. The multiplicity of the corepresentations can be determined with the criterion indicated in Eq. (11.10).

The symmetry analysis of the spin-wave spectrum will now be illustrated by way of the example of magnetic structures of some perovskites (see Section 4 in which the crystal structure of perovskites was described and the symmetry analysis of their phonon spectrum was considered).

Example 1. A triangular antiferromagnetic structure with the wave vector $\mathbf{k} = 0$ was detected in a certain temperature interval in an Mn_3NiN crystal. The magnetic Mn atoms occupy the 3(*c*) positions as indicated in Eq. (4.40); the magnetic moments are aligned with the diagonals of the faces (Fig. 8):

$$\mathbf{S}_1 = (011),\ \mathbf{S}_2 = (\bar{1}01),\ \mathbf{S}_3 = (1\bar{1}0). \tag{11.27}$$

TABLE 9. Transposition of the Magnetic Atoms of a Perovskite Crystal under the Influence of the Elements of the Group $G_{\mathbf{q}}$*

Elements $G_{\mathbf{q}}$	Atoms			Elements $G_{\mathbf{q}}$	Atoms		
	1	2	3		1	2	3
h_1 (H_1)	1	2	3	h_{25} (H_{13})	1 (011)	2 (101)	3 (110)
h_5 (H_5)	3	1	2	h_{29} (H_{17})	3 (110)	1 (011)	2 (101)
h_9 (H_3)	2	3	1	h_{33} (H_{15})	2 (101)	3 (110)	1 (011)
h_{13} (H_8)	2 (101)	1 (011)	3 (110)	h_{37} (H_{20})	2	1	3
h_{17} (H_{12})	1 (011)	3 (110)	2 (101)	h_{41} (H_{24})	1	3	2
h_{21} (H_{10})	3 (110)	2 (i01)	1 (011)	h_{45} (H_{22})	3	2	1

*The elements of the group are given in cubic (*h*) and hexagonal (*H*) arrangements.

One can show [267] that this structure is described by the representation τ_7 with $\mathbf{k} = 0$ of the space group $O_h{}^1$ of perovskites. One can easily establish by direct verification that the following elements of the group leave the structure invariant:

$$h_1,\ h_5,\ h_9,\ h_{13},\ h_{17},\ h_{21},\ h_{25},\ h_{29},\ h_{33},\ h_{37},\ h_{41},\ h_{45}. \tag{11.28}$$

They, together with the translations of the simple cubic lattice, form the space group $D_{3d}{}^5$. Thus, the magnetic group of the crystal

$$M = D_{3d}^5 \tag{11.29}$$

consists only of unitary elements [type I of Eq. (11.1)].

Table 9 lists the transpositions of the magnetic atoms subjected to the effect of the elements (11.28) of the group $D_{3d}{}^5$ and also lists the return translations $\mathbf{a}_p$. For the ensuing calculations we still need the matrices d_i of the transition to the local coordinate system for each of the three magnetic atoms. According to the general equation (11.23), we have

$$d_1 = \begin{pmatrix} 1 & 0 & 0 \\ 0 & -1\sqrt{2} & -1/\sqrt{2} \\ 0 & 1/\sqrt{2} & -1/\sqrt{2} \end{pmatrix},\quad d_2 = \begin{pmatrix} 1/\sqrt{2} & 0 & 1/\sqrt{2} \\ 0 & 1 & 0 \\ -1/\sqrt{2} & 0 & 1/\sqrt{2} \end{pmatrix},$$

$$d_3 = \begin{pmatrix} 1/2 & 1/2 & -1/\sqrt{2} \\ 1/2 & 1/2 & 1/\sqrt{2} \\ 1/\sqrt{2} & -1\sqrt{2} & 0 \end{pmatrix}. \tag{11.30}$$

Now the quantities $O_{ij}(h)$ defined by Eq. (11.13) must be compiled. It suffices to calculate the following matrices for determining the basis functions:

$$O_{11}(h_{17}) = O_{22}(h_{21}) = O_{33}(h_{13}) = \begin{pmatrix} -1 & 0 & 0 \\ 0 & -1 & 0 \\ 0 & 0 & 1 \end{pmatrix}; \tag{11.31}$$

$$O_{12}(h_9) = \begin{pmatrix} 0 & -1 & 0 \\ 1 & 0 & 0 \\ 0 & 0 & 1 \end{pmatrix}, \quad O_{12}(h_{13}) = \begin{pmatrix} 0 & 1 & 0 \\ -1 & 0 & 0 \\ 0 & 0 & 1 \end{pmatrix}; \tag{11.32}$$

$$O_{13}(h_5) = \begin{pmatrix} \frac{-1}{\sqrt{2}} & \frac{-1}{\sqrt{2}} & 0 \\ \frac{1}{\sqrt{2}} & \frac{-1}{\sqrt{2}} & 0 \\ 0 & 0 & 1 \end{pmatrix}, \quad O_{13}(h_{21}) = \begin{pmatrix} \frac{1}{\sqrt{2}} & \frac{1}{\sqrt{2}} & 0 \\ \frac{-1}{\sqrt{2}} & \frac{1}{\sqrt{2}} & 0 \\ 0 & 0 & 1 \end{pmatrix}. \tag{11.33}$$

The quantities of Eq. (11.31) are required for calculating the characters of the spin-wave representation with Eqs. (11.18). According to the table of transpositions of the atoms, one still has to know $O_{11}(h_{41})$, $O_{22}(h_{45})$, and $O_{33}(h_{37})$ but by virtue of the properties of the elements of the group ($h_{41} = h_{17}h_{25}$, $h_{45} = h_{21}h_{25}$, and $h_{37} = h_{13}h_{25}$), expressions with the aid of the matrices (11.31)-(11.32) are possible, e.g.,

$$O_{11}(h_{41}) = O_{11}(h_{17})\, O_{11}(h_{25}) = -O_{11}(h_{17}).$$

Now let us calculate the characters of the spin-wave representation for the wave vector $\mathbf{q} = 0$. The group $G_{\mathbf{q}}$ of the wave vector coincides with the group D_{3d}^5 and comprises the 12 rotational elements (11.28). The characters for them are

$$\chi^{\mathbf{q}}_{\mathrm{SW}}(h_1) = \chi^{\mathbf{q}}_{\mathrm{SW}}(h_{25}) = 6;$$

$$\chi^{\mathbf{q}}_{\mathrm{SW}}(h_{13}) = \chi^{\mathbf{q}}_{\mathrm{SW}}(h_{17}) = \chi^{\mathbf{q}}_{\mathrm{SW}}(h_{21}) = \chi^{\mathbf{q}}_{\mathrm{SW}}(h_{37}) = \chi^{\mathbf{q}}_{\mathrm{SW}}(h_{41}) = \\ = \chi^{\mathbf{q}}_{\mathrm{SW}}(h_{45}) = -2;$$

$$\chi^{\mathbf{q}}_{\mathrm{SW}}(h_5) = \chi^{\mathbf{q}}_{\mathrm{SW}}(h_9) = \chi^{\mathbf{q}}_{\mathrm{SW}}(h_{29}) = \chi^{\mathbf{q}}_{\mathrm{SW}}(h_{33}) = 0.$$

TABLE 10. The Basis Functions of the Spin-Wave Representation for the Magnetic Structure of the Mn_3NiN Perovskite on the **Γ/R** line (including the points **Γ** and **R**), $\varepsilon = \exp(i(2\pi/3))$, $\delta = \exp(i(\pi/4))$.

Wave vectors			Atoms					
Γ	Γ–R	R	1		2		3	
τ_3	τ_2	τ_3	1	0	i	0	$-\delta^*$	0
τ_3'	τ_2'	τ_3'	0	1	0	$-i$	0	$-\delta$
τ_5	τ_3	τ_5	1	0	$i\varepsilon^*$	0	$-\varepsilon\delta^*$	0
			$-\varepsilon^*$	0	$-i$	0	$\varepsilon\delta^*$	0
τ_5'	τ_3'	τ_5'	0	1	0	$-i\varepsilon^*$	0	$-\varepsilon\delta$
			0	$-\varepsilon^*$	0	i	0	$\varepsilon\delta$

The irreducible representations of the group D_{3d}^5 for $\mathbf{q} = 0$ are one-dimensional ($\tau_1, \ldots, \tau_4$) and two-dimensional (τ_5, τ_6). The six-dimensional spin-wave representation is split in the following fashion into irreducible representations:

$$d_{SW}^{\mathbf{q}} = 2\tau_3 + 2\tau_5.$$

The spin-wave basis functions of the representations τ_3 and τ_5 are calculated with Eq. (11.20). When the atom $j = 1$ is used for starting, the matrices of Eq. (11.32) suffice for the calculation. The results are listed in Table 10. This table includes the basis functions for the point **R** of the Brillouin zone [$\mathbf{q} = 1/2(\mathbf{b}_1 + \mathbf{b}_2 + \mathbf{b}_3)$] and on the line **Λ** [$\mathbf{q} = \mu(\mathbf{b}_1 + \mathbf{b}_2 + \mathbf{b}_3)$]. The nomenclature of the irreducible representations at the points **Γ** and **R** coincides with the Kovalev nomenclature. As to the points on the line **Λ**, the group of the wave vector is for them two times smaller than the group D_{3d}^5 and comprises the rotational elements h_1, h_5, h_9, h_{37}, h_{41}, and h_{45}. This group has two one-dimensional (τ_1, τ_2) and one two-dimensional (τ_3) representation. The spin-wave representation is in this case characterized by the relation

$$d_{SW}^{\mathbf{q}} = 2\tau_2 + 2\tau_3, \quad \mathbf{q} = \mu(\mathbf{b}_1 + \mathbf{b}_2 + \mathbf{b}_3). \tag{11.34}$$

It is noteworthy that the atomic components of the basis function for the lines **Λ** and its end points **Γ** and **R** coincide. Since one spin-wave energy corresponds to each irreducible representation entering into the spin-wave representation, no additional splitting of the spin-wave spectrum occurs in the transition from the point **Γ** to the point **R** along the line **λ**. Thus, four branches of spin waves exist in a particular magnetic structure on the entire line **Λ**, namely two nondegenerate of the same symmetry and two double-degenerate of another symmetry.

The results can be confirmed by direct diagonalization of the matrix H(**q**) of the Hamiltonian. To do this, the unitary matrix must be compiled on the basis of Table 10:

$$U'(\mathbf{q}) = \frac{1}{\sqrt{3}} \begin{pmatrix} 1 & 0 & 1 & -\varepsilon^* & 0 & 0 \\ i & 0 & i\varepsilon^* & -i & 0 & 0 \\ -\delta^* & 0 & -\varepsilon\delta^* & \varepsilon\delta^* & 0 & 0 \\ 0 & 1 & 0 & 0 & 1 & -\varepsilon^* \\ 0 & -i & 0 & 0 & -i\varepsilon^* & i \\ 0 & -\delta & 0 & 0 & -\varepsilon\delta & \varepsilon\delta \\ \tau_2 & \tau_2' & \multicolumn{2}{c}{\underbrace{\qquad\qquad}_{\tau_3}} & \multicolumn{2}{c}{\underbrace{\qquad\qquad}_{\tau_3'}} \end{pmatrix}. \tag{11.35}$$

This matrix was stated for $\mathbf{q} = \mu(\mathbf{b}_1 + \mathbf{b}_2 + \mathbf{b}_3)$; the matrix has the same form for the other wave vector: $\mathbf{q} = 0$ and $\mathbf{q} = 1/2(\mathbf{b}_1 + \mathbf{b}_2 + \mathbf{b}_3)$. Further, one must take into account the symmetry relations of the elements of the matrix H(**q**) which result from Eq. (11.24). The matrices of the spin-wave representation for the generators h_5 and h_{13} of the group $G_{\mathbf{q}}$ of the wave vector with $\mathbf{q} = 0$ and $\mathbf{q} = 1/2(\mathbf{b}_1 + \mathbf{b}_2 + \mathbf{b}_3)$ can be easily calculated with Eq. (11.16):

$$d_{SW}^{\mathbf{q}}(h_5) = \left(\begin{array}{ccc:ccc} 0 & -i & 0 & & & \\ 0 & 0 & \delta^* & & & \\ -\delta^* & 0 & 0 & & & \\ \hdashline & & & 0 & i & 0 \\ & & & 0 & 0 & \delta \\ & & & -\delta & 0 & 0 \end{array}\right),$$

$$d_{SW}^{q}(h_{13})=\left(\begin{array}{ccc|ccc} 0 & i & 0 & & & \\ -i & 0 & 0 & & & \\ 0 & 0 & -1 & & & \\ \hline & & & 0 & -i & 0 \\ & & & i & 0 & 0 \\ & & & 0 & 0 & -1 \end{array}\right).$$

Using these matrices in Eqs. (11.24), the explicit form of the matrix H(**q**) is obtained:

$$\left(\begin{array}{ccc|ccc} B_{11}(\mathbf{q}) & B_{12}(\mathbf{q}) & \delta^{*}B_{12}(\mathbf{q}) & A_{11}^{*}(-\mathbf{q}) & A_{12}^{*}(-\mathbf{q}) & \delta A_{12}^{*}(-\mathbf{q}) \\ -B_{12}(\mathbf{q}) & B_{11}(\mathbf{q}) & \delta B_{12}(\mathbf{q}) & A_{12}^{*}(-\mathbf{q}) & -A_{11}^{*}(-\mathbf{q}) & -\delta^{*}A_{12}^{*}(-\mathbf{q}) \\ -\delta B_{12}(\mathbf{q}) & -\delta^{*}B_{12}(\mathbf{q}) & B_{11}(\mathbf{q}) & \delta A_{12}^{*}(-\mathbf{q}) & -\delta^{*}A_{12}^{*}(-\mathbf{q}) & -iA_{11}^{*}(-\mathbf{q}) \\ \hline A_{11}(\mathbf{q}) & A_{12}(\mathbf{q}) & \delta^{*}A_{12}(\mathbf{q}) & B_{11}(-\mathbf{q}) & B_{21}(-\mathbf{q}) & \delta B_{21}(-\mathbf{q}) \\ A_{12}(\mathbf{q}) & -A_{11}(\mathbf{q}) & -\delta A_{12}(\mathbf{q}) & -B_{21}(-\mathbf{q}) & B_{11}(-\mathbf{q}) & \delta^{*}B_{21}(-\mathbf{q}) \\ \delta^{*}A_{12}(\mathbf{q}) & -\delta A_{12}(\mathbf{q}) & iA_{11}(\mathbf{q}) & -\delta^{*}B_{21}(-\mathbf{q}) & -\delta B_{21}(-\mathbf{q}) & B_{11}(-\mathbf{q}) \end{array}\right). \quad (11.36)$$

Thus, in each 3 × 3 block of this matrix there are only two independent elements, and a total of four independent elements, B_{11}, B_{12}, A_{11}, and A_{12}, exist in this matrix.

After the unitary transformation, the matrix H(**q**) consists of two diagonal blocks:

$$U'^{+}(\mathbf{q})\,\mathrm{H}(\mathbf{q})\,U'(\mathbf{q})=\mathrm{H}^{\mathbf{q}\tau_2}\oplus\mathrm{H}^{\mathbf{q}\tau_3}, \quad (11.37)$$

where

$$\mathrm{H}^{\mathbf{q}\tau_2}=\begin{pmatrix} B_{11}(\mathbf{q})+2iB_{12}(\mathbf{q}) & A_{11}^{*}(\mathbf{q})-2iA_{12}^{*}(\mathbf{q}) \\ A_{11}(\mathbf{q})+2iA_{12}(\mathbf{q}) & B_{11}(\mathbf{q})+2iB_{12}(\mathbf{q}) \end{pmatrix}; \quad (11.38)$$

$$\mathrm{H}^{\mathbf{q}\tau_3}=\begin{pmatrix} [B_{11}(\mathbf{q})-iB_{12}(\mathbf{q})]\,\hat{E} & [A_{11}^{*}(\mathbf{q})+iA_{12}^{*}(\mathbf{q})]\,\hat{E} \\ [A_{11}(\mathbf{q})-iA_{12}(\mathbf{q})]\,\hat{E} & [B_{11}(\mathbf{q})-iB_{12}(\mathbf{q})]\,\hat{E} \end{pmatrix}. \quad (11.39)$$

In the second block, Ê is a second-order matrix, and its appearance results from the irreducible representation τ_3 being two-dimensional.

The matrix H(**q**) has two-block form because the symmetry in the system was used. The incomplete diagonalization is associated with the repeated representations, and further diagonalization is no longer a consequence of symmetry. In order to diagonalize the matrix blocks with the repeated representations, superpositions of the basis functions of these representations

$$\sum_s C_{\Lambda s}\,\sigma_s\begin{pmatrix} \mathbf{q}\nu \\ \lambda \end{pmatrix} \tag{11.40}$$

must be used and sets of mixing coefficients $C_{\Lambda s}$ must be chosen accordingly for each eigenvalue $\hbar\omega_\nu$ of a block.

The eigenvalues of the matrices (11.38) and (11.39) can be easily obtained by solving the corresponding quadratic equations. We state the results for the spin-wave spectrum in the exchange approximation, taking into account that the Fourier components of the exchange integral satisfy the general relations:

$$J_{ij}(\mathbf{q}) = J_{ji}(-\mathbf{q}) = J^*_{ij}(-\mathbf{q}). \tag{11.41}$$

In the case of a perovskite crystal, the following symmetry relation between the elements of the exchange matrix exists at $\mathbf{q} = \mu(\mathbf{b}_1 + \mathbf{b}_2 + \mathbf{b}_3)$:

$$J_{11}(\mathbf{q}) = J_{22}(\mathbf{q}) = J_{33}(\mathbf{q});$$
$$J_{12}(\mathbf{q}) = J_{13}(\mathbf{q}) = J_{23}(\mathbf{q}) = J_{21}(\mathbf{q}) = J_{31}(\mathbf{q}) = J_{32}(\mathbf{q}). \tag{11.42}$$

Let us disregard the structure distortions of the magnetically ordered crystal. It follows from the simultaneous conditions (11.41) and (11.42) that the two independent parameters of the exchange relation $J_{11}(\mathbf{q})$ and $J_{12}(\mathbf{q})$ must be real. Using these relations in the explicit notation of the matrix elements in Eqs. (11.38) and (11.39), we obtain the spectrum of the spin waves on the line Λ:

$$\hbar\omega^{\tau_2}(\mathbf{q}) = S\sqrt{[J_{11}(\mathbf{q}) - J_{11}(0) - J_{12}(\mathbf{q}) + J_{12}(0)]\,[J_{11}(\mathbf{q}) - J_{11}(0) + 2J_{12}(\mathbf{q}) + J_{12}(0)]};$$

$$\hbar\omega^{\tau_3}(\mathbf{q}) = S\sqrt{[J_{11}(\mathbf{q}) - J_{11}(0) - J_{12}(\mathbf{q}) + J_{12}(0)]\left[J_{11}(\mathbf{q}) - J_{11}(0) + \frac{1}{2}J_{12}(\mathbf{q}) + J_{12}(0)\right]}. \tag{11.43}$$

At the boundaries of the line $\Gamma - \mathbf{R}$, i.e., at $\mathbf{q} = 0$ and $\mathbf{q} = 1/2(\mathbf{b}_1 + \mathbf{b}_2 + \mathbf{b}_3)$, Eqs. (11.43) render the frequencies of the spin waves at the points Γ and $\mathbf{R}$ but their nomenclature is changed (τ_3 and τ_5 in place of τ_2 and τ_3).

Example 2. A triangular magnetic structure with the wave vector **k** = 0 exists in the Mn_3NiN crystal still in a single temperature interval. As in Example 1, the atomic spins of the magnetic unit cells, i.e.,

$$\mathbf{S}_1 = (2\bar{1}\bar{1}), \quad \mathbf{S}_2 = (\bar{1}2\bar{1}), \quad \mathbf{S}_3 = (\bar{1}\bar{1}2) \tag{11.44}$$

are on a plane which is perpendicular to the [111] axis triple and contains magnetic atoms (see Figure 8), but the atomic spins are oriented toward the point of intersection of this line with the axis triple. The magnetic structure described belongs to the representation τ_9 with $\mathbf{k} = 0$ of the space group $O_h{}'$ of the perovskite crystal [267].

One can convince oneself by direct verification that only a part of the elements of this group, which are listed in (11.28), leave the magnetic structure invariant, namely

$$h_1, h_5, h_9, h_{25}, h_{29}, h_{33}, \tag{11.45}$$

whereas the other elements accomplish the same only in combination with the nonunitary operation θ. The elements (11.45) form the unitary subgroups $S_6{}^2$ so that the magnetic group M has the structure

$$M = S_6^2 + \theta h_{13} S_6^2. \tag{11.46}$$

Let us cite the results of a group theoretical analysis of the spin-wave spectrum for the particular magnetic structure on the line **Λ** with the points **Γ** and **R**. The unitary subgroup $S_6{}^2$ for the points **Λ** and **R** has six one-dimensional representations $\tau_1 - \tau_6$. For these points, we have

$$d_{SW}^{\mathbf{q}} = 2\tau_1 + 2\tau_3 + 2\tau_5; \quad \mathbf{q} = 0, \quad \mathbf{q} = \frac{1}{2}(\mathbf{b}_1 + \mathbf{b}_2 + \mathbf{b}_3). \tag{11.47}$$

On the line **Λ**, the star of the wave vector for $S_6{}^2$ is a two-arm star and there are three one-dimensional irreducible representations τ_1, τ_2, τ_3. The spin-wave representation is characterized by the expressions

$$d_{SW}^{\mathbf{q}} = 2\tau_1 + 2\tau_2 + 2\tau_3, \quad \mathbf{q} = \mu\,(\mathbf{b}_1 + \mathbf{b}_2 + \mathbf{b}_3). \tag{11.48}$$

Thus, the symmetry which can be described by the unitary subgroup renders three branches of spin waves and each of these branches occurs twice. There is no degeneracy in the spectrum on the entire line **Λ**.

For lack of space we will not cite the results of calculating the basis functions and the blocks of the quasidiagonalized matrix of the Hamiltonian, which are analogs to Eq. (11.37). It turns out on the exchange approximation that the frequencies of the spin waves belonging to the one-dimensional representation τ_1 on the line Λ coincide with Eqs. (11.43) for the representation τ_2, whereas in the case of the representations τ_2 and τ_3, they coincide with the expression (11.43) for the representation τ_3. The frequency degeneracy, which was not predicted by the symmetry of the magnetically ordered crystal, is for the representations τ_2 and τ_3 a consequence of the additional symmetry of the exchange Hamiltonian.

Let us note that the spin-wave spectrum with an energy corresponding to the wave vector $\mathbf{q} = \mu(\mathbf{b}_1 + \mathbf{b}_2 + \mathbf{b}_3)$, has an additional degeneracy since a second arm $\mathbf{q} = -\mu(\mathbf{b}_1 + \mathbf{b}_2 + \mathbf{b}_3)$ exists for the group $G_{\mathbf{q}}$. Let us consider as an example the irreducible representation τ_1 of the group $G_{\mathbf{q}}$. It is easy to verify with the Herring criterion that this representation is complex. The complex-conjugate representation is on the arm $-\mathbf{q}$. These particular two representations are merged into a single irreducible representation of the unitary symmetry group of the magnetic structure. This two-dimensional representation defines the particular frequency of the spin wave with the wave vectors $\mathbf{q}$ and $-\mathbf{q}$. Thus, along the line Λ, the spin-wave spectrum of the magnetic structure under consideration has a mirror-like symmetry relative to the point $\mathbf{q} = 0$.

Now let us check how these results are affected by an additional symmetry which can be described by an anti-unitary class. In other words, let us consider the corepresentations of the group M of Eq. (11.46). The calculation of the left side of Eq. (11.10) renders 1 for all irreducible representations contained in Eqs. (11.47) and (11.48). According to the criterion of Eq. (11.10), the corepresentation generated by the cited irreducible representations of the group $G_{\mathbf{q}}$ does not cause additional degeneracy of the spin-wave spectrum. Thus, there are three non-degenerate branches of spin waves of different symmetry on the entire line Λ, including its end points.

12. Scattering of Neutrons at Hybrid Magnon–Phonon Excitations

Hybridization of magnons and phonons in a crystal. When the dispersion curves of phonons and magnons intersect, there arises a random degeneracy which can be removed by certain interactions of the spin

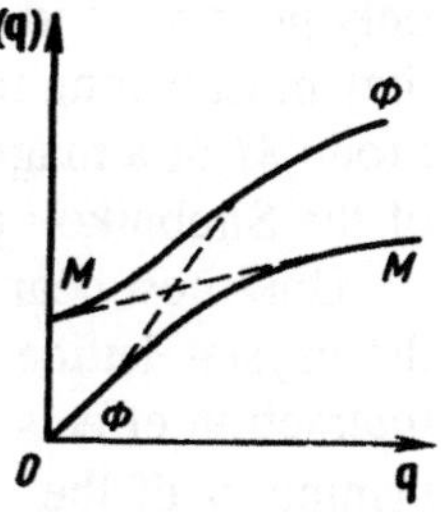

Fig. 9. Hybridization of phonons and magnons in a crystal.

system of the crystal with the lattice. As a consequence, the energy spectrum of the system near the point of intersection is importantly modified so that two new branches of hybridized states representing hybridization of a spin wave and a phonon develop. Such a collectivized branch of excitation can be considered a spin-density wave accompanied by atomic displacements, or, vice versa, an atomic displacement wave which in the crystal is accompanied by a spin-density wave [92, 3]. A typical hybridization pattern is illustrated in Fig. 9.

Hybridization originates from spin-lattice interaction which is linear in the atomic displacements and atomic spins:

$$\mathcal{H}_{LM}=\sum_{lm} D_{lm}^{\alpha\beta}\, u_l^{\alpha}\, S_m^{\beta}. \tag{12.1}$$

A spin-lattice interaction as defined by Eq. (12.1) originates from Hamiltonian terms of the type $\sum_{lmn} D_{lmn}{}^{\alpha\beta\gamma}u_l{}^{\alpha}S_m{}^{\beta}S_n{}^{\gamma}$ stated in the spin-wave approximation, when one of the spin operators is replaced by an eigenvalue in the ground state. The spin operators $S_m{}^{\beta}$ in Eq. (12.1) must be represented by linear combinations of the letter-type operators of the spin deviation so that the problem of the invariance of expression (12.1) relative to a reversal of time must not arise.

Let us consider hybridization in a crystal with some magnetic structure, assuming only that a magnetic unit cell exists in the crystal. All crystals with Shubnikov symmetry belong to this type of crystals with magnetic order. Assume that there are σ_L atoms in a primitive magnetic cell of the crystal and that among them there are σ_M magnetic atoms. Thus, in such a crystal there exist a total of $2\sigma_L$ phonon branches and σ_M spin waves. In an intersection of the dispersion curves of phonons and magnons, hybridization is possible, and this possibility is given by the symmetry of the branches of the spectrum. As has been shown in [152], only phonons and magnons of the same symmetry can be hybridized, i.e., they must belong to a particular irreducible representation of the symmetry group M of a magnetically ordered crystal which can be described by one of the Shubnikov groups.

One must bear in mind that, together with hybridization, alignment of the crystal lattice with the magnetic structure is one of the spin-lattice interaction effects in a crystal with magnetic order and, consequently, the symmetry of the entire crystal must be lowered to that of the group M. Now the phonon modes must be classified in accordance with the irreducible representations of the group M or, more precisely, in accordance with its corepresentations, like the spin waves [151-153]. Thus, the symmetry condition ("selection rule") obtained in [152] must read: phonons and magnons belonging to a single irreducible corepresentation of the group M must hybridize. We will expound below the dynamic theory of hybridization and apply it to the calculation of cross sections of neutron scattering at hybridized states [38, 210].

The Green's functions. The full Hamiltonian of the system must be composed of three terms:

$$\mathcal{H} = \mathcal{H}_L + \mathcal{H}_M + \mathcal{H}_{LM}, \tag{12.2}$$

where $\mathcal{H}_L$ denotes the Hamiltonian of the lattice, $\mathcal{H}_M$ denotes the Hamiltonian of the spin waves, and $\mathcal{H}_{LM}$ denotes the Hamiltonian of the spin-lattice interaction of Eq. (12.1). Let us state them in the following forms:

$$\mathcal{H}_L = \sum_{l\alpha} \frac{(p_l^\alpha)^2}{2M_l} + \frac{1}{2} \sum_{ll'} \sum_{\alpha\beta} \varphi_{ll'}^{\alpha\beta} u_l^\alpha u_{l'}^\beta; \tag{12.3}$$

$$\mathcal{H}_M = \sum_{mm'} \Psi_m^+ \mathrm{H}_{mm'} \Psi_{m'}; \tag{12.4}$$

$$\mathcal{H}_{LM} = \sum_{lm} \sum_{\alpha} \left(u_l^\alpha V_{lm}^\alpha \Psi_m + \Psi_m^+ \overset{+}{V}{}_{ml}^\alpha u_l^\alpha \right). \tag{12.5}$$

Here and in what follows, the subscript l numbers all the atoms of the crystal; m numbers only the magnetic atoms of the crystal. We use the following notation:

$$\Psi_m = \begin{pmatrix} b_m \\ b_m^+ \end{pmatrix}, \quad \Psi_m^+ = (b_m^+, b_m);$$

$$\overset{+}{V}{}_{ml}^\alpha = \begin{pmatrix} \overset{*}{K}{}_{lm}^\alpha \\ K_{lm}^\alpha \end{pmatrix}, \quad V_{lm}^\alpha = \left(K_{lm}^\alpha, \overset{*}{K}{}_{lm}^\alpha\right);$$

$$\mathrm{H}_{mm'} = \begin{pmatrix} B_{mm'} & A^{*}_{mm'} \\ A_{mm'} & B_{m'm} \end{pmatrix}; \tag{12.6}$$

$$K^{\alpha}_{lm} = \sqrt{2S_m} \sum_{\beta} D^{\alpha\beta}_{lm} \left(d^{\beta x}_{m} - i d^{\beta y}_{m}\right). \tag{12.7}$$

The Hamiltonian $\mathcal{H}$ is a quadratic form in displacements and operators of the spin deviations but since a term with the operator of the kinetic energy is present in $\mathcal{H}_L$, the determination of the eigenvalues is not reduced to the diagonalization of the mixed quadratic form in the variables u_l^α and b_m, b_m^+. In this situation one conveniently uses the technique of the two-time Green's functions described in Section 3.

Let us introduce the Green's functions:

$$G^{\alpha\beta}_{ll'} = \sqrt{M_l} \langle\langle u^\alpha_l \mid u^\beta_{l'} \rangle\rangle \sqrt{M_{l'}}; \tag{12.8}$$

$$D_{mm'} = \langle\langle \Psi_m \mid \Psi^+_{m'} \rangle\rangle; \tag{12.9}$$

$$\Gamma^{\beta}_{ml'} = \langle\langle \Psi_m \mid u^\beta_{l'} \rangle\rangle \sqrt{M_{l'}}, \quad \Lambda^{\alpha}_{lm'} = \sqrt{M_l} \langle\langle u^\alpha_l \mid \Psi^+_{m'} \rangle\rangle. \tag{12.10}$$

It is easy to obtain for them a precise equation system:

$$\left.\begin{aligned} G^{\alpha\beta}_{ll'} &= G^{\circ\alpha\beta}_{ll'} + \sum_{l_1 m_1} G^{\circ\alpha\gamma_1}_{ll_1} W^{\gamma_1}_{l_1 m_1} \Gamma^{\beta}_{m_1 l'}; \\ D_{mm'} &= D^{\circ}_{mm'} + \sum_{l_1 m_1} D^{\circ}_{mm_1} \overset{+}{W}{}^{\gamma_1}_{m_1 l_1} \Lambda^{\gamma_1}_{l_1 m'}; \\ \Gamma^{\beta}_{ml'} &= \sum_{l_1 m_1} D^{\circ}_{mm_1} \overset{+}{W}{}^{\gamma_1}_{m_1 l_1} G^{\gamma_1 \beta}_{l_1 l'}; \\ \Lambda^{\alpha}_{lm'} &= \sum_{l_1 m_1} G^{\circ\alpha\gamma_1}_{ll_1} W^{\gamma_1}_{l_1 m_1} D_{m_1 m'}, \end{aligned}\right\} \tag{12.11}$$

where

$$W^{\gamma}_{lm} = \frac{1}{\sqrt{M_l}} \left(K^{\gamma}_{lm}, \overset{*}{K}{}^{\gamma}_{lm}\right), \quad \overset{+}{W}{}^{\gamma}_{ml} = \frac{1}{\sqrt{M_l}} \begin{pmatrix} \overset{*}{K}{}^{\gamma}_{lm} \\ K^{\gamma}_{lm} \end{pmatrix}. \tag{12.12}$$

We number the atoms of the crystal by double subscripts $l = \{ns\}$, where n denotes the number of the cell and s denotes the number of the atom in the cell. After a Fourier expansion in the translations of the crystal with magnetic order, we obtain from Eq. (12.11) closed equations for the phonon and spin-wave Green's function:

$$G_{ss'}^{\alpha\beta} = G_{ss'}^{\circ\alpha\beta} \sum_{s_1 s_2} \sum_{i_1 i_2} G_{ss_1}^{\circ\alpha\gamma_1} \left(W_{s_1 i_1}^{\gamma_1} D_{i_1 i_2}^{\circ} \overset{+}{W}{}_{i_2 s_2}^{\gamma_2}\right) G_{s_2 s'}^{\gamma_2 \beta}; \qquad (12.13)$$

$$D_{ii'} = D_{ii'}^{\circ} + \sum_{s_1 s_2} \sum_{i_1 i_2} D_{ii_1}^{\circ} \left(\overset{+}{W}{}_{i_1 s_1}^{\gamma_1} G_{s_2 s_1}^{\gamma_1 \gamma_2} W_{s_2 i_2}^{\gamma_2}\right) D_{i_2 i'}. \qquad (12.14)$$

Each of the quantities is a function of **q** and ω, with **q** belonging to the first Brillouin zone of the reciprocal lattice of the crystal with magnetic order.

The zeroth Green's functions of the phonons and spin waves are given by the equations

$$G_{ss'}^{\circ\alpha\beta^{-1}}(\mathbf{q}, \omega) = \omega^2 \delta_{\alpha\beta} \delta_{ss'} - \frac{1}{\sqrt{M_s M_{s'}}} \varphi_{ss'}^{\alpha\beta}(\mathbf{q}); \qquad (12.15)$$

$$D^{-1}(\mathbf{q}, \omega) = \hbar\omega\tau_3 - \mathrm{H}(\mathbf{q}), \qquad (12.16)$$

where τ_3 (and also τ_1 in what follows) are Pauli matrices and H(**q**) denotes the spin-wave matrix defined by Eq. (10.8).

Explicit expressions for the zeroth Green's functions result from Eqs. (12.15) and (12.16):

$$G_{ss'}^{\circ\alpha\beta}(\mathbf{q}, \omega) = \sum_{\Lambda} \frac{e_{s\Lambda}^{\alpha}(\mathbf{q}) \overset{*\beta}{e}_{s'\Lambda}(\mathbf{q})}{\omega^2 - \omega_{\Lambda}^2(\mathbf{q})}; \qquad (12.17)$$

$$D^{\circ}(\mathbf{q}, \omega) = \sum_{M} \eta_M \frac{m_M(\mathbf{q}) m_M^{+}(\mathbf{q})}{\hbar\omega - \hbar\omega_M(\mathbf{q})}. \qquad (12.18)$$

where $\omega_{\Lambda}{}^2(\mathbf{q})$ denotes the eigenvalue of the force matrix of Eq. (12.15) and $e_{s\Lambda}{}^{\alpha}(\mathbf{q})$ denotes a component of the eigenvector corresponding to this matrix, $\hbar\omega_M(\mathbf{q})$ denotes the eigenvalue of the matrix τ_3H(**q**) and $m_M(\mathbf{q})$ denotes its eigenvector. They can be determined with the equation

$$\tau_3 \mathrm{H}(\mathbf{q}) m_M(\mathbf{q}) = \hbar\omega_M(\mathbf{q}) m_M(\mathbf{q}),$$

which is another way of stating Eq. (10.9). It is a well known fact that the latter has σ_M positive eigenvalues and the same number of negative eigenvalues $\hbar\omega_M(\mathbf{q})$. The quantity η_M of Eq. (12.18) is equal to +1 for positive, and equal to −1 for negative $\hbar\omega_M(\mathbf{q})$ values. The matrices of the Green's functions G° and D° are $3\sigma_L \times 3\sigma_L$ and $2\sigma_M \times 2\sigma_M$ matrices, respectively.

Let us note some general symmetry properties of the Green's functions we introduced. Owing to the properties (10.4) which for matrices A and B result from the hermiticity of the spin-wave Hamiltonian, the equation

$$D^{\circ}(\mathbf{q},\ \omega) = \tau_1 D^{\circ}(-\mathbf{q},\ -\omega)\,\tau_1,$$

holds and with it the following other important relations can be deduced from Eqs. (12.11).

$$G(\mathbf{q},\ \omega) = \overset{*}{G}(-\mathbf{q},\ -\omega),\ \ D(\mathbf{q},\ \omega) = \tau_1 \overset{*}{D}(-\mathbf{q},\ -\omega)\,\tau_1,$$
$$\Gamma(\mathbf{q},\ \omega) = \tau_1 \overset{*}{\Gamma}(-\mathbf{q},\ -\omega),\ \ \Lambda(\mathbf{q},\ \omega) = \overset{*}{\Lambda}(-\mathbf{q},\ -\omega)\,\tau_1, \qquad (12.19)$$

to establish the values of the Green's functions for the momenta $\mathbf{q}$ and $-\mathbf{q}$.

Symmetry analysis of the equations. The precise Eqs. (12.13) and (12.14) for the Green's functions of the phonons and spin waves can be substantially simplified [38] by making use of the symmetry of a crystal with magnetic order, the symmetry to be described by the group M. We know that the phonon modes of the crystal are characterized by irreducible representations $d^{\mathbf{q}\nu}$ which appear in the composition of the vibrational representation $d_m{}^{\mathbf{q}}$ of Eq. (4.16), and that the magnon modes are characterized by the representations entering into the composition of the spin-wave representation $d_{SW}{}^{\mathbf{q}}$ of Eq. (11.17) of the group $G_{\mathbf{q}}$. Thus, the list of the phonon and magnon modes which for the crystal are possible with a given wave vector $\mathbf{q}$ is determined by the nonzero values of the numbers $n_m{}^{\nu}$ and $n_{SW}{}^{\mu}$ in the equations

$$d_m^{\mathbf{q}} = \sum_{\nu} n_m^{\nu}\, d^{\mathbf{q}\nu},\quad d_{SW}^{\mathbf{q}} = \sum_{\mu} n_{SW}^{\mu}\, d^{\mathbf{q}\mu}. \qquad (12.20)$$

The vibrational representation $d_m{}^{\mathbf{q}} - 3\sigma_L$ has the dimension $3\sigma_L$, and the spin-wave representation has the dimensions $2\sigma_M$; the same number of phonon and magnon branches exists in the particular crystal. The phonon and magnon modes themselves are described by the basis functions of the irreducible representations $d^{\mathbf{q}\nu}$ and $d^{\mathbf{q}\mu}$. The unitary matrices $U_L(\mathbf{q})$ and $U_M(\mathbf{q})$ can be derived from these basis functions, with the matrices diagonalizing the matrices of the force constants for the lattice and for the spin waves and the matrices of the phonon and magnon Green's functions. Obviously, the U_L matrices are $3\sigma_L \times 3\sigma_L$ matrices, and the U_M matrices are $2\sigma_M \times 2\sigma_M$ matrices.

We rewrite Eq. (12.13) in the following form:

$$U_L^+ GU_L = U_L^+ G^\circ U_L + (U_L^+ G^\circ U_L)(U_L^+ WU_M)(U_M^+ D^\circ U_M) \times \times (U_M^+ W^+ U_L)(U_L^+ GU_L), \tag{12.21}$$

where each quantity refers to the same particular **q** value. The transformed quantities are split into blocks which are characterized by the irreducible representations appearing in the expansion of Eq. (12.20). We state this fact in the form of the following expressions:

$$U_L^+ G^\circ U_L = \sum_\nu \oplus G_\nu^\circ, \quad U_M^+ D^\circ U_M = \sum_\mu \oplus D_\mu^\circ,$$
$$U_L^+ WU_M = \sum_\rho \oplus W_\rho, \quad U_L^+ W^+ U_M = \sum_\rho \oplus W_\rho^+. \tag{12.22}$$

Each of the blocks $G_\nu{}^\circ$ has the size $n_m{}^\nu l_\nu \times n_m{}^\nu l_\nu$, and each of the blocks $D_\mu{}^\circ$ has the size $n_{SW}{}^\mu l_\mu \times n_{SW}{}^\mu l_\mu$, where l_μ and l_μ denote the dimensionalities of the irreducible representations $d^{\mathbf{q}\nu}$ and $d^{\mathbf{q}\mu}$. The blocks W_ρ are rectangular matrices of the size $n_m{}^\rho l_\rho \times n_{SW}{}^\rho l_\rho$, and the blocks $W_\rho{}^+$ are rectangular matrices of the size $n_{SW}{}^\rho l_\rho \times n_m{}^\rho l_{\rho_{\mathbf{q}}}$, where ρ denotes the number of the representation contained in $d_M{}^{\mathbf{q}}$ and $d_{SW}{}^{\mathbf{q}}$. It must be noted in addition that all blocks $G_\nu{}^\circ$ and $D_\mu{}^\circ$ are on the main diagonal, whereas the blocks W_ρ and $W_\rho{}^+$ appear at points corresponding to the position of the coincidings irreducible representations $d^{\mathbf{q}\nu}$ and $d^{\mathbf{q}\mu}$. The other elements in the matrices (12.22) are equal to zero.

Since the spin–lattice interaction connects only those phonon and magnon modes which are characterized by the same particular irreducible representation $d^{\mathbf{q}\rho}$, we have $G_\nu = G_\nu{}^\circ$ for $\nu \neq \rho$; $D_\mu = D_\mu{}^\circ$ for $\mu \neq \rho$, but G_ρ and D_ρ are for $\nu = \mu = \rho$ obtained form the matrix equations

$$\left.\begin{aligned} G_\rho &= G_\rho^\circ + G_\rho^\circ W_\rho D_\rho^\circ W_\rho^+ G_\rho; \\ D_\rho &= D_\rho^\circ + D_\rho^\circ W_\rho^+ G_\rho^\circ W_\rho D_\rho, \end{aligned}\right\} \tag{12.23}$$

the dimensionality of which is much smaller than the dimensionality of the initial equations (12.13) and (12.14) so that, in practice, the problem can be solved for a specific crystal.

Similarly, we obtain for the mixed Green's functions:

$$U_M^+ \Gamma U_L = \sum_\rho \oplus \Gamma_\rho, \quad U_L^+ \Lambda U_M = \sum_\rho \oplus \Lambda_\rho; \tag{12.24}$$

$$\Gamma_\rho = D_\rho^\circ W_\rho^+ G_\rho, \quad \Lambda_\rho = G_\rho^\circ W_\rho D_\rho. \tag{12.25}$$

The spectrum of the hybridized magnon phonon modes is given by the equation

$$\det|G_\rho^{\circ -1}(\mathbf{q},\omega)-W_\rho(\mathbf{q})\,D_\rho^\circ(\mathbf{q},\omega)\,W_\rho^+(\mathbf{q})|=0. \tag{12.26}$$

Through the dimensionality of this determinant is $n_m{}^\rho l_\rho$, the order of the algebraic equation (12.26) for the frequencies is greater and equal to $(n_m{}^\rho + n_{SW}{}^\rho)l_\rho$ because of the singularity of the Green's function $D_\rho{}^\circ$.

Let us now determine the structure of the matrix blocks in Eqs. (12.23). In the case of the repeated representation $d^{q\rho}$ $(n_L{}^\rho > 1,\ n_M{}^\rho > 1)$, $G_\rho{}^\circ$ and $D_\rho{}^\circ$ (and also W_ρ and $W_\rho{}^+$) have band structure, i.e., they can be represented as a direct product of some matrices $\tilde{G}_\rho{}^\circ$ and $\tilde{D}_\rho{}^\circ$ times the unit matrix E of dimensionality l_ρ of the irreducible representation (and also of W_ρ and $W_\rho{}^+$). Then the determinant in Eq. (12.26) is factorized and Eq. (12.26) is reduced to the following form:

$$\{\det|\tilde{G}_\rho^{\circ -1}(\mathbf{q},\omega)-\tilde{W}_\rho(\mathbf{q})\,\tilde{D}_\rho^\circ(\mathbf{q},\omega)\,\tilde{W}_\rho^+(\mathbf{q})|\}^{l_\rho}=0.$$

This equation defines the l_ρ degenerate frequencies which can be found from the algebraic equation of the order $n_L{}^\rho + n_M{}^\rho$.

Equations (12.23) split into degenerate matrix equations for the quantities $\tilde{G}_\rho$, $\tilde{G}_\rho{}^\circ$, $\tilde{D}_\rho{}^\circ$, etc.

Further diagonalization of the matrix blocks $G_\rho{}^\circ$ and $D_\rho{}^\circ$ in Eqs. (12.23) has no longer to do with symmetry considerations but, in principle, can be performed with the unitary transformations u_L and u_M. We introduce the notation

$$\begin{aligned}(u_L^+\,G_\rho^\circ\,u_L)_{\lambda\lambda'}&=g_\lambda^\circ\,\delta_{\lambda\lambda'}, & (u_M^+\,D_\rho^\circ\,u_M)_{\mu\mu'}&=d_\mu^\circ\,\delta_{\mu\mu'},\\ (u_L^+\,W_\rho\,u_M)_{\lambda\mu}&=w_{\lambda\mu}, & (u_M^+\,W_\rho^+\,u_L)_{\mu\lambda}&=w_{\mu\lambda}^+.\end{aligned} \tag{12.27}$$

Then the equation for the matrix elements $g_{\lambda\lambda'}$ of the phonon block $(u_L{}^+G_\rho u_L)$ has the form (the subscript ρ will be omitted in what follows):

$$g_{\lambda\lambda'}=g_\lambda^\circ\,\delta_{\lambda\lambda'}+g_\lambda^\circ\sum_{\lambda_1\mu}\left(w_{\lambda\mu}\,d_\mu^\circ\,w_{\mu\lambda_1}^+\right)g_{\lambda_1\lambda'}. \tag{12.28}$$

The equation for the magnon Green's function $d_{\mu\mu'}$ is analogous. This form of the equations is suitable for obtaining the spectrum near the point at which the phonon and magnon branches intersect. Assume that $\omega_\lambda{}^L(\mathbf{q})$ denotes the phonon branch corresponding to $g_\lambda{}^\circ(\mathbf{q}$, and that $\omega_\mu{}^M(\mathbf{q})$

denotes the magnon branch corresponding to $d_u{}^\circ(\mathbf{q}, \omega)$. [The form of $g_\lambda{}^\circ$ and $d_\mu{}^\circ$ is given by Eq. (12.17), where $\lambda = 1, \ldots, n_\mu{}^\rho l_\rho$, and $\mu = 1, \ldots, n_{SW} l_\rho$]. Assume that the first phonon branch and the second magnon branch of the spectral part belonging to the irreducible representation $d^{\mathbf{q}\rho}$ intersect:

$$\omega_1^L(\mathbf{q}) = \omega_2^M(\mathbf{q}) \text{ for } \mathbf{q} = \mathbf{q}_0. \tag{12.29}$$

If in the neighborhood of this point the distances between the branches are much smaller than between them and other branches of the same symmetry as $d^{\mathbf{q}\rho}$, and if the latter are much smaller than $|w_{12}|$, we obtain in the lowest order from Eq. (12.28) and the corresponding equation for $d_{\mu\mu'}$:

$$g_1(\mathbf{q}, \omega) = \hbar\eta^2 \cdot \frac{\omega - \omega_2^M(\mathbf{q})}{[\omega - \omega_1^L(\mathbf{q})][\omega - \omega_2^M(\mathbf{q})] - |w_{12}(\mathbf{q})|^2 \eta^2}; \tag{12.30}$$

$$d_2(\mathbf{q}, \omega) = \frac{1}{\hbar} \cdot \frac{\omega - \omega_1^L(\mathbf{q})}{[\omega - \omega_1^L(\mathbf{q})][\omega - \omega_2^M(\mathbf{q})] - |w_{12}(\mathbf{q})|^2 \eta^2}, \tag{12.31}$$

whereas for the other branches $g_\lambda = g_\lambda{}^\circ$ and $d_\mu = d_\mu{}^\circ$. Besides that, we have

$$\gamma_{21}(\mathbf{q}, \omega) = \eta^2 \frac{w_{12}^*(\mathbf{q})}{[\omega - \omega_1^L(\mathbf{q})][\omega - \omega_2^M(\mathbf{q})] - |w_{12}(\mathbf{q})|^2 \eta^2} = \lambda_{12}^*(\mathbf{q}, \omega), \tag{12.32}$$

where γ_{21} and λ_{12} denote matrix elements from the blocks Γ_ρ and Λ_ρ after a second unitary transformation. The other matrix elements are equal to zero in this approximation. Thus, in the neighborhood of the point of intersection, we have $\eta = 1/\sqrt{2\hbar\omega(\mathbf{q})}$ in Eqs. (12.30)–(12.32), and the expressions for the Green's functions are of the same form as in the case of two isolated interacting Bose branches of the spectrum. Far from the intersection point selected, other branches of the symmetry of $d^{\mathbf{q}\rho}$ may intersect, and, in the general case, the Green's functions must be determined from Eq. (12.23) or (12.28). The coupling parameter $w_{12}(\mathbf{q})$ is expressed through the initial microscopic parameters with the aid of the expressions

$$w_{12}(\mathbf{q}) = \sum_{is} \overset{*}{e}{}_{s1}^{\alpha}(\mathbf{q})\, W_{si}^{\alpha}\, m_{i2}(\mathbf{q}), \tag{12.33}$$

where $e_{s1}{}^{\alpha}(\mathbf{q})$ and $m_{i2}(\mathbf{q})$ denote the atomic components of the polarization vectors of the corresponding phonon and magnon modes, respectively, obtained by diagonalization of the force matrices of the lattice and of the spin system.

Let us apply the mathematical technique which we have developed to investigations of inelastic neutron scattering.

Scattering at hybridized modes. We start from the relation which is given by Eq. (3.10) between the Green's functions and the correlators and express the cross section of inelastic neutron scattering through the Green's functions of the phonons and magnons and through function $\mathbf{\Gamma}$ and $\mathbf{\Lambda}$:

$$\frac{d^2\sigma}{d\Omega d\varepsilon} = \frac{k}{k_0}\frac{\mathrm{i}}{2\pi}\frac{1}{1-\mathrm{e}^{\beta\varepsilon}}\Bigg\{\sum_{ll'} b_l b_{l'}\,\mathrm{e}^{\mathrm{i}\varkappa(\mathbf{R}_l^\circ-\mathbf{R}_{l'}^\circ)}\sum_{\alpha\beta}\varkappa^\alpha\varkappa^\beta\,\overline{G}_{ll'}^{\alpha\beta}(-\omega) +$$

$$+(r_0\gamma)^2\sum_{mm'}F_m(\varkappa)F_{m'}(\varkappa)\frac{1}{2}\sqrt{S_mS_{m'}}\,\mathrm{e}^{\mathrm{i}\varkappa(\mathbf{R}_m^\circ-\mathbf{R}_{m'}^\circ)}\times$$

$$\times\sum_{\gamma\gamma'}\sum_{\alpha\beta}K^{\gamma\gamma'}(\delta_{\alpha\gamma}\quad e_\alpha e_\gamma)(\delta_{\beta\gamma'}-e_\beta e_{\gamma'})\overset{+}{\Delta}{}_m^\alpha\,\overline{D}_{mm'}(-\omega)\,\Delta_{m'}+$$

$$+(r_0\gamma)\sum_{lm}b_lF_m(\varkappa)\sqrt{\frac{S_m}{2}}\sum_{\alpha\beta\gamma}p_0^\gamma\,[\mathrm{i}\varkappa^\beta(\delta_{\alpha\gamma}-e_\alpha e_\gamma)\,\mathrm{e}^{\mathrm{i}\varkappa(\mathbf{R}_m^\circ-\mathbf{R}_l^\circ)}\times$$

$$\times\overset{+}{\Delta}{}_m^\alpha\,\overline{\Gamma}_{ml}^\alpha(-\omega)\quad\mathrm{i}\varkappa^\alpha(\delta_{\beta\gamma}-e_\beta e_\gamma)\,\mathrm{e}^{-\mathrm{i}\varkappa(\mathbf{R}_m^\circ-\mathbf{R}_l^\circ)}\,\overline{\Lambda}_{lm}^\alpha(-\omega)\,\Lambda_m^\beta]\Bigg\}, \qquad (12.34)$$

where one must assume $\omega = -\varepsilon/\hbar$. We have

$$\Delta_m^\alpha = \begin{pmatrix} d_m^{\alpha+} \\ \overset{*}{d}{}_m^{\alpha+} \end{pmatrix}, \quad \overset{+}{\Delta}{}_m^\alpha = (\overset{*}{d}{}_m^{\alpha+},\ d_m^{\alpha+}), \qquad (12.35)$$

and $k^{\gamma\gamma'}$ is defined by Eq. (10.17).

Each of the matrix quantities, which is marked by an upper prime, denotes the difference between the corresponding Green's functions, e.g.,

$$\overline{G}_{ll'}^{\alpha\beta}(\omega) \equiv G_{ll'}^{\alpha\beta}(\omega+i\delta) - G_{ll'}^{\alpha\beta}(\omega-i\delta). \qquad (12.36)$$

After expanding the Green's functions in Fourier series and performing the first unitary transformation [Eq. (12.22)], the expression for the cross section corresponding to scattering at modes belonging to the representation $d^{q\rho}$ can be reduced to the form:

$$\frac{1}{N}\frac{d^2\sigma}{d\Omega d\varepsilon}=\frac{k}{k_0}\frac{\mathrm{i}}{2\pi}\frac{1}{1-e^{\beta\varepsilon}}\sum_{\mathbf{q}}\sum_{\mathbf{b}}\delta_{\varkappa+\mathbf{q},\mathbf{b}}\Big\{\sum_{\Lambda\Lambda'}\varphi^*_{\Lambda}\overline{G}_{\Lambda\Lambda'}(\varepsilon)\varphi_{\Lambda'}+$$

$$+\sum_{MM'}\left((\mathfrak{M}^*_M\mathfrak{M}_{M'})+\mathrm{i}\mathbf{p}_0[\mathfrak{M}^*_M\times\mathfrak{M}_{M'}]\right)\overline{D}_{MM'}(\varepsilon)+$$

$$+\mathrm{i}\sum_{M\Lambda}(\mathbf{p}_0\mathfrak{M}^*_M)\overline{\Gamma}_{M\Lambda}(\varepsilon)\varphi_{\Lambda}-\mathrm{i}\sum_{M\Lambda}\varphi^*_{\Lambda}\overline{\Lambda}_{\Lambda M}(\varepsilon)(\mathfrak{M}_M\mathbf{p}_0)\Big\}. \qquad (12.37)$$

All quantities on the right side must be given the subscript ρ which was omitted; $G_{\Lambda\Lambda'}$ denotes the matrix element of the block G_ρ, etc.

The partial structural amplitudes of inelastic nuclear scattering, φ_Λ, and of inelastic magnetic scattering, $\mathfrak{M}_M$, were introduced in Eq. (12.37):

$$\varphi_\Lambda(\mathbf{q})=\sum_{s\alpha}U^*_{s\alpha,\Lambda}(\mathbf{q})\,b_s\,\varkappa^\alpha; \qquad (12.38)$$

$$\mathfrak{M}_M(\mathbf{q})=\mathbf{f}_M(\mathbf{q})-(\mathbf{e}\mathbf{f}_M(\mathbf{q}))\,\mathbf{e}, \qquad (12.39)$$

where

$$f^\alpha_M(\mathbf{q})=\sum_i a_i\,[d^{\alpha+}_i U_{+i,M}(\mathbf{q})+\overset{*}{d}{}^{\alpha+}_i U^*_{-i,M}(\mathbf{q})]; \qquad (12.40)$$

$$a_s=b_s\frac{1}{\sqrt{M_s}}e^{-\mathrm{i}\varkappa x_s},\quad a_i=r_0\gamma\sqrt{\frac{S_i}{2}}F_i(\varkappa)\,e^{-\mathrm{i}\varkappa x_i}. \qquad (12.41)$$

In stating Eq. (12.40), we represented the unitary matrix U for the spin system in the form $U\equiv\begin{pmatrix}U_+\\U_-\end{pmatrix}$.

Let us calculate the cross section of scattering at elementary excitations of the crystal near the intersection points of the phonon and magnon branches. For this purpose, we perform in Eqs. (12.11) a second unitary transformation [(Eq. (12.27)] and use the expressions for the Green's functions [Eq. (12.30)–(12.32)]. The cross section of scattering at two hybridized excitations of the frequency

$$\Omega_{1,2}(\mathbf{q})=\frac{\omega^L_1(\mathbf{q})+\omega^M_2(\mathbf{q})}{2}\pm\sqrt{\left(\frac{\omega^L_1(\mathbf{q})-\omega^M_2(\mathbf{q})}{2}\right)^2+|w_{12}(\mathbf{q})|^2\eta^2} \qquad (12.42)$$

is stated in the form

$$\frac{d^2\sigma_{12}}{d\Omega d\varepsilon} = \frac{k}{k_0}\sum_{\mathbf{q}}\sum_{\mathbf{b}}\sum_{n=1}^{2}\{S_n^+(\mathbf{q})(1+N_n(\mathbf{q}))\,\delta_{\varkappa+\mathbf{q},\,\mathbf{b}}\,\delta(\varepsilon+\hbar\Omega_n(\mathbf{q}))+ + S_n^-(\mathbf{q})N_n(\mathbf{q})\,\delta_{\varkappa-\mathbf{q},\,\mathbf{b}}\,\delta(\varepsilon-\hbar\Omega_n(\mathbf{q}))\}, \tag{12.43}$$

where

$$S_1^{\pm}(\mathbf{q}) = \hbar^2\eta^2|\varphi_1'|^2\frac{1}{2}(1+\gamma_{\mathbf{q}})+\big((\overset{*}{\mathfrak{M}}{}_2'\mathfrak{M}_2')\pm \pm i\mathbf{p}_0[\overset{*}{\mathfrak{M}}{}_2'\times\mathfrak{M}_2']\big)\frac{1}{2}(1-\gamma_{\mathbf{q}})\pm\hbar\eta^2\,[i\,(\mathbf{p}_0\overset{*}{\mathfrak{M}}{}_2')\,\xi_{\mathbf{q}}^*\varphi_1'-i\,\overset{*}{\varphi}{}_1'\,(\mathbf{p}_0\mathfrak{M}_2')\,\xi_{\mathbf{q}}]; \tag{12.44}$$

$$S_2^{\pm}(\mathbf{q}) = \hbar^2\eta^2|\varphi_1'|^2\frac{1}{2}(1+\gamma_{\mathbf{q}})+\big((\overset{*}{\mathfrak{M}}{}_2'\mathfrak{M}_2')\pm \pm i\mathbf{p}_0[\overset{*}{\mathfrak{M}}{}_2'\times\mathfrak{M}_2']\big)\frac{1}{2}(1-\gamma_{\mathbf{q}})\mp\hbar\eta^2\,[i\,(\mathbf{p}_0\overset{*}{\mathfrak{M}}{}_2')\,\xi_{\mathbf{q}}^*\varphi_1'- -i\overset{*}{\varphi}{}_1'\,(\mathbf{p}_0\mathfrak{M}_2')\,\xi_{\mathbf{q}}]. \tag{12.45}$$

We have

$$\gamma_{\mathbf{q}} = \frac{\omega_1^L(\mathbf{q})-\omega_2^M(\mathbf{q})}{\Omega_1(\mathbf{q})-\Omega_2(\mathbf{q})}, \qquad \xi_{\mathbf{q}} = \frac{w_{12}}{\Omega_1(\mathbf{q})-\Omega_2(\mathbf{q})}. \tag{12.46}$$

The quantities φ_λ' and $\mathfrak{M}_\mu'$ are obtained from φ_Λ and $\mathfrak{M}_M$ by the unitary transformations u_L and u_M and can be expressed through the polarization vectors of the phonons and the spin waves:

$$\varphi_\lambda' = \sum_{S\alpha}\overset{*}{e}{}^{\alpha}_{S\lambda}(\mathbf{q})\,b_S\,\varkappa^\alpha; \tag{12.47}$$

$$\mathfrak{M}_\mu' = \mathbf{f}_\mu'(\mathbf{q})-(\mathbf{e}\mathbf{f}_\mu'(\mathbf{q}))\,\mathbf{e}; \tag{12.48}$$

$$f_\mu'^{\alpha}(\mathbf{q}) = \sum_i a_i\,[d_i^{\alpha+}u_{i\mu}(\mathbf{q})+\overset{*}{d}{}_i^{\alpha+}v_{i\mu}(\mathbf{q})]. \tag{12.49}$$

Let us state the results [38] of calculations of the polarization vector of the scattered beam in accordance with Eq. (10.29):

$$\mathbf{p}\,\frac{d^2\sigma_{12}}{d\Omega d\varepsilon} = \frac{k}{k_0}\sum_{\mathbf{q}}\sum_{\mathbf{b}}\sum_{n}\{\mathbf{p}_n^+(\mathbf{q})(1+N_n(\mathbf{q}))\,\delta_{\varkappa+\mathbf{q},\,\mathbf{b}}\,\delta(\varepsilon+\hbar\Omega_n(\mathbf{q}))+ + \mathbf{p}_n^-(\mathbf{q})N_n(\mathbf{q})\,\delta_{\varkappa-\mathbf{q},\,\mathbf{b}}\,\delta(\varepsilon-\hbar\Omega_n(\mathbf{q}))\}, \tag{12.50}$$

where $\mathbf{p}_n{}^{\pm}(\mathbf{q})$ denotes the polarization vectors of the beam which was coherently scattered with the emission and absorption of a quantum of frequency $\Omega_n(\mathbf{q})$, wherein

$$\mathbf{p}_1^{\pm}(\mathbf{q})\, S_1^{\pm}(\mathbf{q}) = \hbar^2\eta^2 |\varphi_1'|^2 \frac{1}{2}(1+\gamma_{\mathbf{q}})\,\mathbf{p}_0 + \frac{1}{2}\Lambda^{\pm}(1-\gamma_{\mathbf{q}}) -$$

$$-\hbar\eta^2\left([\mathbf{p}_0 \times \overset{*}{\mathfrak{M}}{}_2']\,\xi_{\mathbf{q}}^*\varphi_1' \mp i\overset{*}{\mathfrak{M}}{}_2'\right) - \hbar\eta^2\left([\mathbf{p}_0 \times \mathfrak{M}_2']\,\xi_{\mathbf{q}}\overset{*}{\varphi}{}_1' \pm i\mathfrak{M}_2'\right); \quad (12.51)$$

$$\mathbf{p}_2^{\pm}(\mathbf{q})\, S_2^{\pm}(\mathbf{q}) = \hbar^2\eta^2 |\varphi_1'|^2 \frac{1}{2}(1-\gamma_{\mathbf{q}})\,\mathbf{p}_0 + \Lambda^{\pm}\frac{1}{2}(1-\gamma_{\mathbf{q}}) +$$

$$+\hbar\eta^2\left([\mathbf{p}_0 \times \overset{*}{\mathfrak{M}}{}_2']\,\xi_{\mathbf{q}}^*\varphi_1' \mp i\overset{*}{\mathfrak{M}}{}_2'\right) + \hbar\eta^2\left([\mathbf{p}_0 \times \mathfrak{M}_2']\,\xi_{\mathbf{q}}\overset{*}{\varphi}{}_1' \pm i\mathfrak{M}_2'\right). \quad (12.52)$$

We have

$$\Lambda^{\pm} = \mp i\,[\overset{*}{\mathfrak{M}}{}_2' \times \mathfrak{M}_2'] + \overset{*}{\mathfrak{M}}{}_2'(\mathfrak{M}_2'\mathbf{p}_0) +$$

$$+\mathfrak{M}_2'(\overset{*}{\mathfrak{M}}{}_2'\mathbf{p}_0) - (\overset{*}{\mathfrak{M}}{}_2'\mathfrak{M}_2')\,\mathbf{p}_0. \quad (12.53)$$

It followed form the resulting formulas that an initially unpolarized beam, which was scattered due to emission or absorption of a single excitation quantum, is polarized. The degree of spontaneous polarization due to scattering at excitations of type 1 is

$$\mathbf{p}_1^{\pm} = \mp \frac{i\,[\overset{*}{\mathfrak{M}}{}_2' \times \mathfrak{M}_2']\frac{1}{2}(1-\gamma_{\mathbf{q}}) + \left(-i\overset{*}{\mathfrak{M}}{}_2'\overset{*}{\xi}_{\mathbf{q}}\varphi_1' + i\mathfrak{M}_2'\xi_{\mathbf{q}}\overset{*}{\varphi}{}_1'\right)\hbar\eta^2}{(\overset{*}{\mathfrak{M}}{}_2'\mathfrak{M}_2')\frac{1}{2}(1-\gamma_{\mathbf{q}}) + \hbar^2\eta^2|\varphi_1'|^2\frac{1}{2}(1+\gamma_{\mathbf{q}})}. \quad (12.54)$$

The value of $\mathbf{p}_2{}^{\pm}$ is distinguished from $\mathbf{p}_1{}^{\pm}$ by the sign of the second term in the numerator before $\gamma_{\mathbf{q}}$. This polarization is composed of the polarization due to magnetic scattering and due to the superposition of magnetic and nuclear scattering. At the intersection point proper, $\gamma_{\mathbf{q}} = 0$ so that the relation

$$\mathbf{p}_1^{\pm} - \mathbf{p}_2^{\pm} = \pm \frac{4\left(-i\overset{*}{\mathfrak{M}}{}_2'\overset{*}{\xi}_{\mathbf{q}}\varphi_1' + i\mathfrak{M}_2'\xi_{\mathbf{q}}\overset{*}{\varphi}{}_1'\right)\hbar\eta^2}{(\overset{*}{\mathfrak{M}}{}_2'\mathfrak{M}_2') + |\varphi_1'|^2}, \quad (12.55)$$

holds and the quantity $\xi_{\mathbf{q}}$, i.e., the coupling parameter w_{12}, can be determined by measuring the polarization of a beam scattered under participation of excitation of types 1 and 2. Measurements of the intensities of the peaks of inelastic scattering as a function of the vector of the initial polarization make it possible to determine with Eqs. (12.44) and (12.45) the vector amplitude $\mathfrak{M}_2'$ of inelastic magnetic scattering and

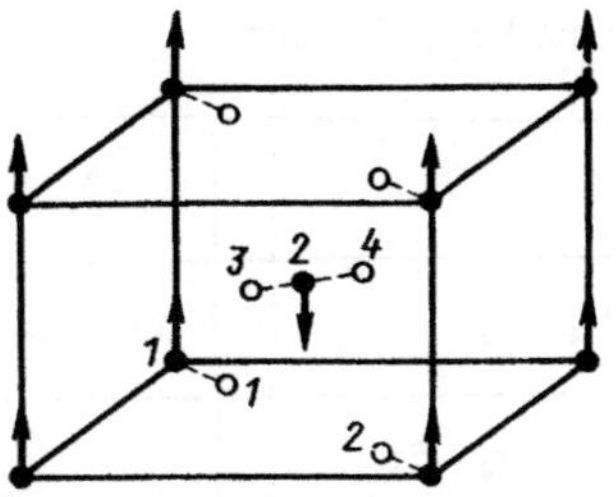

Fig. 10. Unit cell of an FeF_2 crystal with magnetic order.

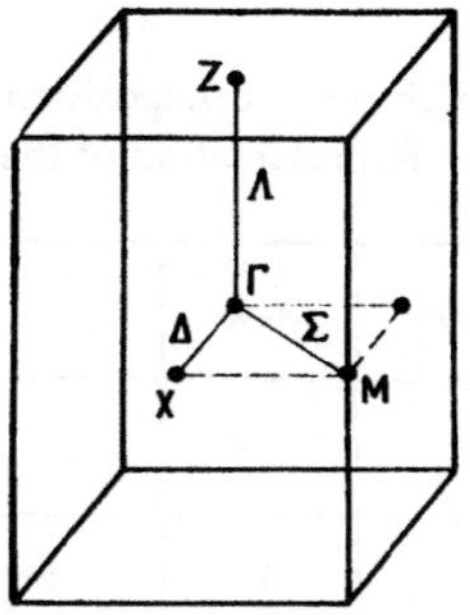

Fig. 11. The Brillouin zone for a primitive tetragonal lattice.

to find with the latter the symmetry (the irreducible representation $d^{q\mu}$ of the spin wave participating in the scattering.

Magnon–phonon hybridization in FeF_2. Let us consider the magnon-phonon magnetization in an antiferromagnetic FeF_2 crystal [152, 210]. This substance has a structure of the rutile type with two iron atoms and four fluorine atoms per unit cell (Fig. 10).

The atoms occupy the positions: Fe(2*a*): *1* (000), *2* $\left(\frac{1}{2}\frac{1}{2}\frac{1}{2}\right)$; F (4*f*): *1* (*uu*0), *2* $(1-u,\ 1-u,\ 0)$, $3\left(\frac{1}{2}+u,\ \frac{1}{2}-u,\ \frac{1}{2}\right)$, $4\left(\frac{1}{2}-u,\ \frac{1}{2}+u,\ 0\right)$ in the $D_{4h}{}^{14}-P4_2/mnm$ space group.

The collinear magnetic structure illustrated in Fig. 10 is characterized by the wave vector $\mathbf{k} = 0$ (the irreducible representation τ_5 of the space group $D_{4h}{}^{14}-P4_2/mnm$ of the crystals) and is described by the Shubnikov group

$$P4_2'/mnm' = D_{2h}^{12} + \theta h_{13} D_{2h}^{12}. \tag{12.56}$$

Thus, the unitary subgroup $G = D_{2h}{}^{12}$ is composed of eight elements in the basis block: h_1-h_4, $h_{25}-h_{28}$; h_{13} means a 180° rotation around the $4(\bar{1}10)$ axis.

The composition of the spin-wave representation $d_{SW}{}^{\mathbf{q}}$ of the group $G_{\mathbf{q}}$ of the wave vector was calculated with Eqs. (11.17) and (11.18) and is listed in Table 11 for points of the Brillouin zone of an orthorhombic lattice corresponding to the $D_{2h}{}^{12}-Pnnm$ group. These points are denoted

TABLE 11. Composition of the Mechanical Representation and the Spin-Wave Representation of the Group $D_{2h}{}^{12}$ for the FeF_2 Crystal*

Point	Atom	d_m^q	d_{SW}^q
Γ	Fe	$\tau_2+2\tau_4+2\tau_6+\tau_8$	$2\tau_3+2\tau_5$
	F	$2\tau_1+\tau_2+\tau_3+2\tau_4+\tau_5+2\tau_6+2\tau_7+\tau_8$	
Δ	Fe	$2\tau_1+\tau_2+\tau_3+2\tau_4$	$2\tau_2+2\tau_3$
	F	$4\tau_1+2\tau_2+2\tau_3+4\tau_4$	
X	Fe	$2\tau_1+\tau_2$	$2\tau_2$
	F	$4\tau_1+2\tau_2$	
Σ	Fe	$4\tau_1+2\tau_2$	$4\tau_2$
	F	$8\tau_1+4\tau_2$	
M	Fe	$2\tau_1+\tau_3+\tau_5+2\tau_7$	$2\tau_2+2\tau_8$
	F	$2\tau_1+\tau_2+\tau_3+2\tau_4+\tau_5+2\tau_6+2\tau_7+\tau_8$	
Λ	Fe	$\tau_1+\tau_2+2\tau_3+2\tau_4$	$2\tau_3+2\tau_4$
	F	$3\tau_1+3\tau_2+3\tau_3+3\tau_4$	
Z	Fe	$\tau_1+2\tau_2$	$2\tau_2$
	F	$3\tau_1+3\tau_2$	

*All representations are one-dimensional except for the points X and Z, where they are two-dimensional.

by the same symbols as in the case of the tetragonal lattice corresponding to the group D_{4h}^{14} of the initial FeF_2 crystal (of rutile structure in the paramagnetic phase (Figure 11). At each of the points **Γ**, **X**, **M**, and **Λ** there are two types of spin waves of different symmetry so that a partial diagonalization of the matrix of the spin-wave Hamiltonian is possible, which matrix must split into two 2 × 2 blocks.

Let us consider the direction **Δ** and **Λ** since we are interested in the spectrum on a line rather than at individual points. The diagonalizing unitary matrices which are composed of the basis functions of the spin-wave representation are for these points

$$U_M(\Lambda)=\frac{1}{\sqrt{2}}\begin{pmatrix}1 & 0 & 1 & 0\\ \varepsilon_{\mathbf{q}} & 0 & -\varepsilon_{\mathbf{q}} & 0\\ 0 & 1 & 0 & 1\\ 0 & \varepsilon_{\mathbf{q}} & 0 & -\varepsilon_{\mathbf{q}}\end{pmatrix};$$

$$\tau_3 \quad \tau_3' \quad \tau_4 \quad \tau_4'$$

$$U_M(\Lambda)=\frac{1}{\sqrt{2}}\begin{pmatrix}1 & 0 & 1 & 0\\ -\varepsilon_{\mathbf{q}} & 0 & \varepsilon_{\mathbf{q}} & 0\\ 0 & 1 & 0 & 1\\ 0 & -\varepsilon_{\mathbf{q}} & 0 & \varepsilon_{\mathbf{q}}\end{pmatrix}, \qquad (12.57)$$

$$\tau_2 \quad \tau_2' \quad \tau_3 \quad \tau_3'$$

where $\varepsilon_{\mathbf{q}}$ is equal to e^{-iqa} for the point **Δ** and equal to e^{-iqc} for the point **Λ**. These matrices reduce the matrix of the spin-wave Hamiltonian and the matrix of the Green's functions of the spin waves to quasidiagonal form.

It follows from Table 11 that there are two branches of spin waves for the direction **Δ**, with the branches related to the one-dimensional irreducible representations τ_2 and τ_3. In the exchange approximation, their energies are degenerate:

$$\hbar\omega^{\tau_2}(\mathbf{q})=\hbar\omega^{\tau_3}(\mathbf{q})=S\sqrt{[J_{11}(0)-J_{11}(\mathbf{q})-J_{12}(0)]^2-|J_{12}(\mathbf{q})|^2}. \qquad (12.58)$$

Similar expressions can be obtained for the direction **Λ** where there are also two branches belonging to one-dimensional representations.

TABLE 12. FeF_2 Crystal Phonon Modes Which Can be Classified by the Irreducible Representations of the Unitary Subgroup $D_{2h}{}^{12}$

Points of the Brillouin zone	Iron			Fluorine				
	Representation	1	2	Representation	1	2	3	4
Δ	τ_1	100	$\varepsilon 00$	τ_1	100		$\bar{\varepsilon}00$	
	τ_1'	010	$0\varepsilon 0$	τ_1'	010		$0\varepsilon 0$	
	τ_2	001	$00\bar{\varepsilon}$	τ_1''		100		$\bar{\varepsilon}00$
	τ_3	001	00ε	τ_1'''		010		$0\varepsilon 0$
	τ_4	100	$\varepsilon 00$	τ_2	001		$00\bar{\varepsilon}$	
	τ_4'	010	$0\bar{\varepsilon}0$	τ_2'		001		$00\bar{\varepsilon}$
				τ_3	001		00ε	
				τ_3'		001		00ε
				τ_4	100		$\varepsilon 00$	
				τ_4'	010		$0\bar{\varepsilon}0$	
				τ_4''		100		$\varepsilon 00$
				τ_4'''		010		$0\bar{\varepsilon}0$
Λ	τ_1	001	00ε	τ_1	100	$\bar{1}00$	$\bar{\varepsilon}00$	$\varepsilon 00$
	τ_2	001	$00\bar{\varepsilon}$	τ_1'	010	$0\bar{1}0$	$0\varepsilon 0$	$0\bar{\varepsilon}0$
	τ_3	100	$\bar{\varepsilon}00$	τ_1''	001	001	00ε	00ε
	τ_3'	010	$0\varepsilon 0$	τ_2	100	$\bar{1}00$	$\varepsilon 00$	$\bar{\varepsilon}00$
	τ_4	100	$\varepsilon 00$	τ_2'	010	$0\bar{1}0$	$0\bar{\varepsilon}0$	$0\varepsilon 0$
	τ_4'	010	$0\bar{\varepsilon}0$	τ_2''	001	001	$00\bar{\varepsilon}$	$00\bar{\varepsilon}$
				τ_3	100	100	$\bar{\varepsilon}00$	$\bar{\varepsilon}00$
				τ_3'	010	010	$0\varepsilon 0$	$0\varepsilon 0$
				τ_3''	001	$00\bar{1}$	00ε	$00\bar{\varepsilon}$
				τ_4	100	100	$\varepsilon 00$	$\varepsilon 00$
				τ_4'	010	010	$0\bar{\varepsilon}0$	$0\bar{\varepsilon}0$
				τ_4''	001	$00\bar{1}$	$00\bar{\varepsilon}$	00ε

Let us now consider the phonon modes of a crystal with magnetic order. They must also be characterized by the irreducible representations of the group $D_{2h}{}^{12}$. The composition of the mechanical representation of the crystal is indicated in Table 12. It follows from a comparison of the mechanical and spin-wave representations that phonon–magnon hybridization is possible on the lines Δ (for the modes τ_2 and τ_3), Λ (for the modes

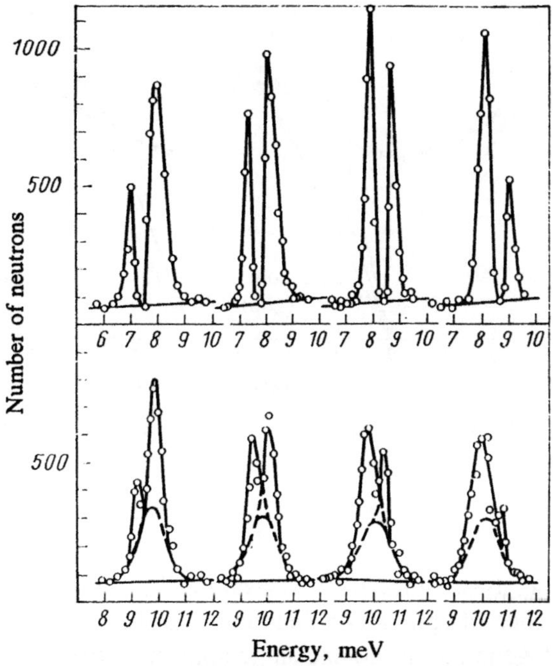

Fig. 12. Intensity of inelastic neutron scattering at FeF_2 as a function of the transferred energy for the wave vectors (00q) at the **q** values 0.35, 0.375, 0.40, and 0.45 [286].

τ_3 and τ_4), and Σ (for the mode τ_2). Since some irreducible representation appears a number of times in $d_M{}^{\mathbf{q}}$ and $d_{SW}{}^{\mathbf{q}}$, we realize that the determinant defining the spectrum of the hybridized modes has the dimension 5 (for modes τ_2 and τ_3 on the line Δ), 7 (for modes τ_3 and τ_4 on the line Λ), and 10 (for the mode τ_2 on the line Σ).

Experimental studies of the spectrum of the elementary excitation in the FeF_2 crystal have been made in [286]. The authors carefully measured the region in which the branches of the magnons and the acoustic phonons should intersect in the three symmetric [001], [100], and [101] directions. As indicated in Fig. 12, two well-resolved peaks caused by scattering at hybridized states were observed in these directions. The dispersion curves are shown in Fig. 13 where the hybridization of all the magnons with the acoustic phonons is well recognizable for all three symmetric directions.

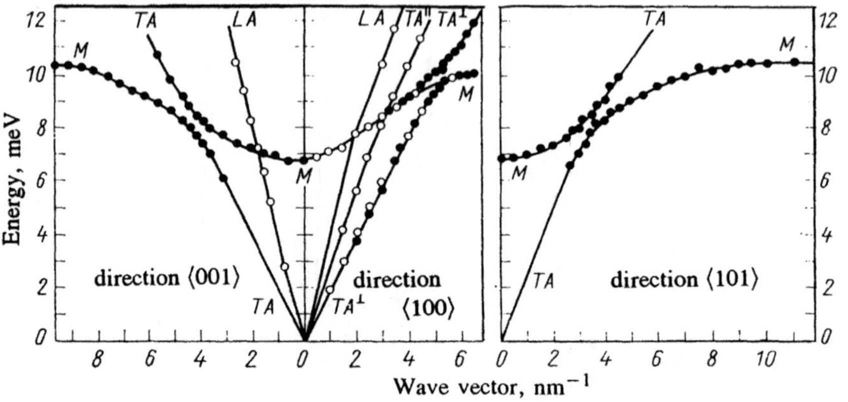

Fig. 13. Dispersion curves for magnons and phonons in an FeF_2 crystal at $T = 4.2$ K [286].

Symmetry analysis in the [001] direction (line Δ) has shown that there exist two branches of spin waves belonging to the representations τ_2 and τ_3 which are degenerate in the exchange approximation. The degeneracy is conserved even when single-ion uniaxial anisotropy is brought into account [235]. The transverse acoustic phonon branch, from which the hybridization of the spin wave originates, belongs to the representation τ_3, as can be readily inferred from Table 12. Consequently, this spin wave also belongs to the irreducible representation τ_3. The other branch of the degenerate spin wave is not hybridized with the acoustic phonon. The other branch is shown in Fig. 13 in the intersection region of the dispersion curves but the experimental points are not shown since the third peak of the spectrum of inelastic scattering along the line Δ is overlapped by stronger peaks corresponding to the scattering at hybridized waves of symmetry τ_3. The parameters of spin–lattice coupling in FeF_2 were determined from a comparison of the theoretical dispersion curves and the experimental data [235].

Hybridization was observed with the methods of neutron spectroscopy also in an other antiferromagnetic crystal, $FeCl_2$, of hexagonal symmetry (space group D_{3d}^5) [122, 368]. The selection rules have been theoretically analyzed in [254]. In the direction $\mathbf{q} \sim [100]$, and "intersection" of the dispersion curves is found at $q = 1.7\ \text{nm}^{-1}$, and the splitting at that point is 33 meV.

13. Experimental Investigations of the Spin-Wave Spectrum of Magnetic Materials

The determination of the dispersion curves of spin waves in crystal with magnetic order significantly helped to establish the interactions accounting for the development of magnetic order in various types of magnetic materials. As far as the microscopic aspect is concerned, materials with magnetic order can be divided into three large groups: dielectrics, rare-earth metals, and transition metals together with their alloys. Localized atomic magnetic moments are preserved in dielectric, and can be described with the effective spin Hamiltonians which account for the exchange and the various forms of anisotropy. The atomic magnetic moments are well-preserved in rare-earth metals but they are coupled by indirect interactions between the conduction electrons. Magnetically ordered $3d$ metals, in which there are dual magnetic moments, namely localized and collectivized moments, are objects of greater complexity.

We consider below research on the spectra of magnetic excitations in each of these groups of magnetic materials; it will be shown how conclusions can be drawn on the atomic magnetic states in these materials and the interatomic interactions on the basis of the form of the dispersion curves for these excitations. The first two classes of magnetic materials, the dielectrics and the rare-earth metals, will be considered in this section, and the $3d$ metals will be treated in Section 14.

Magnetic dielectrics. We cite as examples the results of research on the spin waves measured in the following three substances with the aid of a three-crystal spectrometer: the ferromagnetic materials $ErFe_2$ [291] and MnP [337] and the two-dimensional antiferromagnetic material K_2FeF_4 [341]. Let us begin our discussion with $ErFe_2$ though it is not a dielectric. The atomic magnetic moments of Er are in the crystal close to the moments $\mu(\mathrm{Er}) \approx 8.5\mu_B$ of the free atoms, whereas $\mu \approx 1.6\mu_B$ in the case of Fe instead of the expected value $2\mu_B$ [291]. $ErFe_2$ has the structure of a Laves phase and is characterized by the space group $O_h{}^7$ – $Fd3m$. The primitive cell of the paramagnetic phase comprises six atoms: two Er atoms and four Fe atoms. At $T \approx 600$ K there arises a ferromagnetic structure in which the atomic spins of the Er sublattice are antiparallel to the atomic spins of Fe. The direction of the spins in this collinear structure coincides with an edge of the cube; the Shubnikov group of the crystal turns out to be $I4_1/am'd'$ [276]. Thus, in the

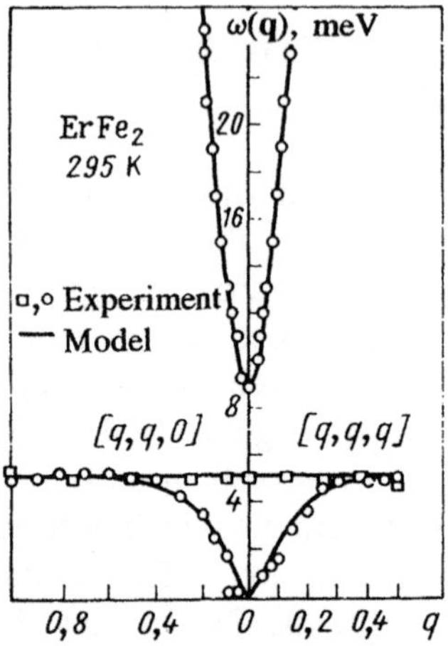

Fig. 14. Dispersion curves of spin waves of a ferromagnetic $ErFe_2$ crystal [291].

primitive magnetic cell there are six magnetic atoms and, consequently, six branches of spin waves must develop.

Figure 14 illustrates the dispersion curves only for three spin waves, an acoustic spin wave and two optical spin waves [291]. The exchange coupling parameters which can be derived from the data help to determine the energy of the three other optical branches. They turn out to be very large, of the order of 200–300 meV, whereas the energies corresponding to the lowest branches amount to about 10 meV. The lower optical branch is practically without dispersion, which means that the Er–Er exchange interaction is very weak. Besides that, the very strong dispersion of the second optical branch points to a very strong Fe–Fe exchange interaction. The solid curves of the figure correspond to the theoretical curves (taking into account only interactions between nearest neighbors) with the following trial parameters:

$$J_{\mathrm{Fe-Fe}} = 30 \text{ meV}, \quad J_{\mathrm{Fe-Er}} = -0{,}32 \text{ meV}, \quad J_{\mathrm{Er-Er}} = 0.$$

This work is a good example of how the parameters of the effective spin Hamiltonian are determined from data relating to inelastic neutron scattering.

Let us now consider another example of research on the spin wave dynamics in a ferromagnetic material. The orthorhombic MnP crystal is a ferromagnet below 291 K but a magnetic phase transition under formation of a helical magnetic structure occurs at the temperature $T_N =$ 47 K. In the ferromagnetic phase, the spontaneous moment is parallel to

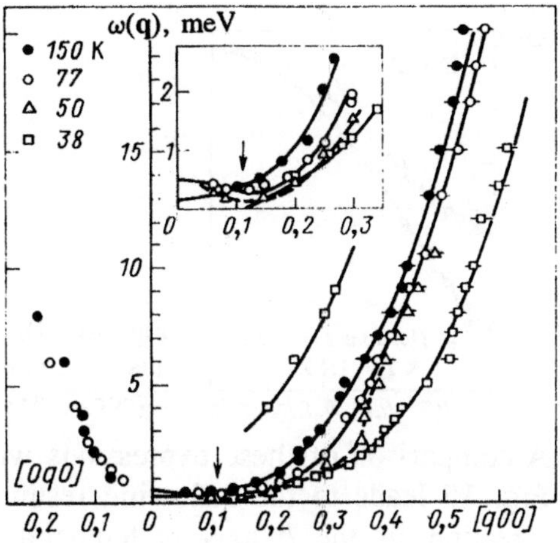

Fig. 15. Temperature dependence of the spectrum of the spin waves in a MnP crystal [337].

the c axis, whereas the spins are rotated in the b–c plane in the helical structure, with the wave vector parallel to the axis a and equal to $\mathbf{k} = 0.117(2\pi/a)$.

The dispersion curves of the spin waves were measured in the direction of the a and b axes at various temperatures above and below T_N (Fig. 15) [337]. In the ferromagnetic state, the spin waves do not exhibit a noticeable temperature dependence in the b direction by contrast to the dispersion curves in the a direction. At 77 K, i.e., 30 K above T_N, in the region $q \approx 0.1\ (2\pi/a)$ there appears a minimum the depth of which increases with decreasing distance from T_N. The position of this minimum corresponds to the wave vector of the developing helical structure. The phase transition to the spiral structure can therefore be interpreted in terms of a soft magnetic mode.

In order to establish which of the parameters change with temperature and cause softening of the spin mode in the ferromagnetic phase, we consider the Hamiltonian of the orthorhombic crystal:

$$\mathcal{H} = -\sum_{ll'} J_{ll'}(\mathbf{S}_l \mathbf{S}_{l'}) + D_1 \sum_l S_l^{z^2} + D_2 \sum_l (S_l^{x^2} - S_l^{y^2}) \tag{13.1}$$

with uniaxial anisotropy: uniaxial ($\sim D_1$) and interplane ($\sim D_2$) anisotropy. The energies of the spin waves in the ferromagnetic and helical phases are given by three respective expressions [337]:

$$\hbar\omega(\mathbf{q}) = 2S\sqrt{[J(0) - J(\mathbf{q}) - 4D_2][J(0) - J(\mathbf{q}) + D_1 - 2D_2]}; \tag{13.2}$$

$$\hbar\omega(\mathbf{q}) = 2S\sqrt{\left[J(\mathbf{k}) - \frac{1}{2}J(\mathbf{k}+\mathbf{q}) - \frac{1}{2}J(\mathbf{k}-\mathbf{q})\right][J(\mathbf{k}) - J(\mathbf{q}) + D_1]}. \tag{13.3}$$

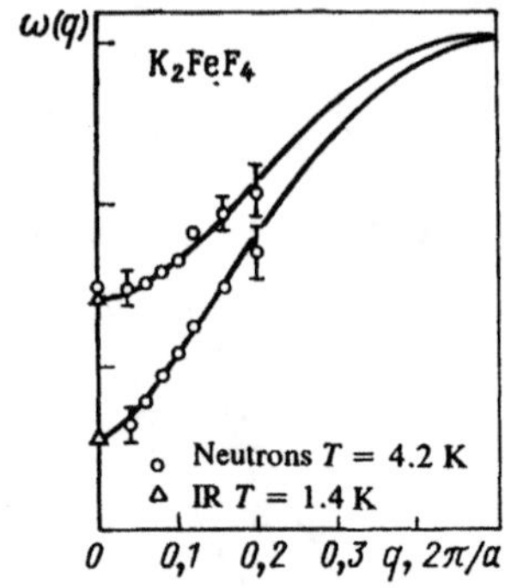

Fig. 16. Dispersion curves of the spin waves of the quasi-two-dimensional antiferromagnetic material K_2FeF_4 [341].

A comparison of these expressions with the dispersion curves shown in Fig. 15 leads to the following result: The exchange parameter in the direction of the *b* axis is temperature-independent but has a certain temperature dependence in the direction of the *a* axis. The nearest-neighbor exchange integral is positive in the direction of the *a* axis and does not change with temperature, whereas it turns out to be negative and monotonically changing with temperature in the interval 38–150 K for the following neighbors, so that the observed magnetic phase transition results.

Finally, let us consider a third example, namely the spin-wave spectrum in the antiferromagnetic material K_2FeF_4 [341]. This orthorhombic crystal is described by the spin wave Hamiltonian Eq. (13.1) with a large positive anisotropy constant D_1 which makes the magnetic system practically two-dimensional. In the Goldstein–Primakov approximation ($S >> 1$), the energies of the two branches of the spin waves in this collinear, two-sublattice two-dimensional antiferromagnetic material are given by the expression

$$\hbar\omega(\mathbf{q}) = 4JS\sqrt{\left[1+\frac{D_1-3D_2}{4J}\right]^2-\left[\gamma_{\mathbf{q}}\pm\frac{D_1+D_2}{4J}\right]^2}, \tag{13.4}$$

where $\gamma_{\mathbf{q}} = \cos(q_x a/2)\cos(q_y a/2)$.

The dispersion curves measured at 4.2 K in the direction of the *a* axis are illustrated in Figure 16 [341]. Additional research work has shown that the energies of the spin waves do not depend upon q_z [in accordance with Eq. (13.4)] so that the two-dimensional form of the antiferromagnetism in this crystal is confirmed. The solid lines of the figure were obtained with Eq. (13.4) and the trial parameters $J = -11.5$ cm^{-1}, $D =$ 3.6 cm^{-1}, $D_2 = -0.37$ cm^{-1}. The two-dimensional form of the antiferromagnetism was also confirmed by measurements of the temperature dependence of the magnetization of the sublattices ($m \approx (T_c - T)^{\beta}$) by way of the temperature dependence of the intensity of the antiferromagnetic peak. It turned out that $\beta = 0.13 \pm 0.01$ in accordance with the accurate value of the critical index for the two-dimensional Ising model, $\beta = 1/8$.

Rare-earth metals and compounds. The magnetic properties of the rare-earth metals are related to the unfilled $4f$ shells of atoms which are well shielded by the closed $5s$–$5p$ shells. Owing to this shielding, the atomic states of the rare-earth ions are preserved in the crystal and the total moment J of the atom is a good quantum number. The magnetic structures and the elementary excitations in the rare-earth systems are given by the combination of three interactions: indirect exchange, crystal field, and magnetoelastic forces [239–241].

The indirect exchange interaction of the $4f$ electrons through the conduction electrons is isotropic in a first approximation and can be described by a Heisenberg-type expression:

$$\mathcal{H}_{ex} = -\sum_{ll'} J(\mathbf{R}_l - \mathbf{R}_{l'}) J_l J_{l'}. \tag{13.5}$$

The Fourier component $J(\mathbf{q})$ of the exchange integral $J(\mathbf{R}_l - \mathbf{R}_{l'})$, which is of the long-range type, is expressed through the susceptibility $\chi(\mathbf{q})$ of the electron gas:

$$J(\mathbf{q}) = 2(g-1)^2 \, |I(\mathbf{q})|^2 \, \chi(\mathbf{q}), \tag{13.6}$$

where $I(\mathbf{q})$ denotes the Fourier component of the s–f exchange integral, the component usually assumed to depend weakly upon the momentum (this corresponds to contact-type s–f coupling) and to be constant; g denotes the Landé factor which arises because the exchange energy of Eq. (13.5) was stated in terms of the total moments instead of the spin moments. The susceptibility of the electron gas is given by the standard equation of the self-consistent field theory:

$$\chi(\mathbf{q}) = \frac{1}{N} \sum_{\mathbf{k}} \sum_{nn'} \frac{n_n(\boldsymbol{\kappa}) - n_{n'}(\boldsymbol{\kappa}+\mathbf{q})}{\varepsilon_{n'}(\boldsymbol{\kappa}+\mathbf{q}) - \varepsilon_n(\boldsymbol{\kappa})}, \tag{13.7}$$

in which $\varepsilon_n(\mathbf{k})$ denotes the energy of the electron in the nth band and $n_n(\mathbf{k})$ denotes the Fermi distribution function. Thus, all features of the electron spectrum of the metal manifest themselves in the behavior of the quantity $J(\mathbf{q})$. In particular, when the Fermi surface of the metal has parallel sections separated in the reciprocal lattice by the momentum $\mathbf{Q}$, the expression for $\chi(\mathbf{q})$ can have a maximum at $\mathbf{q} = \mathbf{Q}$, which must show up in the properties of the elementary magnetic excitations.

The magnetic 4*f* electrons experience the effect of the inhomogeneous electric field from the surrounding ions. Since the intrinsic charge cloud of the ions is highly anisotropic, this (crystal) field tends to align the orbital field and, concurrently, also the total moment of the ion, parallel to some crystallographic directions. The exchange interaction of Eq. (13.5) tends to create some noncommensurate magnetic structure with the wave vector $\mathbf{Q}$ so that the resulting magnetic structure in rare-earth metals and their compounds can be very complicated. The Hamiltonian of the crystal field can be stated in the form

$$\mathcal{H}_{cf} = \sum_{s} \sum_{lm} B_{lm} O_l^m (J_s), \tag{13.8}$$

where O_l^m denotes certain polynomials (like spherical harmonics) of the operators of the atomic spin components; and B_{lm} denotes the parameters of the crystal field. Finally, one must bear in mind the correlation between the atomic magnetic moments and the lattice deformations, both homogeneous and local, which can be very important for rare-earth systems.

First let us consider the case in which the influence of interactions other than the exchange interaction upon the spectrum of the excited states is small. This case is observed in rare-earth metals having a vanishing orbital moment (case of Gd and Eu). In the case of ferromagnetic and simple helical structures, the spin-wave spectrum is defined by the dispersion

$$\hbar\omega(\mathbf{q}) = J\,[J(0) - J(\mathbf{q})]; \tag{13.9}$$

$$\hbar\omega(\mathbf{q}) = J\sqrt{\left[J(\mathbf{Q}) - \frac{1}{2}J(\mathbf{q}+\mathbf{Q}) - \frac{1}{2}J(\mathbf{q}-\mathbf{Q})\right][J(\mathbf{Q} - J(\mathbf{q})]} \tag{13.10}$$

[see Eqs. (9.6) and (9.36)]; in the case of a ferromagnetic structure, $J(\mathbf{q})$ has a maximum at $\mathbf{q} = \mathbf{Q}$. When an orbital moment is present, one must include the energy of the crystal field of Eq. (13.8) which modifies the dispersion curves of Eqs. (13.9) and (13.10) as follows:

$$\hbar\omega(\mathbf{q}) = \sqrt{\{J[J(0) - J(\mathbf{q})] + 6B_{20}J\}\{J[J(0) - J(\mathbf{q})] + 36B_{66}J_5\}}\,; \quad (13.11)$$

$$\hbar\omega(\mathbf{q}) = J \times \quad (13.12)$$

$$\times \sqrt{\left[J(\mathbf{Q}) - \frac{1}{2}J(\mathbf{q}+\mathbf{Q}) - \frac{1}{2}J(\mathbf{q}-\mathbf{Q})\right][J(\mathbf{Q}) - J(\mathbf{q}) + 6B_{20}]}\,,$$

where

$$J_n = \left(J - \frac{1}{2}\right)(J-1)\ldots\left(J - \frac{n}{2}\right).$$

The dispersion curves are more complicated in other magnetic structures observed in rare-earth metals. Equation (13.11) must describe the spin-wave spectrum in the ferromagnetic phase of Tb, and Eq. (13.12) must hold for the helical phases of Tb, Dy, and Ho.

The magnon spectrum was obtained for the majority of rare-earth metals in various magnetic phases when experiments with inelastic neutron scattering were made (see, e.g., the reviews [239, 199]). We cite as an example the dispersion curves obtained for Tb at 4.2 K in the ferromagnetic phase (Fig. 17). The Tb crystal has two atoms in its primitive cell, and therefore there are two branches of spin waves, an acoustic branch and an optical branch. As shown by the drawing, the acoustic branch has an interruption at $\mathbf{q} = 0$ as a consequence of single-ion anisotropy. But the width of the interruption is much smaller than the zone of the magnon branches, which means that the exchange interaction dominates over the anisotropic interaction in Tb. The figure also shows the effects of magnon–phonon hybridization, indicating that the interactions of the magnetic lattice are important. The longitudinal acoustic phonon branch is hybridized with the acoustic magnon branch. Hybridization of optical magnons with acoustic phonons can also be seen.

A comparison of the observed dispersion curves with the theoretical Eqs. (13.11)–(13.12) made it possible to find the amplitude $(g - 1)^{-2} \times [J(\mathbf{q}) - J(0)]$ for some ferromagnetic metals (Fig. 18). According to Eq. (13.6) this function must be the same for all metals, provided that the electron structure of these metals is similar. In reality, there exist substantial differences between the curves depicted but they all have a

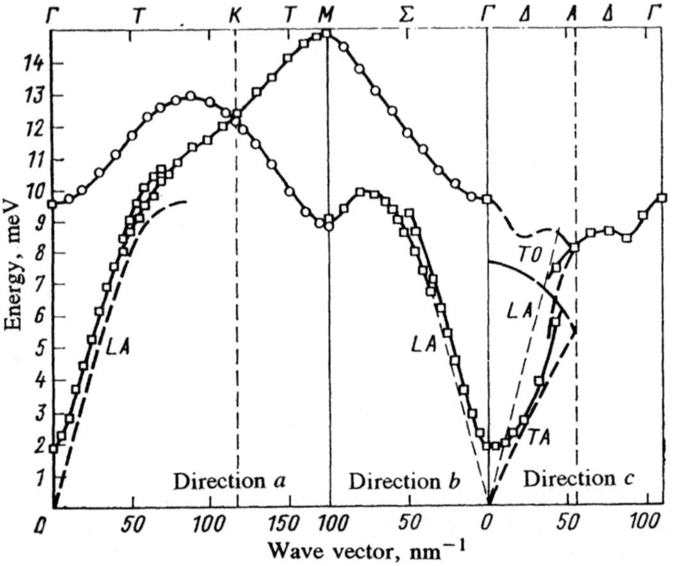

Fig. 17. Magnon dispersion curves for ferromagnetic Tb at 4.2 K [199].

maximum at $\mathbf{q} = 0$, which corresponds to ferromagnetic order in these metals. In the case of metals with helical magnetic structures (Ho, Dy) or with the structure of a "conical spiral" (Er), the Fourier component $J(\mathbf{q})$ has a maximum at a finite $\mathbf{q}$. As indicated above, such a maximum is a necessary condition for structural stability.

These results of neutron investigations show that magnetic order and magnetic excitations in rare-earth metals can be understood when one uses the concept of a dominating isotropic exchange interaction. In light rare-earth metals, the atomic moments J and the parameters of the isotropic exchange are not so large, and anisotropic interaction effects and crystal-field effects appear more conspicuously in them and lead to abnormal features of magnetic order and of magnetic excitations. Pr, in which the energy of the crystal field surpasses the exchange energy, can be taken as an example. The exchange energy of Pr amounts to about 90% of the critical value required for establishing magnetic order above 0.1 K.

The excitations in the paramagnetic Pr phase were carefully studied by the authors of [198]. The ground state of the Pr ion in a crystal field is a singlet $|J_z = 0\rangle$, whereas the first excited states, separated by the

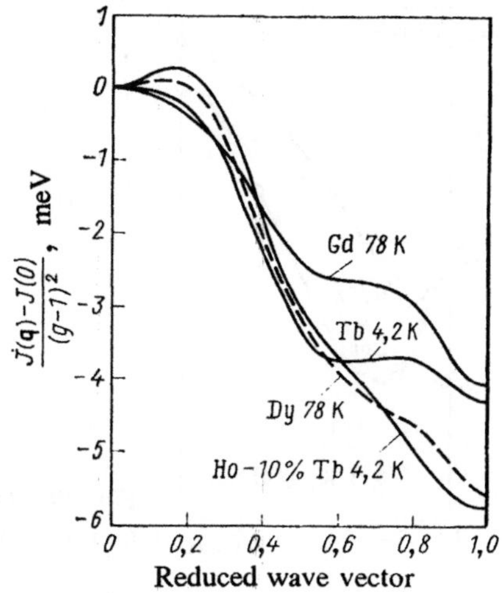

Fig. 18. The Fourier component of the integral of indirect exchange for wave vectors **q** in the *c* direction in the case of certain metals with ferromagnetic or helical structure [199].

interruption Δ, are a doublet. These single-ion states become collectivized states because of the anisotropic exchange interaction which results in a collectivized excitation frequently termed a magnetic excitation; the dispersion law (in the *b* direction) of [232] applies:

$$\hbar\omega_{\pm\alpha}(\mathbf{q}) = \sqrt{\Delta^2 - 2\zeta^2\Delta\,[J^{\alpha\alpha}(\mathbf{q}) \pm |\,J'^{\alpha\alpha}(\mathbf{q})|]}\,. \qquad (13.13)$$

The notation is interpreted as follows: ζ denotes the matrix element of J^x between the ground state and the excited states; $J^{\alpha\alpha}(\mathbf{q})$ and $J'^{\alpha\alpha}(\mathbf{q})$ denote the Fourier components of the anisotropic exchange between the atoms belonging to a single Pr sublattice and various sublattices, respectively. (The Pr crystal has the structure of a double hexagonal close-packed lattice with two atoms per primitive cell.) The projection α assumes two values ($\alpha = x$ and $\alpha = y$) and therefore Eq. (13.13) renders four branches of exciton excitations, which branches are clearly recognizable in inelastic neutron scattering [198, 248]. It follows from a comparison of the theoretical and experimental dispersion laws of excitons that the exchange must be highly anisotropic.

The modes with the subscripts (+) and (−) are conventionally termed optical and acoustic modes, respectively. The optical modes in the *b* direction are illustrated in Fig. 19. At $q = 2.5\ \text{nm}^{-1}$ the energy values of the lower branch are indicated for various temperatures; these values show that this branch of the excitations is a soft magnetic mode whose energy is nonvanishing. This mode has a finite energy at $T = 5$ K even

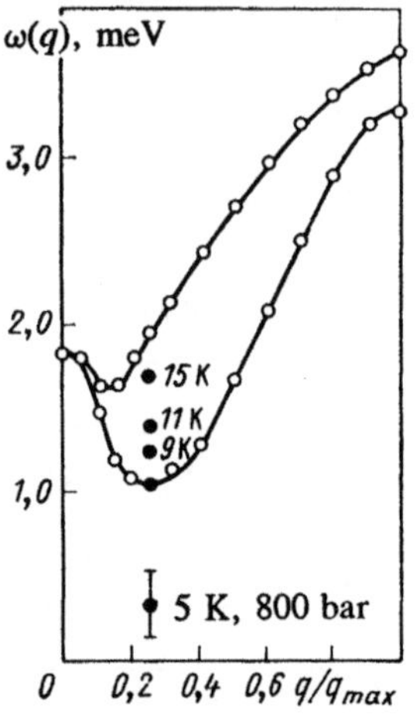

Fig. 19. Magnetic excitations propagating in the *b* direction in paramagnetic praseodymium [248].

when an uniaxial pressure of the order of 1 kbar is applied in the *b* direction though a noncommensurate magnetic structure with the indicated though a noncommensurate magnetic structure with the indicated value $Q = 2.5\ nm^{-1}$ develops in Pr under this pressure. The resulting magnetic structure is like a frozen soft magnetic mode though the fact that the energy of the mode at the point of the magnetic transition is nonvanishing is not understood [248]. A soft magnetic mode was recently detected also in PrSb [350] where an isotropic pressure caused the disappearance of the excitation energy at the symmetric point of the Brillouin zone by generating magnetic order with a noncommensurate structure. The authors of [241] have given a review of unresolved problems in the physics of rare-earth metals and rare-earth compounds.

Quasi-one-dimensional systems. The achievements of neutron physics in research on the condensed state of matter are particularly impressive when magnetic effects in quasi-one-dimensional systems are considered. Such situations arise in certain anisotropic crystals in which the magnetic atoms are arranged in chains separated from each other in the crystal and frequently shielded by other nonmagnetic atoms or molecular groups. The famous $(CH_3)_4NMnCl_3$ (TMMC), which is a magnetic material very close to ideal one-dimensionality, may serve as an example [333].

The Mn^{++} ions are arranged in linear chains composed of $MnCl_3$ molecules and are spaced by 0.325 nm. The chains form a hexagonal crystal. They are separated from each other by $(CH_3)_4N^+$ complexes so

TABLE 13. Quasi-One-Dimensional Magnetic Materials [121]

Crystal	S	T_N, K	T_N/T_{MF}
$CuSO_4 \cdot 5H_2O$	1/2	1,4	0,02
$CuCl_2 2NC_5H_5$ (CPC)	1/2	1,1	0,08
α-CuN	1/2	0,04	0,01
$CsNiCl_3$	1	4,6	0,13
$CsNiF_3$	1	2,6	0,08
$CsMnCl_3 \cdot 2H_2O$	5/2	4,9	0,12
$(CH_3)_4N\ MnCl_3$ (TMMC)	5/2	0,8	0,01

that the distance between the chains is 0.915 nm. This difference in the spacing of the magnetic ions parallel to the chain and on planes perpendicular to it causes a difference of several orders of magnitude in the corresponding exchange integrals and accounts for the quasi-one-dimensional magnetic state in the crystal.

The weak interchain interaction causes a three-dimensional magnetic order in the entire crystal, the order being characterized by the very low Neel temperature $T_N = 0.85$ K. This temperature is lower than the temperature determined form the well-known expression of molecular field theory by two orders of magnitude

$$T_{MF} = \frac{1}{3} \frac{z|J|}{k} S(S+1), \tag{13.14}$$

where J denotes the nearest-neighbor exchange integral. In quasi-one-dimensional systems, the atoms of some chain are the nearest neighbors. Magnetic measurements have rendered for TMMC the value $J = -6.74$ K, which implies $T_{MF} = 76$ K. It has been known for a long time that the molecular field approximation cannot describe a phase transition in one-dimensional systems (see, e.g., [59]) because the fluctuations of the order parameter are of increasing importance. Thus, the magnetic system of the crystal can show a quasi-one-dimensional behavior when structurally separated linear chains of magnetic atoms exist and the temperatures T_N and T_{MF} differ substantially. In addition to TMMC, a large number of quasi-one-dimensional magnetic materials listed in Table 13 are known [121].

It is a specific feature of one-dimensional magnetic systems that no long-range magnetic order exists in them at a finite temperature because the fluctuations have a destructive effect; nevertheless, a considerable degree of short-range order (in the direction of the chains) exists in a wide temperature interval, the order being characterized by the correlation length ξ. For $T \to 0$ the correlation length diverges but for $T = 0$, the long-range order can exist or be missing, depending upon the type of the magnetic system and the form of the exchange interaction. Let us state in the following form the Hamiltonian of the one-dimensional chain of spins, with proper regard for the interaction between the nearest neighbors only:

$$\mathcal{H} = -2J\sum_{l}[\alpha S_l^z S_{l+1}^z + \beta(S_l^x S_{l+1}^x + S_l^y S_{l+1}^y)], \qquad (13.15)$$

where the constants α and β characterize the magnetic anisotropy. Depending upon the values of the constants α and β, there result three very important models of the magnetic material: the Ising model ($\alpha = 1$, $\beta = 0$); the X–Y model ($\alpha = 0$, $\beta = 1$); and the isotropic Heisenberg model ($\alpha = 1$, $\beta = 1$). It follows from the theory that the features of these models differ in a particular measurement in regard to the development of long-range order at $T = 0$ but strong correlation effects at low temperatures can be predicted for all models. There exist excellent reviews [214, 121, 333] of the theoretical and experimental research on one-dimensional magnetic materials, and we therefore restrict our consideration to the general outline of the physics of magnetic quasi-one-dimensional systems.

First of all, let us define the concept of quasi-one-dimensionality. As noted above, a crystal with a quasi-one-dimensional behavior of its magnetic system can have magnetic order in a three-dimensional space because of a weak interaction between the chains at a low temperature $T \leq T_N \leq T_{\mathrm{MF}}$. Above the temperature of three-dimensional establishment of order ($T > T_{\mathrm{MF}}$), the correlators of the spins out of the various chains are negligibly small, whereas the correlators of the spins aligned with the chains are very large. The magnetic chain can become quasi-one-dimensional under these conditions. There is no long-range magnetic order in such a chain but its behavior depends upon the sign of the exchange interaction. Depending upon the sign, one refers to ferromagnetic or antiferromagnetic chains without implying a corresponding long-range order in this definition. TMMC is the most thoroughly studied antiferromagnetic material, whereas $CsNiF_3$ is the corresponding quasi-one-dimensional ferromagnetic material (see Table 13).

The existence of magnon-like excitations with a clearly pronounced dispersion without long-range order is a distinctive feature of quasi-one-dimensional magnetic materials. First, let us consider the predictions which for this reason are provided by the theory. Full order of the spins at $T = 0$ has been established for Heisenberg chains of spins in the classical limit ($S \to \infty$). In this limit, the first excited states correspond to spin waves with an infinitely long lifetime. The first excited states have the following dispersion law in the case of an antiferromagnetic chain [178]:

$$\hbar\omega(q) = 4S\,|J \sin(q_z a)|. \tag{13.16}$$

What happens with these waves at a finite temperature? Though the long-range order is destroyed, there exists short-range order characterized by a large correlation length ξ. For spin waves with a wavelength $1/q_L$, the existing instantaneous magnetic order differs only slightly from the long-range order, and therefore the resulting excitations must many times oscillate on the resulting fluctuations of order, i.e., they must have a rather long lifetime. The energy width of such an excited state can be expressed with the aid of the expression:

$$\Delta\omega(q) \sim \frac{d\omega(q)}{dq} \cdot \frac{1}{\xi} \sim \frac{SJ}{a\xi} \sim \frac{kT}{S}, \tag{13.17}$$

where we used Eq. (13.16) and the estimate $\xi/a \propto S^2J/kT$ for the correlation length of a one-dimensional magnetic material. A similar result is also obtained for a ferromagnetic material. As soon as the length of the spin wave passes through the threshold $\sim 1/\xi$, the wave is suddenly attenuated and in this case ($q > 1/\xi$) the excited states correspond to diffusion modes.

The absence of a magnetic long-range order at $T = 0$ has been established for the one-dimensional quantum system of Heisenberg spins with $S = 1/2$ [142]. The authors of [142] could obtain an accurate expression for the energy of the first excited state at $T = 0$:

$$\hbar\omega(q) = \pi\,|J \sin(q_z a)|. \tag{13.18}$$

These energies coincide with the Eqs. (13.16) of the classical theory except for a numerical parameter. For the X–Y quantum model with $S = 1/2$, there exists a similar precise result [231]. The dispersion law

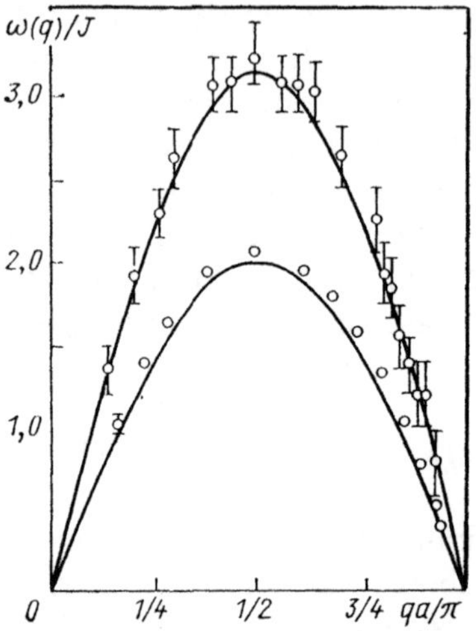

Fig. 20. Dispersion curves for the two quasi-one-dimensional antiferromagnetic materials TMMC and CPC [121].

precisely coincides with expression (13.16) of the classical theory. Thus, the classical one-dimensional systems and the one-dimensional quantum systems behave at low temperatures in a manner which greatly distinguishes one-dimensional systems from three-dimensional systems and implies the existence of excitations with a long lifetime in the absence of any long-range order. These excitations are magnon-like excitations regardless of whether there exists a long-range order in the ground state (in the case of systems with the classical spins $S >> 1$) or no such order exists (in the case of quantum systems).

Experimental research performed with the aid of inelastic neutron scattering has excellently confirmed the conclusions of the theory [119, 204]. Figure 20 shows the experimental data obtained with inelastic neutron scattering for the dispersion curves of two quasi-one-dimensional antiferromagnetic materials, TMMC and CPC (see Table 13). In the case of TMMC, $S = 5/2$ and therefore the experimental data must be compared with the conclusions of the classical theory [see Eq. (13.16)]; in the case of CPC, $S = 1/2$, and the conclusions of quantum theory must be used [see Eq. (13.18)]. In both cases, the experimental curves are precisely situated on the corresponding theoretical curves when the

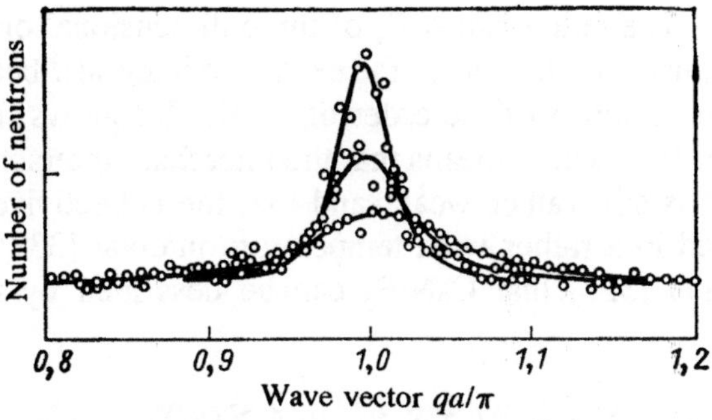

Fig. 21. Temperature dependence of the neutron resonances corresponding to the excitation of a collective spin mode in the quasi-one-dimensional antiferromagnetic material TMMC [121].

exchange integrals have the values $|J| = 6.47$ K in the case of TMMC and $|J| = 13.4$ K in the case of CPC; this means good agreement with the data obtained from measurements of other characteristics.

A similar pattern of collectivized excitations is conserved within a rather wide temperature interval (up to 40 K in the case of TMMC). In the entire interval indicated, one observes rather narrow neutron resonances whose width increases with temperature but remains much smaller than the energy $\hbar\omega(\mathbf{q})$ as long as the correlation length is short. The temperature dependence of the neutron resonances in TMMC is depicted in Fig. 21. In this figure, the wave vector is parallel to the chains (i.e., parallel to the hexagonal axis of the crystal). In the case of a general position of the wave vector **q**, one realizes with a fixed component q_z and varying components q_x and q_y, that the resonance character of the scattering is not affected, which in the pattern of inelastic scattering means that resonance planes (perpendicular to the chains) characteristic of scattering at quasi-one-dimensional systems exist in the reciprocal lattice of the crystal.

The well-known $CsNiF_3$ can serve as an example of a quasi-one-dimensional ferromagnetic system. This substance has hexagonal structure with the space group *P63/mmc*. The structure is composed of NiF_6 complexes which are aligned in chains parallel to the *c* axis and subdivided by Cs ions. The temperature T_N of three-dimensional order establishment is determined as 3.6 K when the exchange integral J between nearest neighbors in the chain is of the order of 10 K. It follows from Table 13 that $T_N/T_{MF} \simeq 0.1$, which means that the interchain interaction is greater in TMMC but is still rather weak, and that the collectivized excitations can be observed in a rather wide temperature interval [332].

It was found [333] that $CsNiF_3$ can be described by the following Hamiltonian:

$$\mathcal{H} = -2J\sum_l S_l S_{l+1} + A\sum_l (S_l^z)^2, \tag{13.19}$$

where $S = 1$; $J/k = 11.8$ K; and $A/k = 9.5$ K. Thus, the anisotropy, which keeps the spins on a plane perpendicular to the chains, is very important in $CsNiF_3$. This means that the dynamics of the spin components parallel to the plane differs considerably from the dynamics of the perpendicular component. The authors of [332] used inelastic neutron scattering to determine two types of collectivized spin modes: one is related to the correlation of the spins within the plane, and the other, to the correlation of the longitudinal components. The inner-plane mode is characterized by a broad peak similar to that observed in TMMC. The mode beyond the plane, however, has an extremely small width, and the intensity of this mode decreases sharply with decreasing temperature.

In conclusion, we note that the basic predictions of the theory of quasi-one-dimensional ferro- and antiferromagnetic systems (existence of magnon-like collectivized modes in linear chains without long-range order) are fully confirmed by experiments involving neutron spectroscopy.

14. Neutron Scattering in a Collectivized Electron System

The model. So far we have expounded the theory of neutron scattering in a localized system of magnetic moments under the assumption that the atomic magnetic moments of the transition elements on the crystal basically conserve their local states. This concept is applicable to nonmetallic compounds of the transition elements and even to rare-earth metals and alloys.

Owing to the extensive overlapping of the wave functions of the d electrons with the wave functions of the neighbors, in d metals there exists a strong delocalization of these electrons and this means, in particular, that there are fractional magnetic moments of the atoms in the metals and great differences in magnitude when compared with the magnetic moments of isolated atoms. A localized moment is a very crude concept in the case of such systems and a consistent approach to neutron scattering in such systems must be based upon the model of collectivized electrons. For the sake of simplicity, we will consider a single-band model of the metal. In *d* metals, the spin part and the space part of the wave function of the electron are separated. It is therefore possible to expand the wave function $\psi_\sigma(\mathbf{r})$ of the electron with the spin $\sigma = \uparrow, \downarrow$ at the point **r** in a series of Bloch functions $\psi_\mathbf{k}(\mathbf{r})$:

$$\psi_\sigma(\mathbf{r}) = \sum_\mathbf{k} a_{\mathbf{k}\sigma}\psi_\mathbf{k}(\mathbf{r}). \tag{14.1}$$

where $a_{\mathbf{k}\sigma}$ donates the secondary quantization operators obeying the Fermi commutation relations:

$$\left.\begin{aligned} a_{\mathbf{k}\sigma}a^+_{\mathbf{k}'\sigma'} + a^+_{\mathbf{k}'\sigma'}a_{\mathbf{k}\sigma} &= \delta_{\mathbf{k}\mathbf{k}'}\delta_{\sigma\sigma'};\\ a_{\mathbf{k}\sigma}a_{\mathbf{k}'\sigma'} + a_{\mathbf{k}'\sigma'}a_{\mathbf{k}\sigma} &= 0. \end{aligned}\right\} \tag{14.2}$$

In the secondary quantization, the energy of the simplest single-band model is stated in the form

$$\mathcal{H} = \sum_{\mathbf{k}\sigma} \varepsilon_\mathbf{k} a^+_{\mathbf{k}\sigma}a_{\mathbf{k}\sigma} + \frac{1}{2}v\sum_{\mathbf{k},\,\mathbf{k}_2\,\mathbf{q}}\sum_{\sigma_1\,\sigma_2} a^+_{\mathbf{k}_1+\mathbf{q}\sigma_1}a_{\mathbf{k}_1\sigma_1}a^+_{\mathbf{k}_2\sigma_2}a_{\mathbf{k}_2+\mathbf{q}\sigma_2}, \tag{14.3}$$

where the first term renders the single-particle energy of the electrons, and the second term relates to the electron interaction; the matrix element of the interaction is assumed to be independent of the momentum. In the coordinate representation, this corresponds to an interaction of electrons only within a single atom so that the Hamiltonian of Eq. (14.3) corresponds to the Hubbard model [200].

Derivation of an expression for the cross section. Obviously, an expression for the differential neutron cross section can be obtained in the collectivized model from the general equation (8.5) of the localized model by successively replacing the spin density **S**(**r**) of the electron system:

$$\sum_l e^{i\varkappa\mathbf{R}_l^\circ}\mathbf{S}_l F_l(\varkappa) \to \int e^{i\varkappa\mathbf{r}}\mathbf{S}(\mathbf{r})\,d\mathbf{r} \equiv \widetilde{\mathbf{S}}_\varkappa. \tag{14.4}$$

Thus, we have the following starting formula for the cross section:

$$\frac{d^2\sigma}{d\Omega d\varepsilon} = (r_0\gamma)^2 \frac{k}{k_0} e^{-2W} \sum_{\alpha\beta} (\delta_{\alpha\beta} - e_\alpha e_\beta) \frac{1}{2\pi\hbar} \int_{-\infty}^{\infty} dt e^{-i\frac{\varepsilon}{\hbar}t} \times$$
$$\times \langle \tilde{S}^{\alpha}_{\varkappa}(t) \tilde{S}^{\beta}_{-\varkappa}(0)\rangle. \tag{14.5}$$

For the actual calculation of a spin correlator we must use the representation of the spin density operator **S(r)** by the operators of the secondary quantization [208]:

$$\mathbf{S}(\mathbf{r}) = \frac{1}{2} \sum_{\sigma\sigma'} \psi^{+}_{\sigma}(\mathbf{r})\, \sigma_{\sigma\sigma'}\, \psi_{\sigma'}(\mathbf{r}), \tag{14.6}$$

where σ donates the Pauli matrices and $\psi_\sigma(\mathbf{r})$ is given by Eq. (14.1). Let us calculate the Fourier component of Eq. (14.6) and state the equation for $\tilde{\mathbf{S}}_{\varkappa}$ in the representation of the secondary quantization:

$$\tilde{\mathbf{S}}_{\varkappa} = \frac{1}{2} \sum_{\mathbf{k}} F(\mathbf{k}, \varkappa)\, a^{+}_{\mathbf{k}+\varkappa\sigma}\, \sigma_{\sigma\sigma'}\, a_{\mathbf{k}\sigma'}, \tag{14.7}$$

where

$$F(\mathbf{k}, \varkappa) = \sum_{l} e^{i\mathbf{k}\mathbf{R}_l} \int W^*(\mathbf{r})\, W(\mathbf{r} + \mathbf{R}_l)\, e^{i\varkappa\mathbf{r}}\, d\mathbf{r}. \tag{14.8}$$

When the latter expression was derived, we expanded the Bloch function $\psi_{\mathbf{k}}(\mathbf{r})$ in a series of Wannier functions $W(\mathbf{r} - \mathbf{R}_l)$:

$$\psi_{\mathbf{k}}(\mathbf{r}) = \frac{1}{\sqrt{N}} \sum_{l} e^{i\mathbf{k}\mathbf{R}_l}\, W(\mathbf{r} - \mathbf{R}_l). \tag{14.9}$$

When the overlapping of the Wannier functions at neighboring lattice points is disregarded (these functions are close to localized atomic functions), the approximation

$$F(\mathbf{k}, \varkappa) \equiv F(\varkappa) = \int e^{i\varkappa\mathbf{r}}\, |W(\mathbf{r})|^2\, d\mathbf{r} \tag{14.10}$$

holds. The quantity $F(\varkappa)$ is the atomic form factor of the collectivized electron state.

Equation (14.7) for the Fourier component of the spin density can be conveniently stated when one introduces the approximation of Eq.

(14.10) in the form $\tilde{\mathbf{S}}_\varkappa \equiv F(\varkappa)\tilde{\mathbf{S}}_\varkappa$. In order to calculate the correlator $\langle S_\varkappa{}^\alpha(t)S_\varkappa{}^\beta(0)\rangle$, which defines the cross section, a retarded Green's function must be introduced by stating it with the aid of the general definition given by Eq. (3.3):

$$G^{\alpha\beta}(\varkappa, t) = \langle\langle S^\alpha_\varkappa(t)\,|\,S^\beta_{-\varkappa}(0)\rangle\rangle = -\frac{i}{\hbar}\theta(t)\langle[S^\alpha_\varkappa(t), S^\beta_{-\varkappa}(0)]\rangle. \quad (14.11)$$

Then, according to the general equations (3.8) and (3.10), the spin density correlator of interest can be expressed through the Green's function with the aid of the equation

$$\langle S^\alpha_\varkappa(t)\,S^\beta_{-\varkappa}(0)\rangle = \frac{1}{2\pi}\int_{-\infty}^{\infty} d\omega e^{-i\omega t}\left\{2\hbar\,\frac{\operatorname{Im} G^{\alpha\beta}(\varkappa, \omega)}{e^{\beta\hbar\omega} - 1}\right\}, \quad (14.12)$$

with which we obtain a convenient expression for calculating the cross section:

$$\frac{d^2\sigma}{d\Omega d\varepsilon} = (r_0\gamma)^2 F^2(\varkappa)\frac{k}{k_0}e^{-2W}\frac{1}{e^{\beta\varepsilon} - 1}\sum_{\alpha\beta}(\delta_{\alpha\beta} - e_\alpha e_\beta) \times$$
$$\times\frac{1}{\pi}\operatorname{Im} G^{\alpha\beta}(\varkappa, \omega)\big|_{\omega=-\varepsilon/\hbar}. \quad (14.13)$$

Scattering at spin waves in a collectivized ferromagnetic material. We consider a ferromagnetic metal in which a spontaneous spin moment arises from an electron interaction which can be described by the energy parameter v. As usual, the longitudinal and transverse spin correlators are separated in the expression (14.13) for the cross section:

$$\frac{d^2\sigma}{d\Omega d\varepsilon} = \frac{1}{4}(r_0\gamma)^2 F^2(\varkappa)\frac{k}{k_0}e^{-2W}\frac{1}{e^{\beta\varepsilon} - 1}\frac{1}{\pi}\{(1 - e_z^2)\operatorname{Im} G^{zz}(\varkappa, \omega) +$$
$$+ (1 + e_z^2)[\operatorname{Im} G^{-+}(\varkappa, \omega) + \operatorname{Im} G^{+-}(\varkappa, \omega)]\}. \quad (14.14)$$

The Green's functions of the transverse components are obtained with Eq. (14.7) for $\mathbf{S}_\varkappa$; we obtain, in particular:

$$S^-_\varkappa = \sum_\kappa a^+_{\kappa+\varkappa\downarrow}\,a_{\kappa\uparrow},\ S^+_\varkappa = \sum_\kappa a^+_{\kappa\uparrow}\,a_{\kappa+\varkappa\downarrow}. \quad (14.15)$$

In order to calculate the functions $G^{+-}(\varkappa, \omega)$, we introduce a delayed Green's function of the form $\langle\langle a_{\mathbf{k}+\varkappa\downarrow}{}^+(t)\,a_{\mathbf{k}\uparrow}(t)\,|\,S_{-\varkappa}{}^+(0)\rangle\rangle$ from which the unknown function $\langle\langle S_\varkappa{}^-(t)\,|\,S_{-\varkappa}{}^+(0)\rangle\rangle$ is obtained by summation over **k**. It is easy to determine for it the equation of motion [according to Eq. (3.6)]:

$$i\hbar\frac{d}{dt}\ll a^{+}_{\kappa+\varkappa\downarrow}(t)\,a_{\kappa\uparrow}(t)\,|\,S^{\pm}_{-\varkappa}(0)\gg=\delta(t)(n_{\kappa+\varkappa\downarrow}-n_{\kappa\uparrow})+$$
$$+[\varepsilon_{\uparrow}(\kappa)-\varepsilon_{\downarrow}(\kappa+\varkappa)]\ll a^{+}_{\kappa+\varkappa\downarrow}(t)\,a_{\kappa\uparrow}(t)\,|\,S^{\pm}_{-\varkappa}(0)\gg+$$
$$+v(n_{\kappa\uparrow}-n_{\kappa+\varkappa\downarrow})\sum_{\kappa}\ll a^{+}_{\kappa'+\varkappa\downarrow}(t)\,a_{\kappa'\uparrow}(t)\,|\,S^{\pm}_{-\varkappa}(0)\gg. \qquad (14.16)$$

The Fermi distribution function $n_{\mathbf{k}\sigma}$ with the spin σ and the quasi-momentum **k** is defined as

$$n_{\kappa\sigma}=\langle a^{+}_{\kappa\sigma}a_{\kappa\sigma}\rangle;$$

$\varepsilon_\sigma(\mathbf{k})$ denotes the single-particle electron energy which was renormalized as a consequence of the electron interaction:

$$\varepsilon_{\sigma}(\kappa)=\varepsilon_{\kappa}-\frac{v}{N}\sum_{\kappa'}n_{\kappa'\sigma}. \qquad (14.17)$$

Let us note that in the deriving Eq. (14.16), the higher Green's functions were split in the form of the usual approximation of random phases:

$$\ll a^{+}_{\kappa\sigma}(t)\,a_{\kappa\sigma}(t)\,a^{+}_{\kappa'+\varkappa\downarrow}(t)\,a_{\kappa'\uparrow}(t)\,|\,S^{\pm}_{-\varkappa}(0)\gg\approx$$
$$\approx<a^{+}_{\kappa\sigma}a_{\kappa\sigma}>\ll a^{+}_{\kappa'+\varkappa\downarrow}(t)\,a_{\kappa'\uparrow}(t)\,|\,S^{\pm}_{-\varkappa}(0)\gg,$$

and, hence, the equation is a closed equation.

Equation (14.16) can be solved in trivial form by the transition to the Fourier representation and renders the following form of the unknown Green's function:

$$G^{-+}(\varkappa,\omega)=-\frac{\chi_0(\varkappa,\omega)}{1-v\chi_0(\varkappa,\omega)}, \qquad (14.18)$$

where

$$\chi_0(\varkappa,\omega)=\frac{1}{N}\sum_{\kappa}\frac{n_{\kappa\uparrow}-n_{\kappa+\varkappa\downarrow}}{\varepsilon_{\downarrow}(\kappa+\varkappa)-\varepsilon_{\uparrow}(\kappa)-\hbar\omega-i\delta} \qquad (14.19)$$

represents the susceptibility of the electron gas magnetized by the exchange field which is generated by the electron interaction. One can show by similar calculations that the other, transverse Green's function $G^{+-}(\varkappa, \omega)$ differs from the functions $G^{-+}(\varkappa, \omega)$ by its sign and by the sign of the argument ω. The cross section of scattering at the spin wave excitations of the system is associated with the transverse terms in Eq. (14.14). We denote this part of the cross section by $d^2\sigma_{SW}/d\Omega d\varepsilon$:

$$\frac{d^2\sigma_{SW}}{d\Omega d\varepsilon} = \frac{1}{4}(r_0\gamma)^2 F^2(\varkappa)\frac{k}{k_0} e^{-2W}[1+(\mathbf{em})^2] \times$$
$$\times (e^{\beta\varepsilon}-1)^{-1}\frac{1}{\pi}\,\mathrm{Im}\left\{G^{-+}\left(\varkappa, -\frac{\varepsilon}{\hbar}\right) - G^{-+}\left(\varkappa, \frac{\varepsilon}{\hbar}\right)\right\}. \tag{14.20}$$

Let us separate the real and imaginary parts in Eq. (14.19) for the susceptibility:

$$\chi_0(\varkappa, \omega) = \mathrm{Re}\,\chi_0(\varkappa, \omega) + i\,\mathrm{Im}\,\chi_0(\varkappa, \omega); \tag{14.21}$$

$$\mathrm{Re}\,\chi_0(\varkappa, \omega) = \frac{1}{N}\sum_{\mathbf{k}} \frac{n_{\mathbf{k}\uparrow} - n_{\mathbf{k}+\varkappa\downarrow}}{\varepsilon_{\mathbf{k}+\varkappa} - \varepsilon_{\mathbf{k}} + 2vn - \hbar\omega}; \tag{14.22}$$

$$\mathrm{Im}\,\chi_0(\varkappa, \omega) = \pi\frac{1}{N}\sum_{\mathbf{k}}(n_{\mathbf{k}\uparrow} - n_{\mathbf{k}+\varkappa\downarrow})\,\delta(\varepsilon_{\mathbf{k}+\varkappa} - \varepsilon_{\mathbf{k}} + 2vn - \hbar\omega). \tag{14.23}$$

Equation (14.17) was used in stating the latter two expressions and the notation $n = (1/N)\sum_{\mathbf{k}}(n_{\mathbf{k}\uparrow} - n_{\mathbf{k}\downarrow})$ was introduced. It follows from Eq. (14.20) that the proportionality

$$\frac{d^2\sigma_{SW}}{d\Omega d\varepsilon} \sim \left.\frac{1}{[1 - v\,\mathrm{Re}\,\chi_0(\varkappa, \omega)]^2 + [v\,\mathrm{Im}\,\chi_0(\varkappa, \omega)]^2}\right|_{\omega = \mp\varepsilon/\hbar}, \tag{14.24}$$

holds, which means that the cross section can have sharp maxima at $\omega = \omega_s$ such that

$$1 - v\,\mathrm{Re}\,\chi_0(\varkappa, \omega_s) = 0, \tag{14.25}$$

and then $\mathrm{Im}\,\chi_0(\varkappa, \omega_s)$ is small in comparison with ω_s.

Equations of the type of Eq. (14.25) are well known in the quantum theory of Fermi systems. These equations determine the spectrum of the collectivized excitations of the electron system.

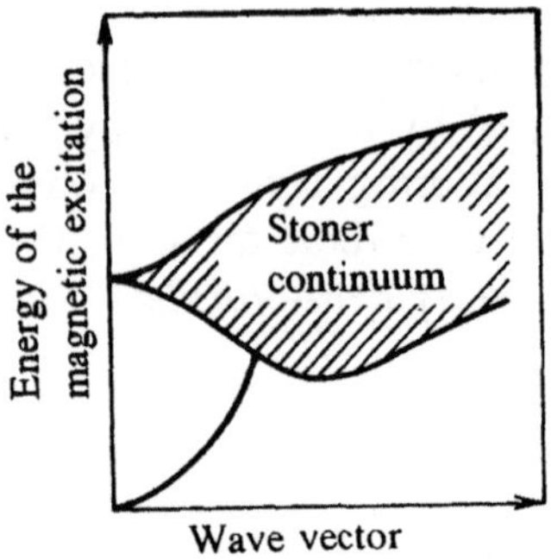

Fig. 22. Spectrum of the magnetic excitations of a collectivized ferromagnet.

Equation (14.25), which describes the behavior of the spectral density of the correlation function of the transverse spin components, defines the spectrum of the collectivized excitations of the system with a total spin change of 1, i.e., the spectrum of the spin waves with the dispersion law $\omega_s(\varkappa)$. These excitations will be nonattenuated, provided that $\mathrm{Im}\,\chi_0(\varkappa, \omega_s)$ is equal to zero, but will be weakly attenuated if $\mathrm{Im}\,\chi_0(\varkappa, \omega_s)$ is small. The range of frequency and momenta for which $\mathrm{Im}\,\chi_0(\varkappa, \omega_s) = 0$ is given by the equation

$$\varepsilon_{\mathbf{k}+\varkappa} - \varepsilon_{\mathbf{k}} + 2vn - \hbar\omega = 0.$$

This region is usually termed the Stoner continuum.

It is a well-known fact that equations of the type of Eq. (14.25) can have nonattenuated solutions beyond the continuum. In particular, for ω and $\varkappa$, they define the usual spin wave spectrum of a ferromagnet with a quadratic dispersion law $\hbar\omega_s = D\varkappa^2$. Figure 22 illustrates the pattern of the solutions to dispersion equation (14.25).

Let us continue the calculation of the cross section of Eq. (14.20). For frequencies ω for which $\mathrm{Im}\,\chi_0(\varkappa, \omega_s) = 0$, the fraction in Eq. (14.24) is expressed through the δ function so that Eq. (14.20) is reduced to the form

$$\frac{d^2\sigma_{\mathrm{SW}}}{d\Omega d\varepsilon} = \frac{1}{4}(r_0\gamma)^2 F^2(\varkappa)\frac{k}{k_0}\, e^{-2W}[1 + (\mathbf{em})^2]\frac{1}{e^{\beta\varepsilon}-1}\cdot\frac{1}{v}\times$$
$$\times\{\delta(1 - v\,\mathrm{Re}\,\chi_0(\varkappa, \omega)) - \delta(1 - v\,\mathrm{Re}\,\chi_0(\varkappa, -\omega))\}_{\omega=-\varepsilon/\hbar}. \quad (14.26)$$

It is readily recognized that near the frequency ω_s, the following expansion of the function in the argument of the δ function holds:

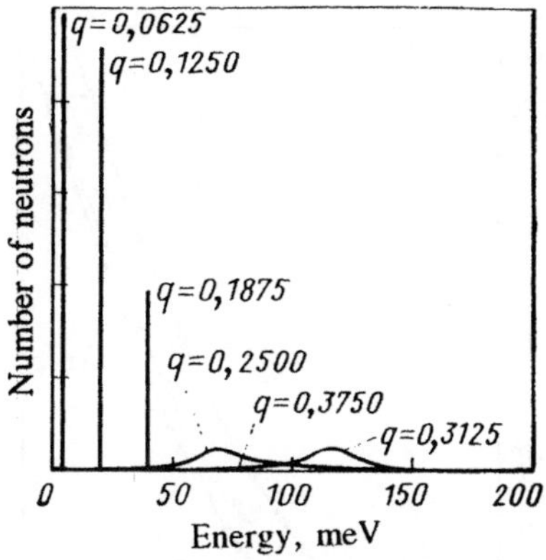

Fig. 23. Intensity of scattering at the spin wave in iron; data of [149] based on the collectivized model of ferromagnetism.

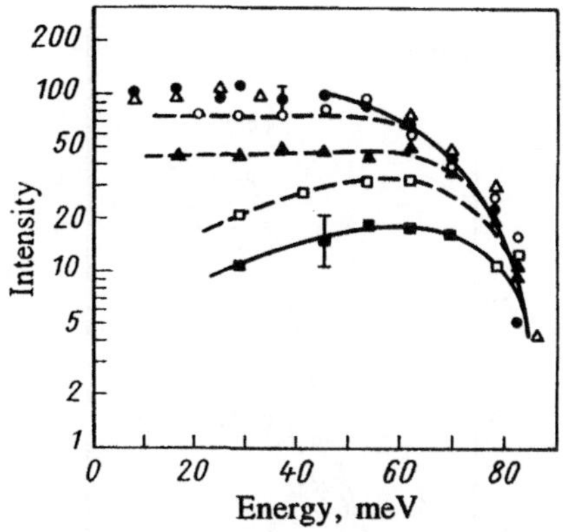

Fig. 24. Intensity of scattering at spin waves in nickel as a function of the energy transferred at the temperature (△) 4.2; (●) 293; (○) 504; (▲) 568; (□) 673; (■) 775 K for wave vectors **q** in the [111] direction [255].

$$1 - v \operatorname{Re} \chi_0(\varkappa, \omega) = \frac{\hbar}{2\pi v}[\omega_s(\varkappa) - \omega] + \ldots \tag{14.27}$$

Making use of this expansion, we write the expression for the cross section:

$$\frac{d^2\sigma_{SW}}{d\Omega d\varepsilon} = (r_0\gamma)^2 F^2(\varkappa)\frac{n}{2}\frac{k}{k_0} e^{-2W}[1+(\mathbf{em})^2] \times$$
$$\times \{n_\varkappa \delta(\varepsilon - \hbar\omega_s(\varkappa)) + (1+n_\varkappa)\delta(\varepsilon + \hbar\omega_s(\varkappa))\}, \tag{14.28}$$

where

$$n_\varkappa = (e^{\beta\hbar\omega_s(\varkappa)} - 1)^{-1}.$$

The resulting formula describes the cross section of inelastic scattering with absorption ($\varepsilon > 0$) and emission ($\varepsilon < 0$) of a spin wave. In regard to accuracy, this formula corresponds to Eq. (9.9) for scattering at spin

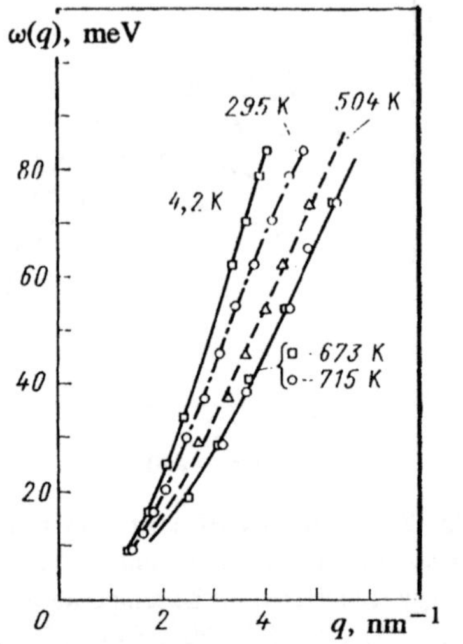

Fig. 25. Dispersion law of the spin waves in nickel at various temperatures [255].

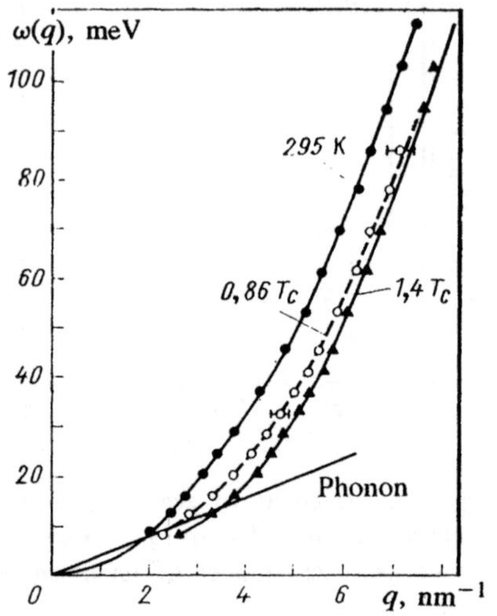

Fig. 26. Dispersion law of the spin waves in the alloy Fe + 12% Si at various temperatures [237].

waves in a ferromagnet in the case of localized spins, provided that we note that $n/2$ coincides with the average spin value referred to a single atom of the metal. Thus, the patterns of inelastic scattering at a ferromagnetic crystal coincide for the localized and collectivized models.

An important difference of the spin waves in the collectivized model is given by the fact that the spin waves are not attenuated (within the approximation of random phases assumed in the calculation) only up to a certain value of the wave vector **q**, with the value differing for the various directions in the Brillouin zone and defined by the boundaries of the Stoner continuum. When they are within the continuum, they are strongly attenuated as a consequence of splitting into electron excitations and hole excitations. The authors of [149] have calculated from the first principles the spin wave spectrum and the intensity of neutron scattering at the spin waves in pure iron. Figure 23 illustrates the results of a calculation for the direction **q** in the [100] direction, though the scattering is isotropic, as has been shown by the authors of [149]. Obviously, the width of the neutron resonances increases sharply at $q \approx 0.2(2\pi/a)$ because the dispersion curve falls into the Stoner continuum.

The spin waves in iron and nickel. This behavior of the spin waves was experimentally established in the two ferromagnetic metals Fe and Ni which *a priori* could be described with the collectivized model. Figure 24 illustrates the dependence of the intensity of neutron scattering in Ni upon the transferred energy ε. We observe that there exists a limit value of the wave vector, and that above that limit, spin waves do not exist. This effect is anisotropic in Ni but more or less isotropic in Fe, with the observed limit value for the wave vectors (and the limit energies of the spin waves) satisfactorily coinciding with the values calculated for Fe [149].

The dispersion curves for Fe [237] and Ni [255] are shown in Figs. 25 and 26. The disappearance of spin waves at large $\mathbf{q}$ values is of principal importance because they are a convincing proof of the fact that the magnetism of the 3*d* transition metals (Ni and Fe) can be treated with the collectivized model.

Let us consider now the temperature effects. We infer from Fig. 24 that the position of the limit wave vectors of spin waves corresponding to the boundaries of the Stoner continuum is temperature-independent. This fact cannot be understood because these boundaries must depend upon the temperature via the quantity n, i.e., the magnetization of the electron gas, with the magnetization being given by the exchange splitting of the electron bands. The existence of spin waves in a paramagnetic phase at $T > T_c$ is another surprising fact. This is the case in the two metals Fe and Ni (see Figs. 25 and 26); pronounced dispersion curves of the spin waves are observed at $T \approx 1.5T_c$. This fact attests to the abnormally strong correlations of the magnetic fluctuations in collectivized ferromagnetic systems. The two temperature effects, namely the temperature independence of the limit energies of spin waves and the existence of spin waves in the paramagnetic region far from the Curie point, are still waiting for their theoretical interpretation.

We direct the reader's attention to a number of papers [223-225] in which the existence of spin wave excitations in Fe and Ni in a wide temperature interval above T_c was theoretically explained. By contrast to the standard theory of collectivized ferromagnetism assuming a single pronounced direction in the system, i.e., the magnetization vector, the authors include in their considerations the effect of fluctuations of the magnetic order upon the states of the band electrons. Near T_c the fluctuations have macroscopic dimensions and their lifetimes are much longer than the characteristic times of the electron motion. Therefore, in the metal's region filled with fluctuations of the magnetic order, the electron states are adiabatically tuned to the fluctuations and are local states throughout the volume of the metal.

If an instantaneous photograph of the entire volume of the metal comprising regions of magnetic order fluctuations were taken, we would recognize that the directions of the magnetic moments of each of these regions are at random. They slowly change their orientation and boundaries in the course of time. Since in a transition through T_c the fluctuations do not disappear, the exchange splitting of the electron spectrum in each of these regions remains and even the collectivized modes are conserved in the electron system in the form of spin waves. The authors of [223] have developed a mathematical calculus in which the local properties of the band structure are taken into account; the mathematics is based on a transition from the laboratory system to a local coordinate system whose z' axis is parallel to the vector of magnetization at a particular point. In calculating specific properties of the entire metal volume, averaging of the expressions obtained over the directions of the local magnetic moment was suggested.

The authors of [223] showed with this approach that near T_c (both above, as well as below T_c), there exist collective spin-wave modes with the dispersion law

$$\hbar\omega\ (q) = Dq^2 - 2D\ \langle |a|^2 \rangle_{\mathrm{av}}, \tag{14.29}$$

where D_q^2 denotes the energy of a usual spin wave in a system with long-range magnetic order (the temperature renormalization of the coefficient D of the spin rigidity was brought into account). Equation (14.29) holds for $q^2 \gg \langle |a|^2 \rangle_{\mathrm{av}}$ so that $\omega(\mathbf{q})$ is positive, as put into evidence by Eq. (14.29). The quantity $a \sim \nabla M$ accounts for the inhomogeneity in the distribution of the magnetization when fluctuations are present; the symbol $\langle |a|^2 \rangle_{\mathrm{av}}$ denotes averaging over the fluctuations. Thus, the parameter $\langle |a|^2 \rangle_{\mathrm{av}}$ is a measure of the energy of the inhomogeneities. The contribution resulting from scattering at fluctuations to the attenuation of the spin waves is proportional to $\langle |a|^2 \rangle_{\mathrm{av}}$ but does not substantially change the extent of the attenuation in the transition through T_c because it turns out for Fe and Ni that $\langle |a|^2 \rangle_{\mathrm{av}} \sim 0.005$ nm and that the temperature dependence is weak [224]. This theory provides a quantitative description of inelastic scattering of neutrons at Fe and Ni near the Curie temperature and in a wide interval above T_c. The authors of [260, 202, 193] have discussed the other problem in a theoretical interpretation of neutron scattering in the ferromagnetic metals Fe, Co, Ni namely the existence of localized atomic moments above T_c.

15. Scattering at Soliton Excitations

Solitons in a one-dimensional ferromagnetic chain. So far we have expounded the theory of spin waves and of neutron scattering at spin waves in a linear approximation in which the spin deviations at a particular atom were assumed to be weak. In this case, the replacement of the operators of the atomic spins by Bose operators of the spin deviations in accordance with the Goldstein–Primakov relations is valid; then the Hamiltonian of a crystal with magnetic order is a quadratic form of Bose operators. The eigenvectors of this quadratic form are the normal modes of the oscillations of the spin system or the spin waves.

We present in this section the results of a nonlinear theory of spin excitations in a ferromagnetic system, the results corresponding to the accurate solutions of nonlinearized equations of spin motion. Individual, isolated solutions of the nonlinear equations have been termed solitons and, accordingly, also the corresponding excitations of a system with magnetic order have been called solitons. By contrast to the spin-wave states of a system, which are described by plane waves, the solitons correspond to localized states in which a disturbance of the ground state of the spin system arises practically at some finite length but propagates without attenuation with a finite velocity v not exceeding some limit value c given by the parameters of the spin interaction. The soliton states usually have an energy below the band of the linear spin-wave excitations, but because of their localization, they are characterized by a specific form factor and, hence, can be observed in neutron experiments.

Soliton solutions of nonlinear equations are know in analytic form only for one-dimensional equations, and we therefore consider soliton excitations first of all in one-dimensional ferromagnetic chains. These results can describe certain three-dimensional crystals in which, by virtue of their crystallochemical structure, there exists a weak interaction between certain chains of atoms with ferromagnetic spin order. Contemplating the application to such quasi-one-dimensional objects, we first consider an ideal one-dimensional chain of the spins in the direction of the unit vector $\mathbf{n}$; this chain can be described by the very simple Hamiltonian

$$\mathcal{H} = -J\sum_l \mathbf{S}_l \mathbf{S}_{l+1} + A\sum_l S_l^{z^2} - g\mu_0 \mathbf{H}\sum_l \mathbf{S}_l. \tag{15.1}$$

where l numbers the atoms in the chain, in which only the exchange interaction of the nearest neighbors is taken into account. The sign of the constant A defines an anisotropy of the type of a "light axis" ($A < 0$) and a "light plane" ($A > 0$).

The equations of motion of the spins have the following general form in the classical limit ($S >> 1$):

$$\hbar \frac{d\mathbf{S}_l}{dt} = [\mathbf{S}_l \times \mathbf{H}_{\text{eff}}], \; \mathbf{H}_{\text{eff}} = -\frac{\partial \mathcal{H}}{\partial \mathbf{S}_l}, \tag{15.2}$$

from which their nonlinearity resulting from the expression

$$\mathbf{H}_{\text{eff}} = J(\mathbf{S}_{l-1} + \mathbf{S}_{l+1}) - 2AS_l^z \mathbf{n} + g\mu_0 \mathbf{H}. \tag{15.3}$$

becomes obvious. The spin vector is conveniently rewritten in a polar coordinate system with the z axis parallel to the vector $\mathbf{n}$ of the chain:

$$\mathbf{S}_l = S\,\{\sin\theta_l \cos\varphi_l,\; \sin\theta_l \sin\varphi_l,\; \cos\theta_l\}, \tag{15.4}$$

where the angles θ_l and φ_l depend upon the time t. In the continuum limit $\theta_l(t) \to \theta(z, t)$, the difference equation system (15.2) is reduced to a pair of differential equations.

First let us consider anisotropy of the "light plane" type and assume the external field to be parallel to the x axis on the plane. The pair of differential equations corresponding to Eq. (15.2) in this case is as follows:

$$\left.\begin{aligned} &\hbar \frac{\partial}{\partial t}\cos\theta = -g\mu_0 H \sin\theta \sin\varphi + Sq^2 J \frac{\partial}{\partial z}\left(\sin^2\theta \frac{\partial \varphi}{\partial z}\right); \\ &\hbar \frac{\partial \varphi}{\partial t} = 2AS\cos\theta. \end{aligned}\right\} \tag{15.5}$$

We disregarded in the second equation terms with derivatives with respect to the space coordinates and assumed that H is small: $\mu_0 H << A << J$. In the case of strong anisotropy, the spins must be situated basically on the plane of the light magnetization, and therefore the angle is $\theta \approx \pi/2$, i.e., $\sin\theta \approx 1$. By combining the two Eqs. (15.5) and replacing $\sin\theta$ by 1, we obtain an approximating equation for the azimuthal angle $\varphi(z, t)$

$$\frac{\partial^2 \varphi}{\partial z^2} - \frac{1}{c^2}\frac{\partial^2 \varphi}{\partial t^2} = \frac{1}{\Delta^2}\sin\varphi, \tag{15.6}$$

where we introduced the constants c having the units of velocity, and Δ having the units of length:

$$c^2 = \frac{2AJ(Sa)^2}{\hbar^2}, \quad \Delta^{-2} = \frac{g\mu_0 H}{SJa^2}. \tag{15.7}$$

Thus, the motion of the vector of the magnetic moment on the plane of anisotropy is given by the well-known nonlinear Sine Gordon equation. Its partial solution of the form $\varphi = \varphi\xi$, where $\xi = z - vt$, can be easily found when Eq. (15.6) is rewritten in the form

$$(1 - v^2/c^2)\frac{d^2\varphi}{d\xi^2} = \Delta^{-2}\frac{dV(\varphi)}{d\varphi}. \tag{15.8}$$

In this particular case, we replaced $\sin\theta$ in Eq. (15.6) by $dV/d\varphi$ and, hence, we have

$$V(\varphi) = 1 - \cos\varphi. \tag{15.9}$$

By integrating the two parts of Eq. (15.8) over $d\varphi$ and introducing the boundary condition

$$\frac{d\varphi}{d\xi} = 0 \text{ for } \varphi = \varphi_1, \tag{15.10}$$

the equation can be rewritten in the form

$$\frac{d\varphi}{d\xi} = \frac{\gamma}{\Delta}\sqrt{2[V(\varphi) - V(\varphi_1)]}, \quad \gamma = (1 - v^2/c^2)^{-1/2}. \tag{15.11}$$

The latter equation can be easily integrated when the "potential" is chosen in the form of Eq. (15.9) to correspond to the sine Gordon equation. We obtain the operation

$$\gamma\left(\frac{\xi - \xi_0}{\Delta}\right) = \int \frac{d\varphi}{\sqrt{2[V(\varphi) - V(\varphi_1)]}} = \ln \mathrm{tg}\frac{\varphi}{4},$$

from which we find

$$\varphi(z, t) = 4\,\mathrm{arctg}\exp[\gamma(z - vt - z_0)/\Delta]. \tag{15.12}$$

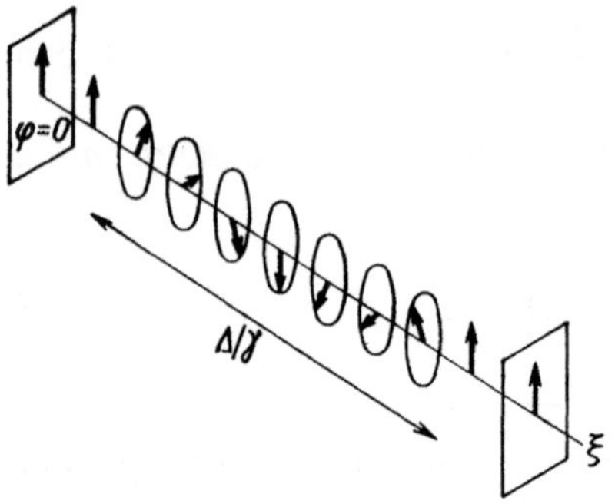

Fig. 27. Soliton in a one-dimensional ferromagnetic chain of spins with "light plane" anisotropy.

where z_0 denotes some coordinate corresponding to the position at the initial moment of time.

The resulting equation (15.12) corresponds to the general condition of Eq. (15.10), which for Eq. (15.6) means

$$\frac{d\varphi}{d\xi} = 0 \text{ for } \varphi = 0,\ 2\pi. \tag{15.13}$$

This boundary condition renders the local behavior of the solution of Eq. (15.12)

$$\varphi(\xi) = \begin{cases} 0, & \xi = -\infty; \\ 2\pi, & \xi = \infty. \end{cases} \tag{15.14}$$

The states at the ends of the linear chain are, in terms of physics, equivalent and correspond to spins which are oriented "upwards". It follows from Eqs. (15.12) and (15.14) that a full 2π rotation of the spins takes place over a length of the order Δ/γ on a plane perpendicular to the axis of the chain (Fig. 27). The soliton obtained connects two states which at the edges of the chain are equivalent in terms of physics and corresponds to a so-called dynamic soliton in contrast to a topological soliton which connects two different states, e.g., states with an upward and downward spin, and describes a domain wall.

Let us calculate the energy E_{sol} of generating in the chain a spin disturbance which can be described by the soliton of Eq. (15.12) [251]:

$$E_{sol}/S^2 Ja = \int_{-\infty}^{\infty} dz \left\{ \frac{1}{2} \left[\left(\frac{\partial\varphi}{\partial z} \right)^2 + \frac{1}{c^2} \left(\frac{\partial\varphi}{\partial t} \right)^2 \right] + \frac{1}{\Delta^2} V(\varphi) \right\}. \tag{15.15}$$

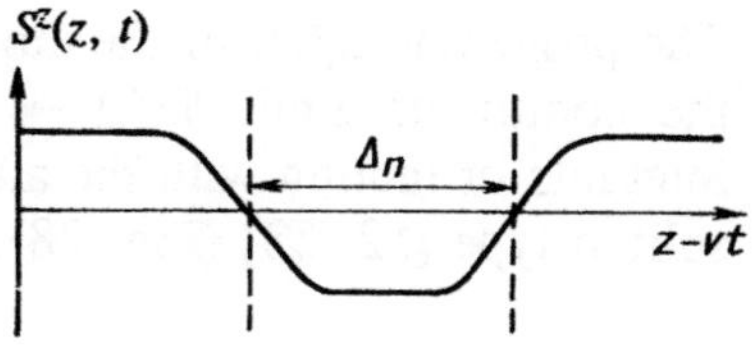

Fig. 28. Soliton in a one-dimensional ferromagnetic chain of spins with "light axis" anisotropy.

This integral can be reduced to the form

$$E_{\text{sol}}/S^2 Ja = \frac{\gamma}{\Delta} \int_{-\infty}^{\infty} d\xi \left(\frac{d\varphi}{d\xi}\right)^2 = \frac{8\gamma}{\Delta}. \tag{15.16}$$

with the aid of Eqs. (15.8)-(15.11).

The velocity v of the soliton cannot exceed the limit velocity c (according to the last result) and, in the "nonrelativistic" limit $v << c$, Eq. (15.16) is reduced to the classical equation

$$E_{\text{sol}} = \varepsilon_s + (m_s v^2/2), \tag{15.17}$$

where

$$\varepsilon_s = 8S^2 Ja/\Delta, \quad m_s = \varepsilon_s/c^2. \tag{15.18}$$

The result stated in Eqs. (15.17) and (15.12) means that a soliton can be considered a quasiparticle which is characterized by a certain mass, capable of moving with a velocity $v < c$, and localized on a length of the order Δ/γ.

Similar results can be obtained for an antiferromagnetic chain [228].

Let us now consider a ferromagnetic chain with a "light axis" anisotropy. The Hamiltonian of Eq. (15.1) with $A < 0$ corresponds to this case. Let us state the result of solving the nonlinear equation of spin motion in this case when the external field is zero [52, 53]. Also in this case the equation has soliton solutions and the result is conveniently expressed by taking into account that the equation of motion has two integrals: the total momentum **p** and the total spin or, more precisely, the total spin deviation

$$n = \sum_l (S - S_l^z). \tag{15.19}$$

The projection $S_l^z(t)$ of the atomic spin onto the direction of the chain in the continuum limit $S_l^z(t) \to S^z(z, t)$ is expressed through the above integrals of motion with the aid of the following localized solution of the soliton type [52, 53] (Fig. 28):

$$S^z(z, t) = S\left\{1 - 2\frac{\mathrm{sh}^2(n/n_0) + \sin^2(\pi q/2q_0)}{\mathrm{ch}^2[(z - vt - z_0)/\Delta_n] + \mathrm{sh}^2(n/n_0)}\right\}. \tag{15.20}$$

where n_0 and q_0 denote the parameters of the ferromagnetic chain

$$n_0 = 4S\sqrt{J/|A|}, \quad q_0 = \frac{2\pi S\hbar}{a}, \tag{15.21}$$

and Δ_n denote the characteristic of space localization (width) of the soliton:

$$\frac{1}{\Delta_n} = \frac{1}{a}\sqrt{\frac{J}{A}}\,\mathrm{th}\frac{n}{n_0}\left[1 + \frac{\sin^2(\pi q/2q_0)}{\mathrm{sh}^2(n/n_0)}\right]. \tag{15.22}$$

Thus, each of the soliton solutions is characterized by the total momentum q and the total spin deviation n. The energy of the corresponding excitation of the system is

$$E_{\mathrm{sol}}(q, n) = 4\sqrt{AJ}\,\mathrm{th}\frac{n}{n_0}\left[1 + \frac{\sin^2(\pi q/q_0)}{\mathrm{sh}^2(n/n_0)}\right]. \tag{15.23}$$

In the case $n << n_0$ for a soliton at rest ($v = 0$) Eq. (15.23) can be stated in the form ($\hbar\omega_0 = 4\sqrt{AJ}$ denotes the energy of the spin wave with momentum zero):

$$E_{\mathrm{sol}}(0, n) = \hbar\omega_0 n\,[1 - (n^2/3n_0^2)], \tag{15.24}$$

which shows that the energy of such a soliton is lower than energy of the n linear spin waves. By comparing the result given by Eq. (15.24) with the result of the analytically solved problem of n interacting spin waves, we observe that the soliton of Eq. (15.20) with its energy of Eq. (15.23) can be treated as a bound state of n spin waves. It follows from Eq. (15.20) that the width of such a quasiparticle decreases when the number n increases and reaches its minimum $a\sqrt{A/J}$ which is smaller than the lattice parameter. One must bear in mind that the continuum consideration which we used is applicable only to lengths which are large in comparison with the lattice parameter.

We obtain from Eq. (15.23) an expression for the energy of a slow soliton with the momentum $q = m_n^* v << q_0$:

$$E(v, n) = \varepsilon_n + (m_n^* v^2/2), \tag{15.25}$$

where ε_n denotes the activation energy and $m_n^* v^2/2$ denotes the kinetic energy of the soliton:

$$\varepsilon_n = 4\sqrt{AJ}\,\mathrm{th}\left(\frac{n}{n_0}\right), \quad m_n^* = m_0^* (n_0/2)\,\mathrm{sh}^2 (n/n_0), \tag{15.26}$$

$m_0^* = \hbar^2/2SJa^2$ denotes the effective mass of the spin wave.

Scattering of neutrons at solitons of a one-dimensional chain. The cross section of magnetic scattering of neutrons is expressed through the Fourier component of the spin correlator. In the continuum approximation, the general expression (8.5) for the scattering cross section of a one-dimensional single-atomic lattice can be restated in the form

$$\frac{d^2\sigma}{d\Omega d\varepsilon} = (r_0\gamma)^2 \frac{k}{k_0} F^2(\varkappa) \sum_{\alpha\beta} (\delta_{\alpha\beta} - e_\alpha e_\beta) \times$$
$$\times \int_{-\infty}^{\infty} \frac{dz}{2\pi a} \int_{-\infty}^{\infty} \frac{dt}{2\pi\hbar} e^{i\left(\varkappa z - \frac{\varepsilon}{\hbar} t\right)} \langle S^\alpha(z, t)\, S^\beta(0, 0)\rangle. \tag{15.27}$$

In order to calculate the cross section of scattering at soliton excitations of a chain, one must use the soliton solution describing the space-time behavior $s^\alpha(z, t)$ of the spin and must average the correlator over the statistical ensemble. The soliton state is characterized by the velocity v, the energy $E_{sol}(v)$, and the initial coordinate z_0. When a single soliton is present in the system, the averaging is understood as integration over z and v, i.e., over the phase space of the soliton with a statistical weight of the Gibbs distribution function [226], i.e.,

$$\langle \ldots \rangle_1 = \frac{1}{Z_1} \int\int \frac{dv dz_0}{B} (\ldots)\, e^{-E_{sol}(v)/kT} \tag{15.28}$$

(the subscript 1 means that a single soliton is present in the chain). We have

$$Z_1 = \int\int \frac{dv dz_0}{B} e^{-E_{sol}(v)/kT}, \tag{15.29}$$

and the normalizing constant B cancels in the calculation of the average values. When a finite but low concentration of solitons exists, $1/L$ must be replaced in Eqs. (15.27) by the equilibrium concentration of the solitons:

$$n_s = \frac{1}{\Delta}\left(\frac{2\varepsilon_s}{\pi kT}\right)^{1/2} e^{-\varepsilon_s/kT}. \tag{15.30}$$

First let us calculate the Fourier component of the correlator $\langle S^x(z, t) \times S^z(0, 0)\rangle$ of a one-dimensional ferromagnetic chain with a "light plane" anisotropy [251]:

$$S_{\|}(\varkappa, \omega) \equiv \frac{S^2}{(2\pi)^2} \int dz dt e^{i(\varkappa z - vt)} \langle \cos\varphi(z, t) \cos\varphi(0, 0)\rangle. \tag{15.31}$$

This quantity is the longitudinal dynamic structure factor because the external field was applied in the direction of the x axis. The soliton solution of Eq. (15.12) must be inserted into the expression indicated, but we must recall that

$$\cos\varphi(z, t) = 1 - \frac{2}{\mathrm{ch}^2[\gamma(z - vt - z_0)/\Delta]}. \tag{15.32}$$

In the case of slow solitons ($v << c$), where Eq. (15.17) can be used for $E_{sol}(v)$, the fourfold integral over the variables z, t, z_0, and v in Eq. (15.31) can be calculated and leads to the result*

$$S_{\|}(\varkappa, \omega) = S^2(1 - 8n_s\Delta)\,\delta(\varkappa)\,\delta(\omega) +$$

$$+ \frac{8n_s}{\pi^2} S^2 \Delta^2 \left(\frac{2\pi\varepsilon_s}{c^2 \varkappa^2 kT}\right)^{1/2} e^{-\frac{\varepsilon_s \omega^2}{2(c\varkappa)^2 kT}} \frac{(\pi\varkappa\Delta/2)^2}{\mathrm{sh}^2(\pi\varkappa\Delta/2)}. \tag{15.33}$$

The first term leads to elastic Bragg scattering with $\varkappa = 0$, with the intensity of the scattering attenuated by the factor $\sim 8n_s\delta$ due to the development of scattering with $\varkappa \neq 0$. This scattering is described by the

---

*The authors of [229, 331] have improved Eq. (15.33) by including contributions resulting from the correlators of the transverse spin components.

second term in Eq. (15.33) and concentrated near the Bragg peak in the region $\varkappa \sim 2/\pi\Delta$. The scattering is inelastic and described by the Gaussian form of a line with the width

$$\delta\omega = \varkappa c\sqrt{kT/\varepsilon_s}, \tag{15.34}$$

whereas the intensity of the inelastic addition is proportional to the Boltzmann factor $e^{-\varepsilon_s/kT}$.

We switch now to an analysis of the cross section of scattering at solitons in a ferromagnetic chain with a "light axis" anisotropy. The dynamic structure factor

$$S(\varkappa, \omega) = \frac{1}{(2\pi)^2}\int\int dz dt e^{i(\varkappa z - \omega t)} \langle S^z(z, t) S^z(0, 0)\rangle, \tag{15.35}$$

must be calculated, with $s^z(z, t)$ defined by the soliton solution of Eq. (15.20). Let us state the final result of calculating this expression [177]:

$$S(\varkappa, \omega) = S^2\left(1 - \sum_n n_s^{(n)} F^{(n)}\right)\delta(\mathbf{q})\,\delta(\omega) + \sum_n n_\varepsilon^{(n)} Q^{(n)}(\varkappa, \omega), \tag{15.36}$$

where

$$F^{(n)} = \frac{4}{\varkappa_n}\left(\frac{\gamma_n - 1}{\gamma_n + 1}\right)^{1/2} |\operatorname{Arch}\gamma_n|, \quad \gamma_n = 1 + 2\operatorname{sh}^2\frac{n}{n_0},$$

$$Q^{(n)}(\varkappa, \omega) = \Delta_n^2 \frac{\gamma_n - 1}{\gamma_n + 1}\frac{\pi}{\varkappa}\left(\frac{m_n^*}{2\pi kT}\right)^{1/2} e^{-\frac{m_n^*\omega^2}{2kT\varkappa^2}} \times$$
$$\times \frac{1 - \cos\left(\frac{\varkappa}{\varkappa_n}|\operatorname{Arch}\gamma_n|\right)}{\operatorname{sh}^2(\pi\varkappa/2\varkappa_n)}. \tag{15.37}$$

The quantities $F^{(n)}$ and $Q^{(n)}$ refer to solitons with a fixed number n of spin deviations; their density is given by the quantity $n_s^{(n)}$. The observed dynamic structure factor is the sum over all solitons with different n. The soliton density is $n_s^{(n)} \sim e^{-\varepsilon_n/kT}$.

It follows from Eq. (15.37) that quasi-elastic scattering at a soliton is characterized by the energy width

$$\delta\omega_n = \varkappa\sqrt{kT/m_n^*} \tag{15.38}$$

and the angular width $\delta\varkappa \sim 2/\pi\Delta_n$.

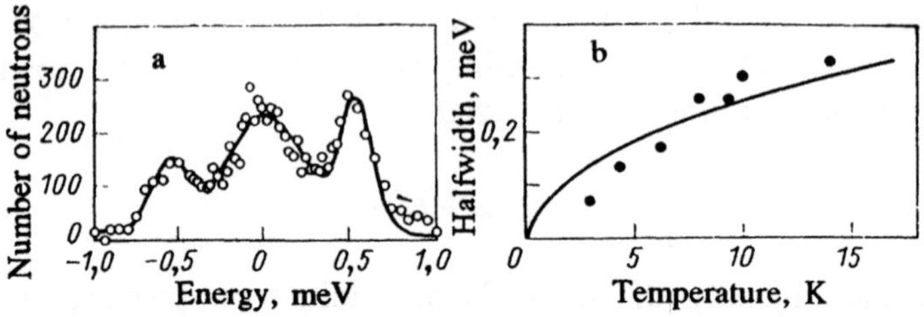

Fig. 29. Inelastic scattering at the quasi-one-dimensional ferromagnetic material $CsNiFe_3$ [219]: a) spectrum at T = 9.3 K; $\varkappa$ = (0; 0; 1.9), H = 0.5 T; b) temperature dependence of the half-width of the quasielastic component.

By comparing the results of Eqs. (15.33) and (15.37), we observe that the cross section of scattering at solitons is of the same form in a ferromagnetic chain with a "light axis" anisotropy or a "light plane" anisotropy. The quasielastic part is described by a Gaussian function of the transferred energy $\varepsilon = \hbar\omega$ with a width dependent on the temperature $\sqrt{T}$.

The authors of [219] have reported on an experimental observation of soliton excitations in ferromagnetic chains in a $CsNiF_3$ crystal which was considered in detail in the preceding section. In the temperature range 3 K $< T \lesssim$ 15 K and in fields of up to 3.6 T, at small angles near the reciprocal (002) lattice point the authors [219] observed two peaks of inelastic scattering at spin waves and, in addition, the central peak of a Gaussian profile, which central peak was interpreted as scattering at a soliton excitation (Fig. 29a). The temperature dependence of the half-width of the quasielastic peak, $\delta\omega \sim \sqrt{T}$, in accordance with the theoretical prediction of Eq. (15.34) corroborates this interpretation of the spectrum (Fig. 29b). However, the behavior of the spectrum as a function of the external field (at a fixed temperature) or of the scattering vector $\varkappa$ agrees only qualitatively with the theoretical equation (15.33) [219]. The soliton activation energy ε_s/k = 27 K was obtained from the temperature dependence of the measured intensity, whereas the value ε_s/k = 34 K in the field H = 0.5 T was calculated with Eq. (15.18) and is somewhat too large. The degree to which the theory and the experiment coincide confirms that in research on $CsNiF_3$, an excitation of the soliton type was detected for the first time. But one must recall the arguments of [288], where it was shown that the observed effects can be explained by a contribution resulting from multimagnon scattering.

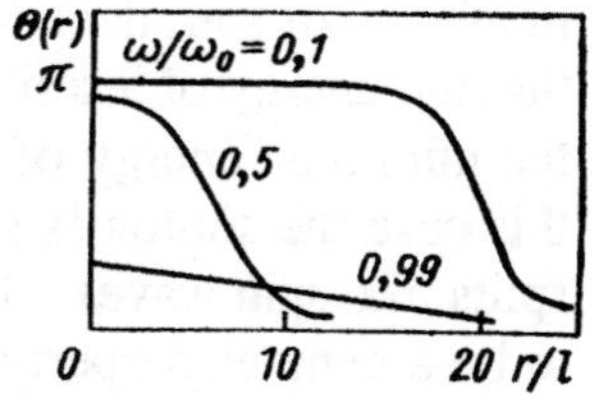

Fig. 30. Distribution of the magnetization in a localized soliton-type oscillation in a three-dimensional ferromagnetic material with axial anisotropy as a function of the parameter ω/ω_0 [53].

It has been reported [130] that in the quasi-one-dimensional antiferromagnetic TMMC (see Section 13), upon application of a strong field onto a plane perpendicular to the chains, there was observed a central peak in inelastic neutron scattering, and this peak was interpreted as scattering at a magnetic soliton (see also [129]).

Three-dimensional solitons. Unfortunately, there do not exist analytic methods of solving nonlinear equations which describe, for example, the movement of a magnetic moment. Therefore, excited states of a ferromagnet in which the deviations of the magnetization from the equilibrium value are fixed in all directions in space are studied mainly with numerical methods. In the case of a ferromagnet with uniaxial localized anisotropy, solutions were obtained [53] under the assumption of radial symmetry of the polar angle $\theta(r)$ (Fig. 30). The states illustrated in Fig. 30 correspond to a soliton at rest ($v = 0$) performing a precessional movement with the frequency ω. The form of the curves depends strongly upon the relation between ω and ω_0 (the minimum frequency of the homogeneous spin wave). At small ω/ω_0 ratios there exists a significant region with spin aligned in a direction opposite to the main direction (angle $\theta = 0$). The radius of this region is $R \sim a\sqrt{J/A}\cdot(\omega/\omega_0)$. A change in magnetization takes place in an interval $\sim a\sqrt{J/A}$ which is small in comparison with R. In this interval, the distribution of the magnetization is close to the distribution of the magnetization on a planar domain boundary

$$\cos\theta(r) \approx \mathrm{th}\,\{\sqrt{A/J}\,[(r - R)/a]\}.$$

When the frequency ω increases, the amplitude of a localized oscillation increases monotonically; at frequencies $(\omega_0 - \omega) << \omega_0$, the amplitude continues to decrease but the size of the region of localization increases. Such soliton vibrations are very much smeared in space and merge with nonlocalized spin waves at $\omega \to \omega_0$.

By contrast to solitons in a one-dimensional ferromagnet, the stability of three-dimensional solitons ("magnon droplets") depends upon the number n of spin deviations. There exists a minimum value $n = n^*$ such that the energy of a soliton per single spin deviation becomes smaller than the minimum energy of the spin-wave spectrum ($E_{sol}(n)/n << \hbar\omega_0$). In this case the soliton is stable. In the opposite case ($n < n^*$), the soliton splits into spin waves. This behavior of a three-dimensional soliton seems to be a general property of Bose particles (with the spin deviations being Bose particles in our case): when the attractive potential between them does not suffice for the formation of bound pairs, bound complexes of n particles with $n > n^*$ may develop. Experimental observations of these nonlinear excitations in three-dimensional ferromagnets by means of neutron diffraction and neutron spectroscopy would naturally be interesting but to date they are difficult because the theory proper has been hardly developed and does not yet allow the determination of specific scattering features. It is only clear that because of the relatively large size of the magnon droplets, the (incoherent) scattering must be concentrated in small regions around the Bragg peaks.

There exists another example of three-dimensional solitons which in a ferromagnet were obtained from the topological properties of the space of the ferromagnet's order parameter rather than from numerical solutions of nonlinear equations [27].

16. Critical Scattering near a Magnetic Phase Transition

Phenomenological description. As in any other phase transition of the second kind, fluctuations of the parameter of magnetic order, which cause a critical magnetic scattering, are important near a magnetic phase transition. The cross section of this type of scattering must be calculated with the method employed in Section 7 in the analysis of critical scattering near a structural phase transition.

Assume that a magnetic phase transition occurs from a paramagnetic phase under formation of a magnetic structure which is characterized by the irreducible representation $\{\mathbf{k}\}\nu$ of the space group G of the crystal. The resulting magnetic structure can be expanded in the basis functions of the relevant irreducible representation. This expansion is in analogy to Eq. (7.1) which was used to describe a structural transition:

$$\mathbf{S}_{ni} = \sum_{L\lambda} C_{L\lambda}\, \mathbf{m}\begin{pmatrix} \mathbf{k}_L & \nu \\ \lambda & \end{pmatrix}\!\Big|\, i\Big)\, e^{i\mathbf{k}_L \mathbf{t}_n}, \tag{16.1}$$

Here the thermodynamic average of the spin of the ith atom in the nth unit cell is represented as a superposition of the pseudovector basis functions $\mathbf{m}\begin{pmatrix}\mathbf{k}_L\nu \\ \lambda\end{pmatrix}\Big| i\Big)$ of the irreducible representation of the group of the wave vector $\mathbf{k}_L$.

As in the case of a structural phase transition, above T_c the average values of the mixing coefficients $C_{L\lambda}$ vanish but their instantaneous values may differ from zero. Let us introduce a wave packet of fluctuations which cause a random configuration of the vectors of the magnetic moments $\mathbf{S}_{ni}$ within the crystal [36]:

$$\mathbf{S}_{ni}=\frac{1}{\sqrt{N}}\sum_{L\lambda}\sum_{\mathbf{q}}C_{L\lambda}(\mathbf{q})\mathbf{m}\left(\begin{matrix}\varkappa_L\,\nu \\ \lambda\end{matrix}\Big| i\right)e^{i\varkappa_L\mathbf{t}_n}. \tag{16.2}$$

At $T > T_c$, the free energy of the crystal is described by the same Ginzburg–Landau functional (7.4) and leads to the same correlation functions (7.8) as in a structural phase transition, except for the fact that the term $\mathcal{H}_{\text{int}}$ must be constructed only from the even invariants in the parameters $C_{L\lambda}(\mathbf{q})$.

In order to calculate the cross sections at $T > T_c$, we substitute the expansion of Eq. (16.2) in the general formula (1.5) of the differential cross section of magnetic scattering. We obtain

$$\frac{d^2\sigma}{d\Omega d\varepsilon}=N(r_0\gamma)^2\sum_{L\lambda}\sum_{\alpha\beta}(\delta_{\alpha\beta}-e_\alpha e_\beta)\overset{*}{f}{}^{\alpha}_{\nu L\lambda}(\varkappa)f^{\beta}_{\nu L\lambda}(\varkappa)\times$$
$$\times\sum_{\mathbf{q}}\langle C^*_{L\lambda}(\mathbf{q})C_{L\lambda}(\mathbf{q})\rangle\sum_{\mathbf{b}}\delta_{\varkappa-\varkappa_L-\mathbf{q},\mathbf{b}}\,\delta(\varepsilon), \tag{16.3}$$

wherein $\mathbf{f}_{\nu L\lambda}$ has the meaning of a structure amplitude of magnetic scattering:

$$\mathbf{f}_{\nu L\lambda}(\varkappa)=\sum_i F_i(\varkappa)e^{-W_i}e^{-i\varkappa\mathbf{x}_i}\mathbf{m}\left(\begin{matrix}\varkappa_L\,\nu \\ \lambda\end{matrix}\Big| i\right). \tag{16.4}$$

We obtain a final expression for the cross section of critical scattering near a magnetic Bragg reflection $\varkappa_0 = \mathbf{k}_L + \mathbf{b}$ from Eq. (16.3) when we include Eq. (7.8) for the correlator of the fluctuations into our considerations:

$$\frac{1}{N}\frac{d\sigma_{\varkappa}^{\text{crit}}}{d\Omega}=(r_0\gamma)^2\sum_{\lambda}|\mathbf{M}_{\nu L\lambda}(\varkappa_0)|^2\frac{kT}{a(T-T_c)+\sum_{\alpha\beta}q^{\alpha}q^{\beta}T_{\nu}^{\alpha\beta}}, \tag{16.5}$$

where

$$\mathbf{M}_{\nu L\lambda}(\varkappa)=\mathbf{f}_{\nu L\lambda}(\varkappa)-(\mathbf{e}\mathbf{f}_{\nu L\lambda}(\varkappa))\,\mathbf{e}. \tag{16.6}$$

Let us note that the vectors $\mathbf{f}_{\nu L\lambda}(\varkappa)$ and $\mathbf{M}_{\nu L\lambda}(\varkappa)$ which characterize the critical magnetic scattering, are analogs to the vectors of Eqs. (8.13) and (8.12) which characterize the elastic scattering of neutrons at a magnetic structure. They are partial contributions of the latter in regard to subscripts of the basis functions of the relevant irreducible representation.

Symmetry analysis of the cross section. The final equations (16.5) for the cross section of critical magnetic scattering and (7.13) for critical nuclear scattering are in a very clear way identical: a correlation function of the Ornstein–Zernicke type is multiplied by some factor (sum over λ) which depends upon the basis functions of the corresponding relevant irreducible representation. In the case of magnetic scattering, this factor can be expressed through the tensor $B'^{\alpha\beta}$ which is an analog to the Elliott–Thorpe tensor of Eq. (5.15):

$$\sum_{\lambda}|\mathbf{M}_{\nu L\lambda}(\varkappa)|^2=\sum_{\alpha\beta}(\delta_{\alpha\beta}-e_{\alpha}e_{\beta})\,B_{\nu L}'^{\alpha\beta}(\varkappa); \tag{16.7}$$

$$B_{\nu L}'^{\alpha\beta}(\varkappa)=\sum_{\lambda}f_{\nu L\lambda}^{\alpha}(\varkappa)\,f_{\nu L\lambda}^{\beta}(\varkappa). \tag{16.8}$$

We represent the tensor $B'^{\alpha\beta}$ in a form which is an analog to Eq. (5.15):

$$B_{\nu L}^{\alpha\beta}(\varkappa)=\sum_{i}\left(A_{\nu L}'(i,\varkappa)\,A_{\nu L}'^{+}(i,\varkappa)\right)^{\alpha\beta}\left(F_i(\varkappa)\,e^{-W_i}\right)^2, \tag{16.9}$$

where

$$A_{\nu L}'^{\alpha\beta}(i,\varkappa)=\frac{l_{\nu}}{\|G_{\kappa L}^{\circ}\|}\sum_{g\in G_{\kappa L}^{\circ}}\overset{*}{\chi}{}^{\kappa L\nu}(g)\,e^{i\varkappa g x_i}\,\delta_g\,R_g^{\alpha\beta}. \tag{16.10}$$

The formula for $A_{\nu L}'^{\alpha\beta}$ is distinguished from the analogous Eq. (5.10) describing the Elliott–Thorpe tensor by the factor $\delta_g=\pm 1$ in Eq. (16.10), with the factor accounting for the pseudovector form of the basis functions of the magnetic representation. It is another difference that the summation is extended only over the magnetic atoms in Eq. (16.10). As in the case of critical nuclear scattering, the calculation of the components of the tensor $B'^{\alpha\beta}$ makes it possible to determine the extinction laws for magnetic scattering. The extinctions of critical scattering at various vectors **b** of the reciprocal lattice make it possible to determine, in principle, the irreducible representation involved in the magnetic phase transition. Making use of Eqs. (16.5) and (16.7), one can solve still another problem, if the irreducible representation relevant for the magnetic phase transition is known, i.e., one can determine the magnetic Bragg reflections near which the cross section of critical scattering has its maximum. This problem is of importance when the determination of the critical indices for correlation lengths is the objective in investigations of critical scattering.

When Eq. (16.5) for the cross section of the critical magnetic scattering and the corresponding Eqs. (7.13) for the critical nuclear scattering were derived, we assumed that the relevant irreducible representation of the group G appears only once in the composition of the magnetic or mechanical representations. When it appears several times, there exists a corresponding number of sets of basis functions and the summation in Eqs. (16.2) and (7.2) must be extended over all of them. For example, in the case of magnetic transition, we then must write in place of Eq. (16.2):

$$\mathbf{S}_{ni} = \frac{1}{\sqrt{N}} \sum_{s} \sum_{L\lambda} \sum_{\mathbf{q}} b^{s}_{L\lambda}(\mathbf{q})\, \mathbf{m}_{s} \begin{pmatrix} \mathbf{k}_L \nu \\ \lambda \end{pmatrix}\Bigg| i \Bigg) e^{i(\mathbf{k}_L+\mathbf{q})\mathbf{t}_n}, \tag{16.11}$$

where the sum over s also accounts for all sets of a repeated representation $\{\mathbf{k}\}\nu$. This complication leads to additional sums over the subscripts s in the expressions for the cross section

$$\frac{1}{N}\frac{d\sigma_{\varkappa}}{d\Omega} = (r_0\gamma)^2 \sum_{ss'} \left\{ \sum_{\lambda} \overset{*}{\mathbf{M}}{}^{s}_{\nu L\lambda}(\varkappa)\, \mathbf{M}^{s'}_{\nu L\lambda}(\varkappa) \right\} \langle \overset{*}{b}{}^{s}_{L\lambda}(\mathbf{q})\, b^{s'}_{L\lambda}(\mathbf{q}) \rangle, \tag{16.12}$$

where $\mathbf{M}_{\nu L\lambda}{}^{s}(\varkappa)$ is defined by the same Eqs. (16.6) and (16.4) but with $\mathbf{m}$ replaced by $\mathbf{m}_s$.

The Ginzburg–Landau functional comprises a quadratic form of the mixing coefficients, $b_{L\lambda}{}^{s}(\mathbf{q})$, and this quadratic form is nondiagonal in the subscripts s and s'. Diagonalization of this form makes it possible to calculate the correlators appearing in Eq. (16.12). The expression for the cross section must comprise the parameters of the Ginzburg–Landau functional. In the particular case of a double degenerate representation, diagonalization is easily performed and leads to the following expression for the critical magnetic scattering [44]:

$$\frac{1}{N}\frac{d\sigma_{\varkappa}^{\text{crit}}}{d\Omega}=(r_0\gamma)^2\sum_{\lambda}\left|\mathbf{M}^1_{\nu L\lambda}(\varkappa_0)\frac{G}{\sqrt{G^2+D^2}}-\mathbf{M}^2_{\nu L\lambda}(\varkappa_0)\frac{D}{\sqrt{G^2+D^2}}\right|^2\times$$
$$\times\frac{kT}{a(T-T_c)+\sum\limits_{\alpha\beta}q_\alpha q_\beta T_\nu^{\alpha\beta}}. \tag{16.13}$$

where D and G denote the parameters of the Ginzburg–Landau functional, with D representing the coefficient before the mixed terms $b_{L\nu\lambda}{}^1 b_{L\nu\lambda}{}^2$. We realize that also in this case in Eq. (16.13) there develops a term of the Ornstein–Zernicke type but the factor before this term depends not only upon the basis functions of the relevant representation but also upon the parameters of the Hamiltonian. A comparison of the theoretical expressions of the type of Eq. (16.13) with the experimentally measured cross section of critical scattering facilitates the determination of the Ginzburg–Landau parameters.

Microscopic description. Based on the microscopic theory, the cross section of critical neutron scattering can be obtained from the general equation (8.5), provided that the fluctuation part is separated in the spin correlators. We first disregard the time dependence of the correlators and leave the problem of the dynamics of the spin fluctuations to the following section. Thus, the scattering becomes elastic and can be expressed in the form

$$\frac{d^2\sigma^{\text{crit}}}{d\Omega d\varepsilon}=(r_0\gamma)^2\sum_{ij}F_i(\varkappa)F_j(\varkappa)\,e^{-W_i-W_j}e^{i\varkappa(x_i-x_j)}\times$$
$$\times\sum_{\alpha\beta}(\delta_{\alpha\beta}-e_\alpha e_\beta)\,\chi_{ij}^{\alpha\beta}(\varkappa)\,\delta(\varepsilon). \tag{16.14}$$

We have introduced the quantity

$$\chi_{ij}^{\alpha\beta}(\mathbf{q})=\langle(S_i^\alpha(\mathbf{q})-\langle S_i^\alpha(\mathbf{q})\rangle)(S_j^\beta(-\mathbf{q})-\langle S_j^\beta(-\mathbf{q})\rangle)\rangle, \tag{16.15}$$

which is proportional to the magnetic susceptibility of the system; $S_i^\alpha(\mathbf{q})$ denotes the Fourier component of the projection of the atomic spin $S_{mi}{}^\alpha$ upon the magnetic lattice of the crystal:

$$S_i^{\alpha}(\mathbf{q}) = \frac{1}{\sqrt{N}} \sum_m e^{i\mathbf{q}\mathbf{t}_m} S_{mi}^{\alpha}, \tag{16.16}$$

so that $\mathbf{t}_m$ is the translation vector of the magnetic lattice and the subscripts i and j number the magnetic atoms in the magnetic unit cell of the crystal. The average fluctuations and holds at temperatures below, as well as above, T_c.

Equation (16.15) is a periodic function of $\mathbf{q}$ with the period of the basis vectors of the reciprocal magnetic lattice. Near some lattice point $\mathbf{b}_m$ of the reciprocal lattice, the angular distribution of the cross section of Eq. (16.14) is given by the expression

$$\frac{d\sigma^{\text{crit}}_{\varkappa=\mathbf{b}_m+\mathbf{q}}}{d\Omega} = N(r_0\gamma)^2 \sum_{ij} F_i(\varkappa) F_j(\varkappa) \times$$
$$\times e^{-W_i - W_j} e^{i\varkappa(x_i - x_j)} \sum_{\alpha\beta} (\delta_{\alpha\beta} - e_\alpha e_\beta) \chi_{ij}^{\alpha\beta}(\mathbf{q}). \tag{16.17}$$

The form of the function $\chi_{ij}^{\alpha\beta}$ was obtained in [182] (see also [43]) in the self-consistent field approximation; we briefly cite the results for a ferromagnetic crystal with a single magnetic atom per cell. For the paramagnetic phase ($T > T_c$), the following expression for the cross section results in this case from Eq. (16.17):

$$\frac{d\sigma^{\text{crit}}_{\varkappa=\mathbf{b}+\mathbf{q}}}{d\Omega} = N\frac{2}{3}(r_0\gamma)^2 F^2(\varkappa) e^{-2W} \chi(\mathbf{q}), \tag{16.18}$$

where $\chi(\mathbf{q})$ at $q << b$ is the well-known Ornstein–Zernicke function:

$$\chi(\mathbf{q}) = S(S+1)\frac{1}{r_1^2(k_1^2+q^2)}, \tag{16.19}$$

which is the Fourier image of the spin correlator of the self-consistent field:

$$\langle S_0 S_R \rangle = \frac{v_0 S(S+1)}{4\pi r_1^2} \frac{e^{-\kappa_1 R}}{R}. \tag{16.20}$$

where $\varkappa_1 = \xi^{-1}$ denotes the inverse correlation length; v_0 denotes the volume of the unit cell of the crystal; r_1 denotes a weakly temperature-dependent parameter with the dimension of a length; and **b** denotes a vector of the reciprocal lattice of the crystal.

In the case of a Heisenberg ferromagnet with an exchange interaction between the nearest neighbors, explicit expressions for the parameters r_1 and $\varkappa_1$ are easily obtained:

$$r_1^2 = (a^2/z)(T_c/T), \quad \kappa_1^2 = (z/a^2)[(T - T_c)/T_c], \tag{16.21}$$

where z denotes the number of the nearest neighbors.

In a ferromagnet phase with a spontaneous moment parallel to the z axis, Eq. (16.17) is reduced to the expression

$$\frac{d\sigma^{\text{crit}}_{\varkappa=\mathbf{b}+\mathbf{q}}}{d\Omega} = N(r_0\gamma)^2 F^2(\varkappa)\, e^{-2W}[(1 - e_z^2)\chi_{\parallel}(\mathbf{q}) + (1 + e_z^2)\chi_{\perp}(\mathbf{q})], \tag{16.22}$$

where $\chi_{\parallel}$ and $\chi_{\perp}$ denote the longitudinal and transverse components of the susceptibility tensor, respectively:

$$\chi_{\parallel}(\mathbf{q}) = \frac{S(S+1)}{r_1^2(\kappa_1'^2 + q^2)}, \quad \chi_{\perp}(\mathbf{q}) = \frac{S(S+1)}{r_1^2 q^2}. \tag{16.23}$$

Investigation of the critical indices in ferromagnets. The approximation of an average field works only far from the point of the phase transition, where the fluctuations of the parameters of order are real. In the critical region and at the point proper, the three-dimensional behavior of the correlation functions is no longer described by the result [Eq. (16.19)] of the Ornstein–Zernicke theory. The form of the correlation function is under these conditions derived with the hypothesis of static scaling [86] according to which the double correlation function $\chi(\mathbf{q})$ depends upon both the wave vector and the temperature not separately but as a generalized homogeneous function of the variable q and $\varkappa_1$ (the inverse correlation length). $\chi(\mathbf{q})$ is usually stated in the form [190]

$$\chi(q) = q^{-2+\eta}\,\varphi\left(\frac{\kappa_1}{q}\right), \tag{16.24}$$

where η is termed the Fischer parameter. For $T = T_c$ we have $k_1 = 0$ and therefore $\chi(\mathbf{q}) \sim \mathbf{q}^{-2+\eta}$ follows from Eq. (16.24); this agrees with the notation given in Eq. (7.17) for the correlation function in the lattice space:

In the Ornstein–Zernicke theory,

$$\varphi\left(\frac{\kappa_1}{q}\right)=\frac{1}{1+\kappa_1^2/q^2}\,, \tag{16.25}$$

and $\eta = 0$. Numerical calculations based on various models show that the parameter η is very small. For example, in the case of the Heisenberg model, $\eta \approx 0.04$. When this parameter is disregarded, the Ornstein–Zernicke correlation function satisfied static scaling and can be used to describe neutron scattering in the critical region. There is the other feature, namely the temperature dependence of the correlation length, or of the parameter $\varkappa_1$, which is characterized by the critical index ν:

$$\kappa_1 \sim \tau^{\nu}, \tag{16.26}$$

where τ denotes the reduced temperature defined by the expression

$$\tau = (T - T_c)/T_c. \tag{16.27}$$

β is an other critical index defining the temperature dependence of the magnetization (the order parameter) at $T < T_c$:

$$m \sim |\tau|^{\beta}, \tag{16.28}$$

The definitions given by Eqs. (16.26) and (16.28) for a magnetic phase transition correspond to the definitions of Eqs. (7.15) for the quantities characterizing a structural phase transition. Since the critical behavior of the systems is universal, the relations between the critical indices in the magnetic case remain the same as in the theory of structural transitions and are given by Eqs. (7.19).

Investigations of the critical scattering of neutrons make it possible to determine experimentally two critical indices, namely ν and β. The index is determined from measurements of the cross section near magnetic Bragg peaks at various temperatures above T_c; the index β is obtained by measurements of the temperature dependence of the magnitude of the magnetic Bragg peaks below T_c because their intensity is proportional to $m^2 \sim |\tau|^{2\beta}$. Such investigations were carefully made on the ferromagnetic metals Fe, Co, Ni selected first of all because all of them have a high Curie point and a rather wide temperature interval in which the critical phenomena manifest themselves.

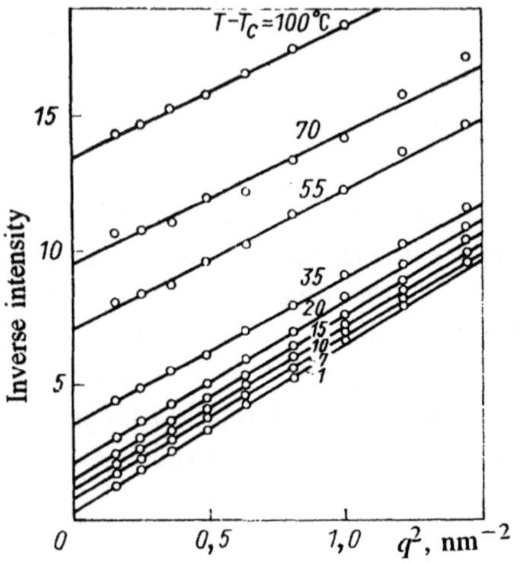

Fig. 31. Inverse intensity of scattering at cobalt as a function of the square of the deviation of the scattering vector from the Bragg peak (corrections for the resolution and the inelasticity of scattering were introduced in [187].

Figure 31 shows data which were obtained in regard to critical scattering at Co [187]. At various temperatures exceeding T_c. It follows from the figure that the experimental results in an appropriate approximation confirm the Ornstein–Zernicke relation

$$\frac{d\sigma}{d\Omega} \sim \frac{1}{\kappa_1^2 + q^2} \tag{16.29}$$

within a wide temperature interval. The corresponding method has been described in detail in [343].

In the critical region, the double correlation function of the spins depends upon the critical index η, and the cross section should be described by the expression

$$\frac{d\sigma}{d\Omega} \sim \frac{1}{(\kappa_1^2 + q^2)^{1-\eta/2}} . \tag{16.30}$$

rather than by the Ornstein–Zernicke function of Eq. (16.29). Since η is small, the above expression differs only slightly from Eq. (16.29) but the evaluation of experimental data with Eq. (16.30) makes it possible to determine η.

When the curves are extrapolated to $\mathbf{q} = 0$, $\varkappa_1^2$ can be determined at each temperature. The experimental $\varkappa_1^2$ points are illustrated in Fig. 32 on a log scale in dependence upon the reduced temperature τ. At

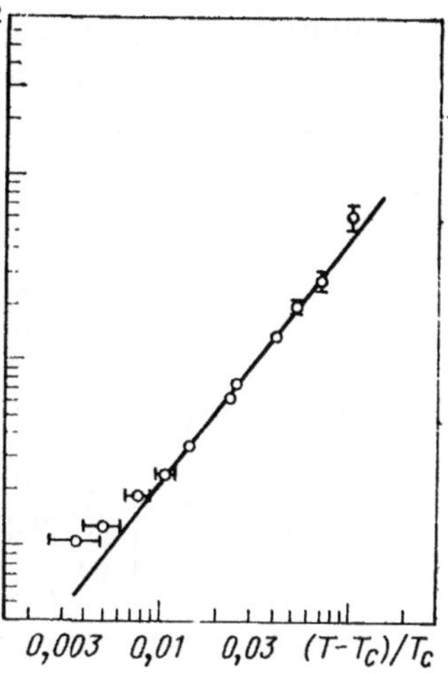

Fig. 32. Temperature dependence $\varkappa_1^2$ for cobalt; logarithmic scale [187].

temperatures from $\tau = 0.01$ and higher, the points are situated on a straight line whose slope determined the critical index $\nu = 0.65 \pm 0.04$. The critical index values obtained with various experimental methods are listed in Table 14 for the ferromagnetic metals of the 3*d* group. The asterisks denote data obtained from neutron scattering. The table gives an idea of the precision with which the experimental data were obtained and of the spread of the results.

The calculated values of the indices are in the case of the Heisenberg model [293]:

$$\nu = 0{,}70;\ \gamma = 1{,}38;\ \beta = 0{,}38;\ \eta = 0{,}04, \tag{16.31}$$

and in the case of the three-dimensional Ising model [179]

$$\nu = 0{,}64;\ \gamma = 1{,}24;\ \beta = 0{,}34;\ \eta = 0{,}05.$$

It follows from a comparison with Table 14 that the measured critical indices are in the case of Fe and Ni rather close to the theoretical values of the Heisenberg model. Deviations exist in the case of Co toward a merging with the classical critical indices of Eq. (7.18). The reason for these deviations is unclear.

The measured values of the critical indices rather properly satisfy the scaling relations. For example, it follows from Eqs. (7.19) that $\beta - [(3\nu - \gamma)/2] = 0$. When the more accurate data for the critical indices listed

TABLE 14. The Critical Indices γ and β Determined with Various Experimental Techniques (see the review of [187])

Metals	γ	β	ν	η
Fe	1,33±0,015 1,33±0,03 1,30±0,04* 1,34±0,04*	0,389±0,005 0,34±0,04 0,35±0,06	0,70±0,015*	0,07±0,05*
Co	1,21±0,04 1,20±0,04 1,32±0,02 1,20±0,05 1,19* 1,23±0,05*	0,42±0,01 0,41	0,625*	0,10±0,05*
Ni	1,34±0,01 1,32±0,02 1,29±0,04 1,30*	0,378±0,004 0,375±0,02 0,35±0,01	0,65±0,04*	—

in Table 14 are inserted, it turns out that the left side of the above-noted relation is 0.005 ± 0.02 in the case of Fe and 0.06 ± 0.07 in the case of Co.

Let us also state the results of neutron diffraction work on the critical indices in nonmetallic ferromagnetic materials [109]:

$$\begin{aligned} &\text{EuO: } \nu = 0{,}681 \pm 0{,}017,\ \beta = 0{,}36 \pm 0{,}01; \\ &\text{EuS: } \nu = 0{,}702 \pm 0{,}022,\ \beta = 0{,}36 \pm 0{,}01. \end{aligned} \tag{16.32}$$

There is excellent agreement with the values of Eq. (16.31) calculated with the Heisenberg model.

Critical indices in antiferromagnets. Let us switch to the description of the critical behavior of antiferromagnets. By contrast to antiferromagnets in which the number n of components of the order parameter is not greater than three (the number of components of the vector of the spontaneous moment), antiferromagnets have physical systems with $n \geq 4$ because n coincides with the dimensionality of the irreducible representation of the space group of the crystal, according to which the order parameter of the phase with magnetic order is transformed, but the dimensionality of these representations is often greater than three.

TABLE 15. Substances Experiencing an Antiferromagnetic Phase Transition, and the Number n of Components of the Order Parameter

Substance	n	G ($T > T_c$)	Magnetic structure ($T < T_c$)
TbAu, DyC_2	4	$I4/mmm$	Sinusoidal
Tb, Dy, Ho	4	$P6_3/mmc$	Helical
NbO_2	4	$P4_2/mnm$	Missing (structural phase transition)
TbD_2	6	$Fm\,3m$	Sinusoidal
Nd	6	$P6_3/mmc$	Sinusoidal
K_2IrCl_6	6	$Fm\,3m$	Type III antiferromagnet
MnS_2	6	$Pa3$	Type III antiferromagnet
MnO, MnSe, NiO, ErSb	8	$Fm\,3m$	Type II antiferromagnet
TbAs, TbP, CeS, TbSb	4	$Fm\,3m$	Type II antiferromagnet
UO_2	6	$Fm\,3m$	Type I antiferromagnet
Cr	12	$Im\,3m$	Sinusoidal
Eu	12	$Im\,3m$	Sinusoidal

The authors of [115, 116, 261–263] studied in a large number of antiferromagnetic crystals of different symmetry their critical behavior at $T > T_c$ with the group reform technique and the ε expansion [15]. A list of these substances is given in Table 15 which indicates for each of them the number n of components of the order parameter. This quantity is determined from the magnetic structure arising from a phase transition at $T < T_c$. To this end, the number of arms of the star of the wave vector must be multiplied by the irreducible representation of the group of the wave vector describing the magnetic structure.

The authors of [115, 116, 261–263] constructed for each of the antiferromagnets listed in Table 15 the effective Ginzburg–Landau Hamiltonian which represents the expansion of the free energy in powers of the order parameter up to terms of the fourth power. For example, in tetragonal $TbAu_2$ and DyC_2 crystals with the space group $I4/mmm$, the TSW structure exists (transverse spin wave with wave vectors on the base plane). There are in each case four arms of the star and the irreducible representations of the Group $G_{\mathbf{k}}$ are one-dimensional so that the number n of components of the order parameter is equal to 4. The Ginzburg–Landau Hamiltonian has in this case the form [263]:

$$\mathcal{H}_{GL}=\int dr\left\{r\sum_{\Lambda=1}^{n}C_{\Lambda}^{2}+\sum_{\Lambda=1}^{n}(\nabla C_{\Lambda})^{2}+u_{1}\left(\sum_{\Lambda=1}^{n}C_{\Lambda}^{2}\right)^{2}+u_{2}\sum_{\Lambda=1}^{n}C_{\Lambda}^{4}\right\} \quad (16.33)$$

with $n = 4$. In hexagonal Tb, Dy, and Ho crystals, a helical magnetic structure with a wave vector parallel to the hexagonal axis and with spins on the base plane exists at $T < T_c$. The number of arms in the star is equal to two, and the irreducible representation of the group $G_{\mathbf{k}}$, which describes the order of the spins on the base plane, is two-dimensional so that the number of components of the order parameter is $n = 2 \times 2 = 4$. Despite the different symmetry of the crystals, the Hamiltonian $\mathcal{H}_{GL}$ is also in this case given by Eq. (16.33) up to terms of the fourth power. Table 15 includes the NbO_2 crystals which do not have magnetic order but experience a structural phase transition of the second kind from the tetragonal $P4_2/mnm$ phase into a phase with the space group $I4_1/a$. The wave vector of the new structure is $\mathbf{k} \sim [(1/4)(1/4)(1/2)]$ and belongs to a four-arm star so that in this case $n = 4$. Noteworthy is the fact that $\mathcal{H}_{GL}$ is again given by (Eq. (16.33) (with an additional term describing the tetragonal anisotropy which, as it turn out, does not affect the critical behavior of the system).

Thus, for our six substances of different symmetry, with different forms of magnetic order, and even with different types of phase transitions (NbO_2), the critical behavior is described by the same Ginzburg–Landau Hamiltonian with $n = 4$. It turns out that the critical behavior of all these systems must be the same and must be described by the critical indices [116]:

$$\nu = 0{,}70;\ \eta = 0{,}02;\ \gamma = 1{,}39;\ \beta = 0{,}39;\ \alpha = -0{,}17. \quad (16.34)$$

The indices ν and η were calculated with the ε expansion, the other indices were obtained from similarity relations. This example demonstrates that the critical behavior of the systems is universal.

A similar analysis, which was made for the antiferromagnets TbD_2, Nd, MnS_2, and K_2InCl_6 [116] shows that the magnetic phase transitions in them can be described by the same Ginzburg–Landau Hamiltonian. Since the number of components of the order parameter is the same ($n = 6$), this implies the same critical behavior which can be characterized by the critical indices

$$\nu = 0{,}69;\ \eta = 0{,}02;\ \gamma = 1{,}38;\ \beta = 0{,}38;\ \alpha = -0{,}14. \quad (16.35)$$

The effect of the anisotropy can be assessed when one compares these values with those obtained from the isotropic vector model with $n = 6$ [see Eq. (7.20)]:

$$\nu = 0{,}73;\ \eta = 0{,}015;\ \gamma = 1{,}45;\ \beta = 0{,}41;\ \alpha = -0{,}27. \quad (16.36)$$

There exists definite interest in an experimental determination of the critical indices of the antiferromagnets listed, as well as of other antiferromagnets, to further verify the ε expansion and the similarity hypothesis.

A breakaway of a phase transition of the second kind (if the Landau theory predicts such a phase transition) toward a phase transition of the first kind is one of the possible effects of the strong interaction of fluctuations in the critical region. Research on the phase transition with the ε expansion method can reveal this possibility if it turns out that the nonlinear equations of the reform group do not provide a stable fixed point [115, 26]. In the case of the antiferromagnets listed in Table 15 below the separating line, such a situation is encountered. Accordingly, it is assumed that a magnetic phase transition in them must be of the first kind. It is generally accepted that the transitions in Eu, Cr, UO_2, and MnO are transitions of the first kind. Special experiments which were made to check the predictions of the theory [115] in CeSe and CeTe [278] (isomorphous with CeS) have, however, shown that a transition of the second kind occurs in them.

$RbMnF_3$ is the first antiferromagnet in which the critical behavior was studied in detail [343]. This is a cubic crystal which at $T_c = 83$ K experiences a magnetic phase transition to a structure with a double magnetic cell. The following values of the critical indices were obtained:

$$\nu = 0{,}701 \pm 0{,}011;\ \eta = 0{,}055 \pm 0{,}010;\ \gamma = 1{,}366 \pm 0{,}024;$$
$$\beta = 0{,}32 \pm 0{,}02.$$

17. The Critical Dynamics of Ferromagnets

The energy dependence of the cross section. The dynamics of the spin system near the Curie point (both above and below T_c) is characterized by energy values $\varepsilon >> kT_c$. This makes the scattering process in this temperature interval quasi-elastic. Nevertheless, modern methods of inelastic neutron scattering facilitate detailed studies of the energy distribution of the scattering cross section in the critical region and make it possible to obtain reliable information on the dynamics of the spin system in it.

In order to study inelastic scattering processes at spin systems, we must start from the general expression [Eq. (8.5)] for the cross section and take into account the time dependence of the spin correlators. The scattering cross section is conveniently related to the retarded Green's spin function which was defined with the aid of general expression (3.3):

$$G_{ij}^{\alpha\beta}(\mathbf{q}, t) = -\frac{i}{\hbar}\,\theta(t)\,\langle[S_i^{\alpha}(\mathbf{q}, t),\ S_j^{\beta}(-\mathbf{q}, 0)]\rangle, \tag{17.1}$$

where $S_i^{\alpha}(\mathbf{q})$ donates the Fourier component of the spin operator. Taking into account the general relation [Eq. (3.10)] between the correlator and the Green's function corresponding to it, we rewrite Eq. (8.5) for the cross section in the following form:

$$\frac{d^2\sigma}{d\Omega d\varepsilon} = N(r_0\gamma)^2\,\frac{k}{k_0}\,\frac{1}{e^{\beta\varepsilon}-1}\sum_{ij} F_i(\varkappa)\,F_j(\varkappa)\,e^{-W_i-W_j}\,e^{i\varkappa(x_i-x_j)}\times$$
$$\times\sum_{\alpha\beta}(\delta_{\alpha\beta}-e_\alpha e_\beta)\,\frac{1}{\pi}\,\mathrm{Im}\,G_{ij}^{\alpha\beta}(\varkappa, -\varepsilon), \tag{17.2}$$

where $\mathrm{Im}\,G_{ij}^{\alpha\beta}$ is the spectral density of the spin correlator.

Furthermore, we restrict our considerations to the simplest case of a ferromagnetic crystal with a single magnetic atom per cell. We also recall that $|\varepsilon| << kT$ in the critical region. Then we rewrite Eq. (17.2) for the above conditions at $T > T_c$

$$\frac{d^2\sigma}{d\Omega d\varepsilon} = N(r_0\gamma)^2\,\frac{2}{3}\,S(S+1)\,e^{-2W}F^2(\varkappa)\times$$
$$\times\frac{k}{k_0}\,\frac{kT}{\varepsilon}\,\frac{1}{\pi}\,\mathrm{Im}\,G(\varkappa, -\varepsilon), \tag{17.3}$$

where $\mathrm{Im}\,G(\mathbf{q}, \omega)$ denotes the spectral density of the spin correlator $\langle S(\varkappa, t)\,S(-\varkappa, 0)\rangle$. When this expression is integrated over the energies, we obtain the angular distribution of the critical scattering in the form of Eq. (16.18):

$$\frac{d\sigma^{\mathrm{crit}}_{\varkappa=\mathbf{b}+\mathbf{q}}}{d\Omega} = N(r_0\gamma)^2\,\frac{2}{3}\,S(S+1)\,e^{-2W}F^2(\varkappa)\,\widehat{\chi}(\mathbf{q}), \tag{17.4}$$

where

$$\widehat{\chi}(\mathbf{q}) = \int_{-\infty}^{\infty} d\varepsilon \frac{kT}{\varepsilon} \frac{1}{\pi} \operatorname{Im} G(\mathbf{q}, -\varepsilon) \frac{k}{k_0} \tag{17.5}$$

must be close to the $\chi(\mathbf{q})$ expression for the static susceptibility, provided that the spectral density of the spin correlator is concentrated near $\varepsilon = 0$.

Equation (17.3) is usually stated as

$$\frac{d^2 \sigma_{\varkappa=\mathbf{b}+\mathbf{q}}}{d\Omega d\varepsilon} = N (r_0 \gamma)^2 \frac{2}{3} S(S+1) e^{-2W} F^2(\varkappa) \times \\ \times \frac{k}{k_0} \chi(\mathbf{q}) F(\mathbf{q}, \varepsilon), \tag{17.6}$$

where $\chi(\mathbf{q})$ donates the static magnetic susceptibility of the system and the function $F(\mathbf{q}, \varepsilon)$, which accounts for the energy dependence of the cross section, was normalized so that

$$\int_{-\infty}^{\infty} F(\mathbf{q}, \varepsilon)\, d\varepsilon = 1. \tag{17.7}$$

the factor $F(\mathbf{q}, \varepsilon)$ is proportional to the spectral density of the spin correlator and we will term it the spectral density function in what follows below.

In a phase with magnetic order ($T < T_c$), the general equation (17.2) for the cross section leads to the following form of the cross section as a generalized of Eq. (17.6):

$$\frac{d^2 \sigma_{\varkappa=\mathbf{b}+\mathbf{q}}}{d\Omega d\varepsilon} = N (r_0 \gamma)^2 S(S+1) e^{-2W} F^2(\varkappa) \times \\ \times \{(1 - e_z^2) \chi_{\parallel}(\mathbf{q}) F_{\parallel}(\mathbf{q}, \varepsilon) + (1 + e_z^2) \chi_{\perp}(\mathbf{q}) F_{\perp}(\mathbf{q}, \varepsilon)\}, \tag{17.8}$$

where $\chi_{\parallel}(\mathbf{q})$ and $\chi_{\perp}(\mathbf{q})$ denote the static magnetic susceptibilities and $F_{\parallel}(\mathbf{q}, \varepsilon)$ and $F_{\perp}(\mathbf{q}, \varepsilon)$ denote the spectral density functions of the longitudinal and transverse spin components, respectively.

As we observed in the preceding section, in the static case the quantity $\chi(\mathbf{q})$ depends upon the temperature through the parameter $\varkappa_1$, i.e., through the temperature dependence of the correlation length. We will see in what follows that this behavior is observed also in the spectral density function. Thus, the cross section of critical scattering depends upon two parameters which can be varied in the experiment, namely q and $\varkappa_1$ (disregarding the third parameter, the energy of the transition). Depending upon the relation between them, hydrodynamic (for $\mathbf{q} << \varkappa_1$) and critical (for $\mathbf{q} >> \varkappa_1$) conditions are to be distinguished. In the critical mode (region), the fluctuations strongly interact because of their high density. In the hydrodynamic case, the interaction of the fluctuations can brought into account with perturbation theory. Let us start the discussion with the latter case. First let us consider the form which the spectral density $F(\mathbf{q}, \varepsilon)$ must have in the phase with order and in the paramagnetic phase of the crystal in the hydrodynamic region. For $T < T_c$, the fluctuations of the transverse component of the magnetic moment are spin waves which in isotropic ferromagnets have a quadratic dispersion law for small q:

$$\hbar\omega(\mathbf{q}) = Dq^2. \tag{17.9}$$

The spectral function which can be obtained from the linear theory of the spin waves at low temperatures has two sharp peaks at the energies $\pm\hbar\omega(\mathbf{q})$. The interaction of spin waves with each other becomes important at high temperatures, and this leads to their renormalization and to an attenuation which increases when T_c is approached. The form of the spectral density for the transverse correlators near T_c can be expected to be of the same form as at low T, namely

$$F_\perp(\mathbf{q}, \varepsilon) = \frac{1}{2\pi}\left[\frac{\Gamma(\mathbf{q})}{[\varepsilon - \hbar\omega(\mathbf{q})]^2 + \Gamma^2(\mathbf{q})} + \frac{\Gamma(\mathbf{q})}{[\varepsilon + \hbar\omega(\mathbf{q})]^2 + \Gamma^2(\mathbf{q})}\right], \tag{17.10}$$

where $\Gamma(\mathbf{q})$ is in the general case some function of $\mathbf{q}$.

Based on the hydrodynamic theory, the authors of [190] obtained another expression for the spectral density

$$F_\perp(\mathbf{q}, \varepsilon) = \frac{1}{\pi}\frac{\gamma(\mathbf{q})(\hbar\omega(\mathbf{q}))^2}{[\varepsilon^2 - (\hbar\omega(\mathbf{q}))^2]^2 + \gamma^2(\mathbf{q})(\hbar\omega(\mathbf{q}))^2}, \tag{17.11}$$

which also renders the peaks at the energies $\pm\hbar\omega(\mathbf{q})$. When these peaks are spaced far apart, Eqs. (17.10) and (17.11) differ only slightly; differences arise only when the attenuation $\Gamma(\mathbf{q})$ in Eq. (17.10) or $\gamma(\mathbf{q})$ in Eq. (17.11) becomes comparable with the energy $\hbar\omega(\mathbf{q})$. But one must bear in mind that Eq. (17.11) is valid only at $\gamma(\mathbf{q}) << \hbar\omega(\mathbf{q})$.

The hydrodynamic theory leads to the following attenuation dependence upon the wave vector and the temperature:

$$\gamma(\mathbf{q}) = \gamma' q^{1}, \quad \gamma' \sim |\tau|^{u}. \qquad (17.12)$$

Also a temperature dependence in the form of a power function has been suggested for the constant D of spin wave rigidity:

$$D \sim |\tau|^{x}. \qquad (17.13)$$

Order parameter fluctuations, which determine the structure of the spectral density of a longitudinal correlations, are assumed to be important near T_c from the side of the phase with magnetic order. These fluctuations are expected to have diffusion character and lead to a central peak in $F_{\parallel}(\mathbf{q}, \varepsilon)$ at $\varepsilon = 0$. However, we will find out below that quasi-elastic scattering which could be ascribed to longitudinal fluctuations is not observed in experiments on many ferromagnets.

Now let us consider the behavior of $F(\mathbf{q}, \varepsilon)$ in the paramagnetic phase. Van Hove [349] has shown for the hydrodynamic region that the dynamics of the spin fluctuations has diffusion character. The magnetization $\mathbf{M}(\mathbf{r}, t)$ obeys at each point the diffusion equation

$$\frac{d\mathbf{M}}{dt} = \Lambda \Delta \mathbf{M}, \qquad (17.14)$$

where the constant Λ of spin diffusion depends upon temperature. The evolution of a plane magnetization wave with the wave vector $\mathbf{q}$ (wave given at the initial time $t = 0$) is obviously determined by the function $e^{-\Lambda q^2 t}$, and this to the following time dependence of the spin correlator:

$$\langle S(\mathbf{q}, t)\, S(-\mathbf{q}, 0)\rangle = \langle S(\mathbf{q})\, S(-\mathbf{q})\rangle\, e^{-\Lambda q^2 t}. \qquad (17.15)$$

The Fourier transformation of this expression with respect to the variable t defines the spectral density function in the form of a Lorentz function:

$$F(\mathbf{q}, \varepsilon) = \frac{1}{\pi} \frac{\Lambda q^2}{\varepsilon^2 + (\Lambda q^2)^2}. \qquad (17.16)$$

This form of the spectral distribution is also obtained in microscopic theories [256] for high temperatures. The constant Λ of spin diffusion is assumed to be temperature-dependent according to a power law:

$$\Lambda \sim \tau^{y}. \qquad (17.17)$$

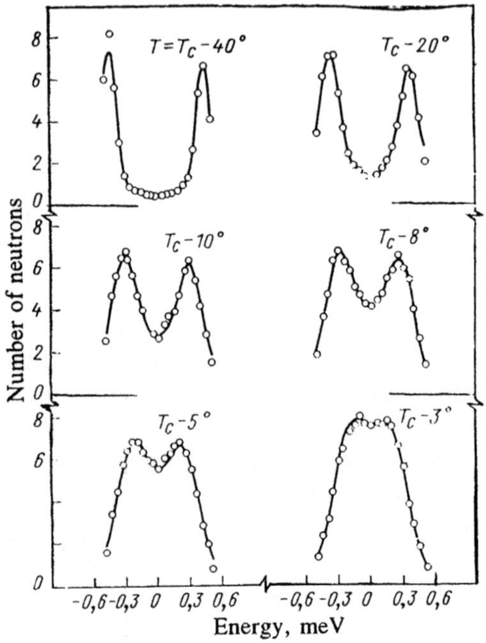

Fig. 33. Spectrum of inelastic scattering at spin waves in cobalt at $\varkappa = 0.5$ nm^{-1} for a number of temperatures below T_c [187].

We have cited the basic relations of the hydrodynamic approximation theory both above and below T_c, with $q << \varkappa_1$. In the other limit ($q >> \varkappa_1$), which corresponds to the critical region, the form of the spectral function can be established by the hypothesis of dynamic scaling (see below). First let us consider the experimental results concerning the dynamics of the spin system in a phase with order.

The dynamics of the spin system below T_c. Inelastic scattering experiments made with a three-crystal spectrometer have shown that there exist long-wave spin waves in a ferromagnetic phase very close to T_c. A typical spectrum which of inelastic magnetic scattering was obtained in the ferromagnet Co is illustrated in Fig. 33.

The solid curves were constructed by fitting to the form of the spectral Halperin–Hohenberg function with the least-square method. Two inelastic scattering peaks appear clearly up to the temperature $T_c - 3$ K. The spacing of the peaks (which is proportional to the energy (**q**) of the spin wave) decreases when T_c is approached but the width of the peaks grows, indicating increasing attenuation. The comparison of the two forms of the spectral function of Eqs. (17.10) and (17.11) show optimal agreement with the experiment in the case of Eq. (17.11). Similar spectra have been

TABLE 16. Values of the Dynamic Critical Indices Determined from Inelastic Neutron Scattering [187].

Physics parameter	Asymptote	Scaling law	Calculation	Fe	Co [187]	Ni [252]
$\Omega(q)$	$\sim q^z$	$z = \frac{5}{2} - \frac{\eta}{2}$	2,48	2,7±0,3[146] 2,8±0,3[127]	2,4±0,2	2,46±0,25
D	$\sim \lvert\tau\rvert^x$	$x = \nu' - \beta$	0,32	0,37±0,03[146] 0,36±0,03[127]	0,39±0,05	0,39±0,04
Λ	$\sim \tau^y$	$y = \frac{\nu}{\nu'}(\nu' - \beta)$	0,32	0,38±0,06[279]	—	—

obtained in Co for wave vectors $0.3 \leq q \leq 0.8\ \text{nm}^{-1}$ of the spin waves and in the temperature interval $0.002 \leq 1 - (T / T_c) \leq 0.03$. The data obtained confirm a quadratic dispersion law but the temperature dependence of the constant D of spin rigidity follows a power law (Figure 34) with the critical index $x = 0.39$. According to the hydrodynamic theory and the hypothesis of scaling, x can be expressed through the critical indices: $x = \nu' - \beta$ [190]. When the results of static scaling $\nu' = \nu$ and the ν and β values for Co are used, the value $x = 0.23$ is obtained, and this is much lower than the observed value. The discrepancy may be related to violation of the relation $\nu' = \nu$ which was not verified since critical scattering at $T < T_c$ was not observed in any of the ferromagnetic metals (scattering at the longitudinal fluctuations of the parameter of order) [187]. When the ν and β values obtained by the high-temperature expansions in the Heisenberg model are used, the result is $x = 0.32$. The values of the critical index are listed in Table 16 for D referring to the metals Fe, Co, and Ni.

The attenuation of the spin waves near T_c was intensively studied in iron [127] in the wave vector interval $1.4 \leq q \leq 2.6\ \text{nm}^{-1}$ and in the temperature interval $5.3\ \text{K} \leq T_c - T \leq 60\ \text{K}$. The dependence of the attenuation upon the wave vector is in the hydrodynamic theory given by Eqs. (17.12) according to which $\gamma_{\mathbf{q}} \sim q^4$. As mentioned above, at $T < T_c$, the spectra of inelastic scattering at Co agree with the spectral density of Eq. (17.11) in a better way than with that of Eq. (17.10). The same holds for Fe [127]. This makes it possible to use the form given by Eq. (17.11) for determining the attenuation of the spin waves. For example, the dependence $\gamma_{\mathbf{q}} \propto q^4$ was established in Fe up to wave vectors $q <$ 2.6 nm^{-1}, whereas the temperature variations of the coefficients γ' enabled the determination of the corresponding critical index which is given by the expression $\gamma' \sim |\tau|^u$. The value $u = -0.96 \pm 0.10$ which was found agrees within the experimental error limits with the value $u = -1.09 \pm 0.10$ predicted by the hypothesis of dynamic scaling [190]. The spin-wave spectrum was analyzed in detail near T_c not only for the three ferromagnetic dielectrics (or semiconductors). Thus, the corresponding results were obtained for the isotropic Heisenberg ferromagnet EuO, except for the dependence of the attenuation on the wave vector, which dependence turned out to be much weaker than q^4.

In research on inelastic scattering $T < T_c$, a diffuse central peak which could be ascribed to critical scattering at longitudinal spin components was detected in each of the ferromagnetic metals. This can be recognized in Fig. 33 for Co. Special investigations made on Fe [127] have shown that if a central peak exists, its intensity must be less than 3% of the spin-wave scattering peaks which are associated with the correlation of the transverse spin components. The reason for this is so far unclear though there exist theoretical arguments which can provide some explanation of the absence of critical scattering related to longitudinal correlations. The dynamic order parameters interact with the spin waves and this may cause a large attenuation of the same and a very broad diffusion peak [12]. However, as has been shown in the same paper [12], the spectral density of the longitudinal correlation function comprises two peaks of inelastic scattering at spin waves with the energy $\hbar\omega(\mathbf{q}) = Dq^2$. When a large part of the spectral density is concentrated at these peaks, the absence of a central peak could be accounted for. Of course, this problem still requires careful experimental and theoretical studies.

Dynamic scaling. The similarity hypothesis for static phenomena in the critical region confirms that a double correlation function $\chi(\mathbf{q})$ of the spins (susceptibility) must be a generalized homogeneous function of the wave vector $\mathbf{q}$ and of the inverse correlation length $\varkappa_1 \equiv \xi^{-1}$ so that $\chi(q) \sim q^{-2+\eta}\varphi(\varkappa_1/q)$. This conclusion is the mathematical expression of the physical idea that near T_c any length must be measured in fractions of the correlation length ξ which defines the scale of all lengths in the system.

Halperin and Hohenberg [190, 196] have expanded the hypothesis of scaling to dynamic phenomena by assuming that the correlation function must be a homogeneous function of the energy ε and of some characteristic energy $\Omega(\mathbf{q})$ which defines the energy scale in the critical region.

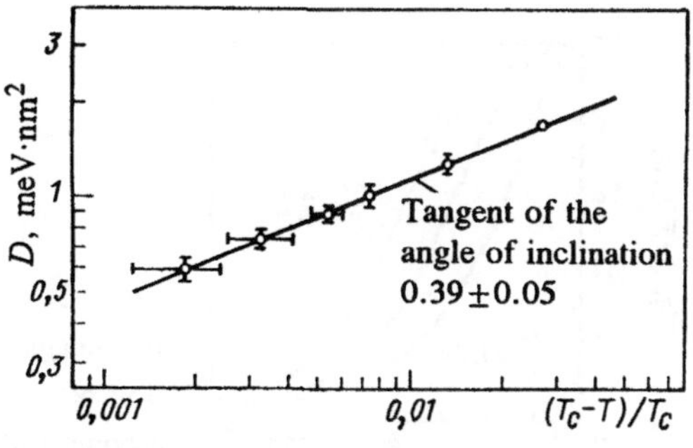

Fig. 34. Temperature dependence of the constant of spin rigidity in the case of cobalt (on a log scale) [187].

Thus, the spectral density $F(\mathbf{q}, \varepsilon)$ must be a function of the form

$$F(\mathbf{q}, \varepsilon) = F(\kappa_1/q, \varepsilon/\Omega(q)), \tag{17.18}$$

wherein $\Omega(q)$ must be defined as the width of the peak of the energy distribution with the expression

$$\int_{-\Omega(q)}^{\Omega(q)} F(\mathbf{q}, \varepsilon)\, d\varepsilon = \frac{1}{2}. \tag{17.19}$$

The second assumption of the hypothesis of dynamic scaling relates to the form of the characteristic energy Ω as a generalized homogeneous function q and $\varkappa_1$:

$$\Omega(q) = \sigma q^z f(\varkappa_1/q). \tag{17.20}$$

The number z is termed the critical dynamic index. This number defines the q dependence of ω at the transition point proper because it is assumed that $f(0) = 1$. However, in the hydrodynamic mode ($q << \varkappa_1$) at $T < T_c$, the quantity $\Omega(\mathbf{q})$ corresponds to the spin-wave energy Dq^2, with $D \sim \sqrt{\varkappa_1}$ in this limit [196]. In this limit, the result can be restated in the form $\Omega(q) \sim q^{5/2}\sqrt{\varkappa_1/q}$, from which we obtain $z = 5/2$ in a comparison with Eq. (17.20). Thus, at the transition point proper we have

$$\Omega(q) = \sigma q^{5/2} \; (T = T_c). \tag{17.21}$$

The hypothesis of scaling does not predict the form of the correlation function $F(\mathbf{q}, \varepsilon)$ and the quantity $f(\varkappa_1/q)$. They can be obtained only from microscopic calculations by numerical solution of the equations describing

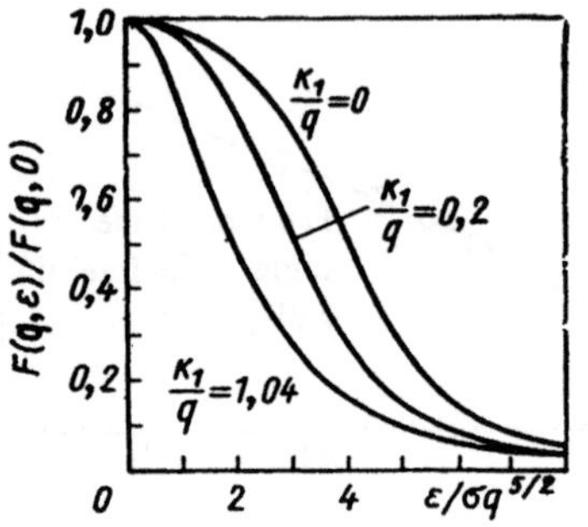

Fig. 35. Spectral function for a Heisenberg-type ferromagnet with a simple cubic lattice; the interaction between the nearest neighbors was taken into account [201].

the spin dynamics. Figure 35 illustrates the results of calculations [201] made for the isotropic Heisenberg model with nearest neighbors. The form of the normalized function $F(\mathbf{q}, \varepsilon)$ of the argument $\varepsilon/\sigma q^{5/2}$ is given exclusively by the ratio of the quantities $\varkappa_1/q$, and this by itself confirms the hypothesis of dynamic scaling [Eqs. (17.18) and (17.20)]. It follows also from the figure that when the temperature rises (increases in $\varkappa_1/q$), the spectral function is deformed and changes from a Gaussian form to a Lorentzian form.

The scaling function [289] was obtained for a Heisenberg ferromagnetic by numerical solution of the kinetic equation of Kawasaki [218] (Fig. 36). The minimum corresponds to the boundary separating the hydrodynamic conditions ($\varkappa_1/q \gtrsim 2$) from the critical conditions. A scaling function of this type is often approximated by a two-term expression:

$$f(\varkappa_1/q) = \exp[-A(\varkappa_1/q)] + B\sqrt{\varkappa_1/q}, \qquad (17.22)$$

so that the constants of fitting are $A = 1.83$ and $B = 0.43$ for the theoretical curve (dashed curve of Fig. 36), whereas considerable deviations from the theory are found with the experimental curve in the case of Co (solid line of Fig. 36): $A = 2.74$ and $B = 0.47$.

In experiments, the scaling function is defined by the width of the energy distribution at $T > T_c$; the energy distribution is determined at fixed q as a function of the temperature and, vice versa, as a function of q at a fixed temperature. Typical energy distributions are illustrated in Fig. 37 for Co. In the transition through the Curie point, the two inelastic peaks, which reflect the scattering at the spin waves, merge into a single broad peak. When T increases, the width of the peak first decreases and thereafter begins to increase again, leading to the temperature-induced

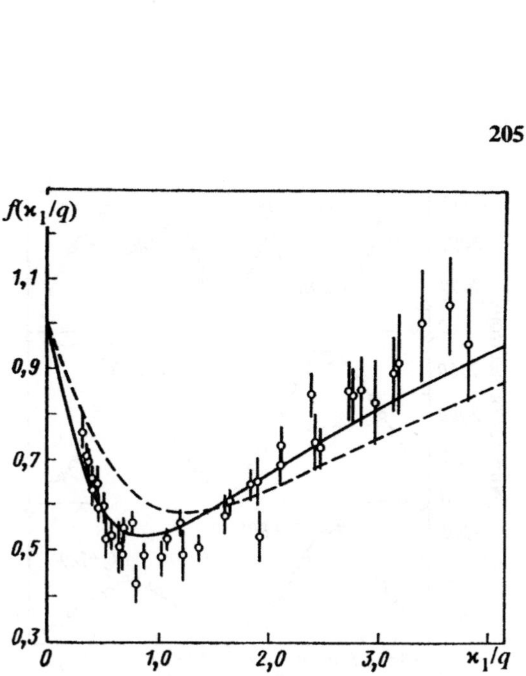

Fig. 36. The scaling function $f(\varkappa_1/q)$ for a ferromagnet ($T > T_c$) [289]; the dashed line corresponds to calculations made with the Heisenberg model; the experimental points are the results of calculations made with Eq. (17.20) and data on neutron scattering at cobalt [187].

change in the scaling function as indicated in Fig. 36. The dependence of the width of the distribution $\Omega(q)$ upon q was obtained from such curves, particularly for the case $T = T_c$ (Fig. 38). The experimental points place themselves on a straight line, and this means that $\Omega(q)$ follows a power law. The slope of the curves renders for Co the critical index $z = 2.4$ which is very close to the value $z = 2.5$ predicted by the hypothesis of scaling. The coefficient σ in the expression $\Omega(q) = \sigma q^{5/2}$ is at $T = T_c$ a measure of the inelasticity of the critical scattering. In the case of Co, $\sigma = (300 \pm 30)$ nm$^{5/2}$ meV [187], whereas in the case of Fe, $\sigma = 130$ nm$^{5/2}$ meV [127, 146, 279]; in the case of Ni, $\sigma = 330$ nm$^{5/2}$ meV [252]. This shows that the critical scattering in Ni is the most inelastic. The values of the dynamic critical indices determined from critical neutron scattering for the three ferromagnetic metals are listed in Table 16. Obviously, within the experimental errors, the dynamic critical index z agrees with the predictions of the hypothesis of dynamic scaling ($z = 2.5$) though a definitely excessive value is obtained for iron. The constant D of spin rigidity renders for the three metals values which are close but significantly exceed the calculated value of the Heisenberg model. The experimental results for the constant Λ of spin diffusion are very small. A number of experimental data obtained for Λ in other research work were not included in the table because the region of the hydrodynamic approximation in $\varkappa_1/q$ was not reached; thus, the hydrodynamic region in the case of Co and Ni must be additionally investigated. We realize that, as a whole, the available experimental data on the critical

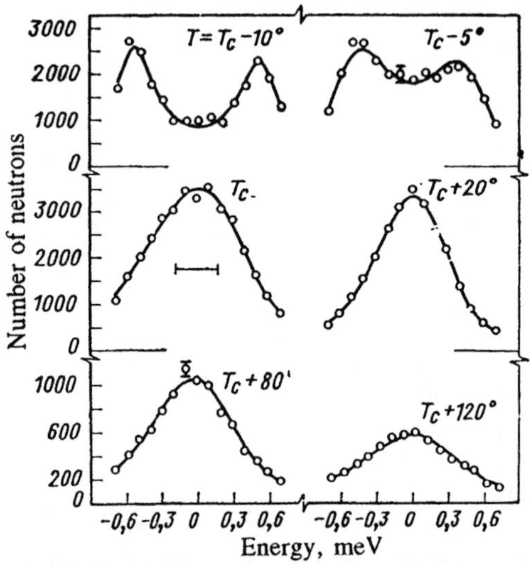

Fig. 37. Energy distribution of scattering at cobalt in the transition through the Curie point (scattering vector $\varkappa$ = 0.7 nm^{-1} [187].

dynamics of magnetic materials can be understood when one accepts the hypothesis of dynamic scaling. The same holds for the results of research on the spin dynamics in EuO [109].

The ferromagnets Fe, Co, Ni, and EuO which have been studied are crystals with a weak magnetic anisotropy. The substance MnP is an example of a highly anisotropic ferromagnet. Research on the inelastic scattering at the transition point T_c = 291 K rendered the value z = 2.57 ± 0.11 which is very close to the theoretical value z = 2.5. Thus, the anisotropy does not significantly affect the dynamics of the spin fluctuations at T_c. At the same time, the experimental dependence of the scattering cross section at $T < T_c$ (upon the spin waves) in MnP is greatly at variance when a comparison with the cubic ferromagnets is made.

Polarization effects. The above experimental data on the dynamics of spin fluctuations were obtained in research on inelastic scattering at three-crystal spectrometers. The authors of [79, 80, 88, 275, 342] have developed a new line of research on the critical dynamics by analyzing the polarization of a scattered beam of originally polarized neutrons. They used the setup described in [79], which makes it possible to perform a three-dimensional polarization analysis, i.e., to measure the cross section $d\sigma/d\Omega$ and the polarization vector **p** of neutrons scattered into a given solid angle. Though the measured quantities are integrals (over the energies), they help to obtain unique information on the spin dynamics and provide, in particular, three-particle correlations of the spins.

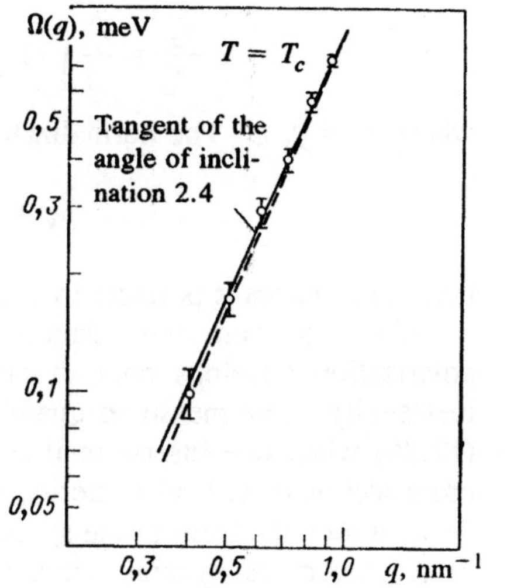

Fig. 38. Width of the energy distribution $\Omega(\mathbf{q})$ in cobalt at $T = T_c$ as a function of the wave vector **q** [187].

The method is based on the expression for the polarization vector of the neutron beam scattered at a paramagnetic crystal [70]:

$$\mathbf{p} = -\mathbf{e}\,(\mathbf{e}\mathbf{p}_0). \tag{17.23}$$

When the scattering takes place into a small angle ϑ with an energy transfer ε, approximations for the projections of the unit vector **e** of scattering follow from the energy and momentum conservation laws [70]:

$$e_y = \frac{2E_0\,\vartheta}{\sqrt{(2E_0\,\vartheta)^2+\varepsilon^2}}, \quad e_z = \frac{\varepsilon}{\sqrt{(2E_0\,\vartheta)^2+\varepsilon^2}} \tag{17.24}$$

and we obtain for the scattering vector $\varkappa$ proper:

$$\varkappa = k_0\left(\vartheta^2 + \frac{\varepsilon^2}{4E_0^2}\right)^{1/2}. \tag{17.25}$$

These expression hold for $\vartheta << 1$, $\varepsilon << E_0$, where E_0 and $\varkappa_0$ denote the energy and the wave vector of the incident neutron beam, respectively. The z axis was chosen parallel to the direction of the incident beam, whereas the y axis is on the plane of scattering.

In two experiments in which $\mathbf{p}_0$ was aligned with y and z, respectively, the polarization measured in scattering under the angle ϑ must be given by the integrals of Eq. (17.23) over the energy transferred:

$$\frac{p_\alpha}{p_\alpha^\circ} = -\int e_\alpha^2 \frac{d^2\sigma}{d\Omega d\varepsilon}\, d\varepsilon \Bigg/ \int \frac{d^2\sigma}{d\Omega\, d\varepsilon}\, d\varepsilon, \tag{17.26}$$

where $\alpha = y, z$. The normalization

$$p_y + p_z = -p^\circ, \tag{17.27}$$

holds and makes it possible to monitor the experimental conditions.

We infer from the relations (17.24) that a z component of the polarization develops only in elastic scattering and is a measure for inelasticity. The measured quantities $p_\alpha/p_\alpha{}^\circ$ can be calculated with Eq. (17.26) when one inserts in it Eqs. (17.24) and the expression for the cross section of critical scattering in the form of Eq. (17.3) or Eq. (17.6). By assuming the form of the spectral density function $F(\mathbf{q}, \varepsilon)$ on the basis of the theory of dynamic similarity, the magnitude of the polarization effect can be estimated in the critical region. It was shown in [74] that scattering events with an energy transfer many times greater than the characteristic energy $\Omega(\mathbf{q})$ of the spectral distribution are important in the analysis of the polarization with the aid of Eq. (17.26). Thus, the analysis of the polarization of the scattered neutrons provides information on the asymptotic behavior of the spectral function; this is the important distinction of such experiments over research made with three-crystal spectrometers.

The proposed method was tested on ^{57}Fe. In the interval $16' < \vartheta < 3°$ of scattering angles and the temperature interval $10^{-4} < \tau < 10^{-1}$, the angular and temperature dependencies of the experimentally determined parameter $R = p_z/(p_z + p_y)$ were carefully measured. A comparison was made with the values calculated with Eq. (17.26) where the expression of Eq. (17.20) with the scaling function $f(\varkappa_1/q)$ given in Eq. (17.22) was used as characteristic energy $\Omega(\mathbf{q})$. Processing of the experimental data with the least-square method led to the following values of the critical indices and parameters which characterize the spin dynamics:

$$\left.\begin{aligned} \nu &= 0{,}67 \pm 0{,}01; \quad z = 2{,}617 \pm 0{,}004; \\ A &= 3{,}7 \pm 0{,}3; \quad B = 1{,}013 \pm 0{,}05. \end{aligned}\right\} \tag{17.28}$$

The critical indices ν and z are in good agreement with the theoretical values and also with the experimental values obtained from measurements made with the three-crystal spectrometer (Tables 14 and 16). The parameters A and B are approximately 2 times greater than the experimental values for the Heisenberg model [289]. This difference may be associated with the fact that the critical dynamics differs at large transferred energies from the predictions of the hypothesis of dynamic scaling. For example, in the paramagnetic region at temperatures noticeably exceeding F_c, scattering at states of the type of spin waves is observed. However, processes of this type were not taken into account in Eq. (17.26) which was used to process the experimental data.

The development of a spontaneous polarization in the scattered peak is another polarization effect which can be used in research on the critical spin dynamics [57]. When no magnetic field is applied in the paramagnetic state, polarization does not develop in the first Born approximation [see Eq. (17.23)]. However, polarization appears in second-order perturbation theory and, as has been shown in [57], is related to the appearance of the three-particle spin correlator $S_1^\alpha(t_1)\ S_2^\beta(t_2)\ S_3^\gamma(t_3)\rangle$ in the expression for the cross section of scattering. The resulting polarization has the direction of the pseudovector $[\mathbf{k}_0 \times \mathbf{k}]$ which is composed of the wave vectors of the incident and scattered neutrons. In the static approximation, the triple correlator is equal to zero and polarization develops only when an energy exchange between the neutron and the scatterer takes place. The observed developing polarization provides direct information on the dynamics of the three-spin correlations.

The authors of [57] obtained theoretical expressions for the developing polarization when they used the asymptotic form of the spin Green's functions in the critical region. The critical region is usually defined by the inequality $q >> \varkappa_1$ under the assumption that all interactions between the spins, except for the exchange interactions, can be disregarded. In the near neighborhood of T_c, the long-range dipole–dipole interactions change the critical dynamics of magnetic materials [71–73] by causing anisotropy of the critical fluctuations. The dipole forces must result in a reduction of the dynamic index z, i.e., the characteristic energy of the fluctuations must increase. The dipole region (within the critical region) is characterized by the inequalities

$$4\pi\chi > 1 \text{ or } \varkappa_1 < q_0, \tag{17.29}$$

where χ denotes the static susceptibility at $q = 0$ and q_0 (and ω_0) denote the characteristic momentum (energy) of the dipole interaction:

$$q_0 = \frac{1}{a}\left(\frac{\omega_0}{kT_c}\right)^{1/2}, \quad \omega_0 = \frac{4\pi (g\mu_0)^2}{v_0} \tag{17.30}$$

(v_0 denotes the volume of a unit cell of the crystal). The exchange region, which is characterized by the inequality

$$q \gg \varkappa_1 > q_0 \tag{17.31}$$

is beyond the limits of the dipole region.

Since the asymptotes of the Green's functions differ in the exchange region and the dipole region, different values of the spontaneous polarization of the scattered beam are obtained:

in the exchange region

$$\mathbf{p} \sim \mathbf{m}\frac{r_0}{a}\frac{1}{a^2}\begin{cases} q_c^{3/2}\varkappa_1^{-1}(k_0\vartheta)^{-5/2}, & k_0\vartheta \gg \varkappa_1, \\ q_c^{-3/2}\varkappa_1^{-3/2}(k_0\vartheta), & \varkappa_1 \gg k_0\vartheta, \end{cases} \tag{17.32}$$

and in the dipole part of the critical region:

$$\mathbf{p} \sim \mathbf{m}\frac{r_0}{a}\frac{1}{a^2}\begin{cases} q_0^{1/2}(k_0\vartheta)^{-5/2}; & k_0\vartheta \gg q_0; \\ q_0^{-3/2}\varkappa_1^{-3/2}(k_0\vartheta), & \varkappa_1 \gg k_0\vartheta. \end{cases} \tag{17.33}$$

where $\mathbf{m}$ denotes the following unit pseudovector:

$$\mathbf{m} = \frac{[\boldsymbol{\kappa}_0 \times \boldsymbol{\kappa}]}{|[\boldsymbol{\kappa}_0 \times \boldsymbol{\kappa}]|}, \tag{17.34}$$

and $q_c \sim 1/a(2E_0k_0a/kT_c)^{2/3}$ denotes some characteristic momentum which depends upon the geometry of scattering. The estimates expressed by Eqs. (17.32) and (17.33) are valid for $q_0 \sim q_c$. The principal dimensionless parameter determining the order of magnitude of the polarization is $r_0/a \sim 10^{-4}$. The estimates expressed by Eqs. (17.32) and (17.33) define the dependences of the polarization upon the temperature and the angle ϑ of scattering; these dependencies can be experimentally verified.

A corresponding experiment was made on Fe [57]. It would be more convenient to measure the cross section of scattering (the intensity) ϑ of initially polarized neutrons scattered under a particular angle rather than to directly measure the spontaneous polarization of the scattered beam.

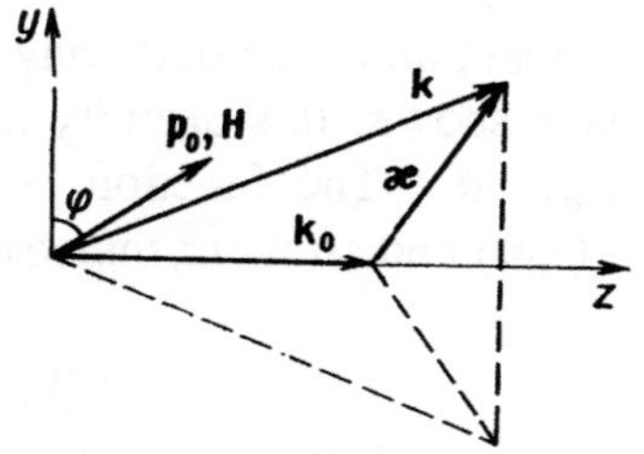

Fig. 39. Kinematic scheme of scattering at critical fluctuations [56].

Then the intensity of scattering of such a beam must be described by an expression of the form

$$I = I_0 (1 + A\mathbf{m}\mathbf{p}_0). \tag{17.35}$$

The vector **m** is perpendicular to the plane of scattering. When two measurements are made (one at $p_0 \parallel \mathbf{m}$ and the other with the inverse sign of the initial polarization $\mathbf{p}_0$), the intensity ratio (the intensities are denoted by I_+ and I_-, respectively) is

$$p = (I_+ - I_-)/(I_+ + I_-) \tag{17.36}$$

and, obviously, determines the degree of polarization developing in the scattering process.

Similar measurements were made on Fe in the temperature interval 3.5 K $< (T - T_c) <$ 35 K. The order of magnitude and the angular dependence of the effect agree qualitatively with the predictions of the theory. In particular, it was found that in critical scattering at $T - T_c =$ 6 K, the degree of spontaneous polarization averaged over the scattering angles is $p = (7.8 \pm 0.9)\cdot 10^{-5}$. Though this value is very small, the polarization was rather accurately determined as a difference effect [Eq. (17.36)], wherein nonmagnetic scattering and background were eliminated. Still another possibility of using polarized neutrons for research on the critical dynamics of magnetic materials in an external magnetic field was considered in [56, 78]. The authors suggested measuring the small-angle scattering intensity at various orientations of the magnetic field relative to the vector of the initial polarization. The kinetic scheme of the scattering is shown in Fig. 39. The polarization vector $\mathbf{p}_0$ and the magnetic field vector **H** are on the plane of scattering and are parallel to the scattering vector $\varkappa$. The intensities I_+ and I_- of scattering are measured in parallel and anti-parallel positions of $\mathbf{p}_0$ and **H**. Then Eq. (17.36) is used to determine the quantity $p = p(k_0\vartheta)$ which depends upon the quasi-elastic momentum of the scattering. As the experiment [78] and the theory [56] have shown, this quantity is asymmetric with respect to the sign of the angle ϑ. The function $p(k_0\vartheta)$ can be expanded in a symmetric part $p_s(k_0\vartheta)$ and an antisymmetric part $p_a(k_0\vartheta)$:

$$p(k_0\vartheta) = p_s(k_0\vartheta) + p_a(k_0\vartheta),$$
$$p_s(k_0\vartheta) = [p(k_0\vartheta) + p(-k_0\vartheta)]/2, \quad p_a(k_0\vartheta) = [p(k_0\vartheta) - p(-k_0\vartheta)]/2. \tag{17.37}$$

This asymmetry is related to peculiar features of the dynamics of spin fluctuations in a magnetic field. The intensity of the critical scattering of polarized neutrons in a weak magnetic field is in the first Born approximation described by the expression [56]:

$$I(\vartheta) \sim \int \frac{d\varepsilon}{\varepsilon} \frac{k}{k_0} \operatorname{Im} \{G(\varkappa, -\varepsilon) - \mu (\mathbf{e}\,\mathbf{H})(\mathbf{e}\,\mathbf{p}_0)\, G^{(3)}(\varkappa, \varepsilon)\}, \tag{17.38}$$

in which $G(\varkappa, \varepsilon)$ denotes the usual Green's function corresponding to paired correlators, and $G^{(3)}(\varkappa, \varepsilon)$ denotes a three-particle function associated with the three-particle spin correlators and with special values of the arguments such that there appears a function of $\varkappa$ and ε alone.

The factor before the three-particle Green's function can be restated in the following form for the selected geometry of scattering (see Figure 39):

$$(\mathbf{eH})(\mathbf{ep}_0) = [e_y^2 \cos^2\varphi + e_z^2 \sin^2\varphi + e_z e_y \sin 2\varphi]\, H p_0. \tag{17.39}$$

Taking into account Eq. (17.24), we note that the first two terms are quadratic in the quasielastic momentum $k_0\vartheta$ and result in a symmetric contribution to the function $p(k_0\vartheta)$, whereas the last term is linear in $k_0\vartheta$ and leads to an antisymmetric contribution. The values can be calculated for $p_s(k_0\vartheta)$ and $p_a(k_0\vartheta)$ in explicit form with the aid of Eqs. (17.38), (17.39), and (17.24). The authors of [56] used the asymptotic form of the three-particle Green's function to obtain asymptotic expressions for the quantities $p_s(k_0\vartheta)$ and $p_a(k_0\vartheta)$, with the asymptotic expression defining the angular and temperature dependences under various conditions.

Measurements were made on Fe with the following parameters: $1\text{ K} < T - T_c < 7\text{ K}$, $10 < H < 600$ Oe, $(0.075 < k_0\vartheta < 0.7)\ \text{nm}^{-1}$.

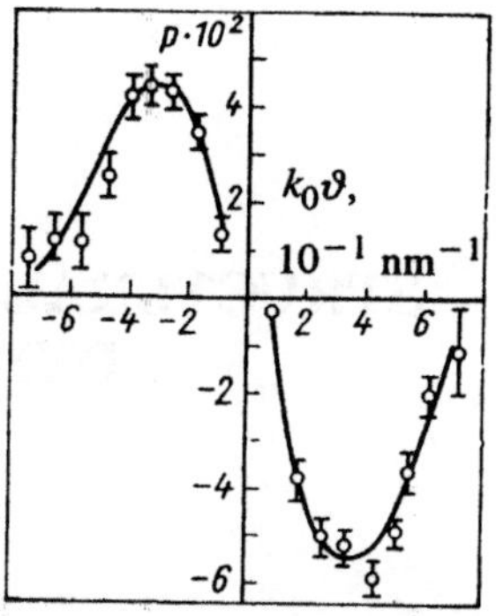

Fig. 40. Dependence of the quantity $p = p(k_0\vartheta)$ upon the quasielastic momentum $k_0\vartheta$ at $\tau = 3.8 \cdot 10^{-3}$, H = 440 Oe, and $\varphi = 20°$ [56].

Figure 40 depicts one of the typical dependences $p(k_0\vartheta)$ at $\tau = 3.8 \cdot 10^{-3}$ and H = 440 Oe; it is recognized that $p(k_0\vartheta)$ reaches $(4\text{–}6) \cdot 10^{-2}$ in the maximum at $k_0\vartheta \approx 0.35$ nm^{-1} and is clearly asymmetric. It was detected in experiment that in weak fields, the asymmetry of scattering (asymmetry characterized by the quantity p_a) increases proportionately to the field strength [this follows from Eq. (17.38)], whereas the increase in asymmetry is reduced at H larger than some critical value H_{crit}. The experimental H_{crit} values determined by extrapolation turned out to be the same for scattering vectors $k_0\vartheta$ ranging from 0.75 to 4.5 nm^{-1}, and the temperature dependence follows a power law $H_{\text{crit}} \sim \tau^{\mu}$ with $\mu = 1.8 \pm 0.2$. This value of the index μ is in good agreement with the theoretical value $\mu = 5/3$, which can be obtained from the following considerations. The critical value H_{crit} is given by the condition that the energies of spin interaction with the magnetic field and the characteristic energy of the critical fluctuations must be equal [see Eq. (17.20)]:

$$\mu H_{\text{crit}} \sim \varkappa_1^{\,z}. \tag{17.40}$$

We obtain $H_{\text{crit}} \sim \tau^{5/3}$ with $\varkappa_1 \sim \tau^{2/3}$ and $z \approx 5/2$.

Thus, the asymmetry observed in the above-described geometry of scattering has a satisfactory theoretical explanation. In principle, research on the temperature and angular dependences of the observed effect can provide information on the three-particle spin correlations [56]. This asymmetry of scattering in a magnetic field is to be preferred over the above-described effect without field, with the effect manifesting itself only in the second Born approximation. The asymmetry effect in the presence of a field is of the order of magnitude of about 10^{-2}, whereas the asymmetry effect without a field is of the order of about 10^{-4}.

Chapter 3

STRUCTURAL PHASE TRANSITIONS AND THE DYNAMICS OF CRYSTALS

We develop in this chapter the basic theoretical ideas about the structural transitions of the distortion type and also about order–disorder transitions and the results of research made with neutron spectroscopy techniques. The theoretical basis of analyzing the scattering cross section at a soft mode was compiled in Section 6. We consider in the present chapter various types of structural phase transitions in crystals, namely in cubic perovskites, terbium molybdate, Nb_3Sn, $NaNO_2$, hydrogen-containing compounds of the type of KH_2PO_4, etc. The investigations of soft modes and of the critical dynamics in structural phase transitions are reviewed. The problem of the central peak in the fluctuation spectrum of the order parameter is discussed. The experimental results are presented on the basis of the theory of symmetry and the modern ideas of phase transitions of the second kind.

18. Structural Phase Transitions in Perovskites

Symmetry analysis of the structural transitions. Crystals with perovskite structure were selected in Section 4 as an example for the application of the theory of symmetry of crystal-lattice oscillations. The phonon modes were calculated for a number of symmetry points and symmetric directions of the Brillouin zone. We use in the present section

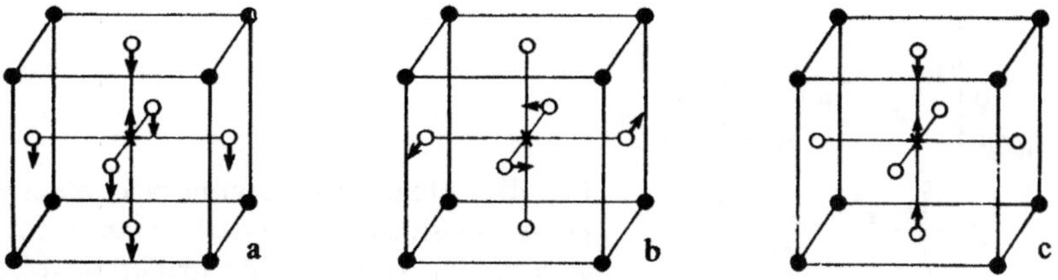

Fig. 41. Displacements of the atoms of the unit cell of ABX_3 crystals with perovskite structure; displacements observed in structural phase transitions.

the previous calculations for analyzing the atomic displacements which arise in numerous structural phase transitions in such crystals.

The unit cell of an ABX_3 crystal with perovskite structure is illustrated in Figure 2 where it is noted that the X atoms occupy the 3(*c*) positions and the A and B atoms the 1(*a*) and 1(*b*) positions of the space group $O_h{}^1$. The atomic displacements which are observed within the unit cell of initial perovskite crystals are indicated in Figure 41. We cannot infer form Figure 41*a*, *b* whether the unit cell is preserved in a phase transition, but in the last case (Figure 41*c*) the unit cell is obviously enlarged. In each of the three cases, the X atoms, which form an octahedron, are shifted either relative to the central atom of the cell or relative to the atoms at the apices of the cube. The distortion structure of the entire crystal can be understood when the atomic displacements in the octahedra alone are plotted. Thus, the following four types of elementary structural distortions in perovskites develop (Figure 42).

In case *a*, the phase transitions take place while the unit cell is preserved [this corresponds to the wave vector $\mathbf{k} = 0$ of the superlattice or to the Γ of the Brillouin zone (see Figure 3)]. The associated mode of the atomic displacement is polar, i.e., electric dipole moments are possible. Case *b* represents the antipolar mode and corresponds to a doubling of the unit cell in the *x* and *y* directions. The superlattice is described by the wave vector $\mathbf{k} = 1/2(\mathbf{b}_1 + \mathbf{b}_2)$ (point **M**), and the primitive cell has the dimensions $(\sqrt{2}, \sqrt{2}, 1)a$, where *a* denotes the parameter of the initial cubic lattice. Obviously, case *c* corresponds to a doubling of the cell in the *x* and *y* directions and, as in the preceding case, is described by wave vectors **k** ending at the point **M**. The last case *d* describes a superlattice which is characterized by doubling of the cell in each of the three directions; this corresponds to the wave vector $\mathbf{k} = 1/2(\mathbf{b}_1 + \mathbf{b}_2 + \mathbf{b}_3)$ (point **R**), with the primitive cell having the dimensions $(\sqrt{2}, \sqrt{2}, 2)a$. The atomic displacements indicated in Figure 42

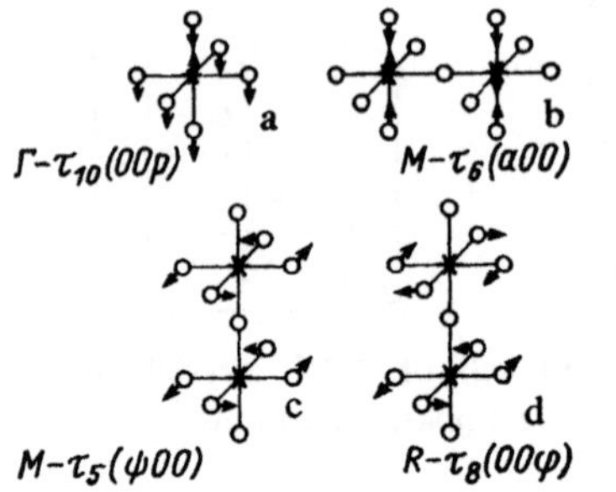

Fig. 42. Atomic displacement types observed in phase transitions in crystals with perovskite structure (the symbols of the irreducible representations of the group O_h^1, which describe the corresponding vibration modes of the initial crystal, are indicated).

reduce the symmetry of the initial crystal, and the dyssymmetric phases of reduced symmetry are described by the space groups C_{4v}^1, C_{4h}^7, D_{4h}^5, and D_{4h}^{18}, respectively. Structure distortions of this type have been detected, particularly in the following perovskites:

$$\begin{aligned} &\text{a: } BaTiO_3,\ PbTiO_3,\ KNbO_3;\\ &\text{b: } Na_{0,8}WO_3,\ PrAlO_3,\ CsPbCl_3;\\ &\text{c: } KMnF_3;\\ &\text{d: } SrTiO_3,\ LaAlO_3,\ RbCaF_3,\ KMnF_3. \end{aligned} \tag{18.1}$$

The structural distortions usually appear in a phase transition of the second kind of compensation of the soft modes indicated in Figure 42. The symbol of the irreducible representation of the particular phonon mode is indicated under each of the drawings in Figure 42; the symbol is determined by comparing the displacements of Figure 42 with Tables 1, 2, and 3 of the basis functions of the mechanical representation of the perovskite crystal for the points **Γ**, **M**, and **R** of the Brillouin zone [185].

The representation τ_{10} for the point **Γ** and τ_8 for the point **R** are three-dimensional, and the corresponding stars are single-arm stars. The small representations τ_6 and τ_5 for the point **M** are one-dimensional but since the star **M** is a three-arm star, the total representation of the group O_h^1 is three-dimensional. Thus, all four types of soft modes illustrated in Figure 42 belong to the three-dimensional representations of the group O_h^1. In addition to the symbols of the irreducible representation, Figure 42 shows the displacement coefficients denoted by the letters p and a for polar and antipolar modes and ψ and φ for the modes described by bending of the octahedron around the z axis [105, 106]. The quantities φ and ψ are very

TABLE 17. Structural Phase Transitions in Perovskites; Participation of Soft Modes in the Form of Bending [106]

Crystal	T_1, K	Фаза	T_2, K	Phase	T_3, K	Phase
$SrTiO_3$	110	00φ	—	—	—	—
$LaAlO_3$	700	$\varphi\varphi\varphi$	—	—	—	—
$SmAlO_3$	2100	$\varphi\varphi\varphi$	1100	$\varphi\psi\varphi$	—	—
$PrAlO_3$	1320	$\varphi\varphi\varphi$	205	$\varphi 0 \varphi$	151	$\varphi_1 0 \varphi_2$
$Na_{0.8}WO_3$	433	00ψ	343	$\psi\psi 0$	290	$\psi\psi\psi$
$KMnF_3$	185	00φ	91	$00(\varphi+\psi)$	—	—
$CsPbCl_3$	320	00ψ	314	$\varphi 0 \psi$	310	$\varphi_1\varphi_2\psi$
$CsPbBr_3$	403	00ψ	361	$\varphi\psi\varphi$	—	—

small (on the order of 10^{-3}) and they are therefore obviously proportional to the angle of bending.

The structural distortions indicated in Figure 42 correspond to condensation of a pure soft mode belonging to the three-dimensional representation. The condensation of two or three soft modes of a particular representation leads to a further decrease in symmetry. Thus, the following chains of phase transitions can result from successive condensations [105]:

$$\Gamma-\tau_{10}: O_h^1 \xrightarrow[(p00)]{} C_{4v}^1 \xrightarrow[(pp0)]{} C_{2v}^{14} \xrightarrow[(ppp)]{} C_{3v}^5;$$

$$M-\tau_6: O_h^1 \xrightarrow[(a00)]{} D_{4h}^7 \xrightarrow[(aa0)]{} D_{4h}^{18} \xrightarrow[(aaa)]{} T_d^3;$$

$$M-\tau_5: O_h^1 \xrightarrow[(00\psi)]{} D_{4h}^5 \xrightarrow[(\psi\psi 0)]{} D_{4h}^{17} \xrightarrow[(\psi\psi\psi)]{} T_h^5;$$

$$R-\tau_8: O_h^1 \xrightarrow[(00\varphi)]{} D_{4h}^{18} \xrightarrow[(\varphi\varphi 0)]{} D_{2h}^{28} \xrightarrow[(\varphi\varphi\varphi)]{} D_{3d}^6. \qquad (18.2)$$

Some of these chains were indeed observed in real crystals. The succession of phase transitions developing in $PrAlO_3$ when the temperature is decreased is a good example (Table 17).

Besides the phases obtained by mixing of the modes indicated in (18.2), in principle there may develop phases corresponding to the types $(\psi_1\psi_2 0)$, (p_1, p_2, p_3), etc. of mixing. They have even lower symmetry than phases of the type $(\psi\psi 0)$, (ppp), etc. Of course, the theory of symmetry does not provide conclusions about the relative stability of the phases and we must therefore return to experiments.

We also note that when the temperature (or the pressure) changes, simultaneous condensation of two or more soft modes with different

TABLE 18. Types of Distortions and Space Groups of the Phases Observed in the System (Na, K)NbO_3 [106].

Bending of octahedron		Polar phase	Resulting phase		Bending of octahedron		Polar phase	Resulting phase	
			G	Cell volume a^3				G	Cell volume a^3
00ψ	D_{4h}^5	—	D_{4h}^5	$\sqrt{2}\cdot\sqrt{2}\cdot 1$	$0\varphi\psi$	D_{2h}^{17}	$p_1 p_2 0$	C_s^1	$2\cdot 2\cdot 2$
$\varphi 0\psi$	D_{2h}^{17}	—	D_{2h}^{17}	$2\cdot 2\cdot 2$	00ψ	D_{4h}^5	$00p$	C_{4v}^2	$\sqrt{2}\cdot\sqrt{2}\cdot 1$
$\varphi\psi_1\psi_2$	D_{2h}^{13}	—	D_{2h}^{13}	$2\cdot 2\cdot 2$	$0\psi 0$	D_{4h}^5	$p_1 0 p_2$	C_s^1	$\sqrt{2}\cdot 1\cdot\sqrt{2}$
$\varphi\psi_1\psi_2$	D_{2h}^{13}	$p00$	C_{2v}^1	$2\cdot 2\cdot 2$	—	O_h^1	$00p$	C_{4v}^1	$1\cdot 1\cdot 1$
$\varphi_1\psi\varphi_2$	C_{2h}^2	$p_1 0 p_2$	C_s^1	$\sqrt{2}\cdot 2\cdot\sqrt{2}$	—	O_h^1	$00p$	C_{2v}^{14}	$1\cdot 1\cdot 1$
$\psi_1\psi_2 0$	D_{2h}^{25}	$00p$	C_{2v}^{20}	$2\cdot 2\cdot 2$					

symmetries is possible. The space groups of the phases developing from the superposition of condensed soft modes of bending at the points **R** and **M** were theoretically determined in [1, 105, 106]. The experimental results of investigations of structural transitions in which soft modes of this type participate are listed in Table 17.

Occasionally one observes phases developing in the superposition of condensed modes of bending with polar modes [see Figures 41*a*, *b*)]. An outstanding example is the system (Na, K)NbO_3 in which all phases listed in Table 18 were observed at various ratios of Na and K concentrations and also at various temperatures. We have not cited all the experimental data on the types of the distortions observed in perovskites (additional information and references can be found in the review [106]), but the results cited show that the patterns of structural phase transitions in these crystals are very dissimilar. However, the relation between the observed phases can be precisely established with the theory of symmetry and any phase can be explained as condensation of one or several soft modes indicated in Figure 41.

Ferro-distortion soft modes. Soft phonon modes were experimentally studied on many perovskites undergoing structural phase conversions. Comprehensive data can be found in review papers [6, 28, 155, 307, 304]; the authors of [6, 28] list the perovskite crystals in which soft modes were studied with the aid of neutron scattering or light scattering. These investigations have confirmed the general pattern of phase transitions involving the soft-mode mechanism in this class of structure.

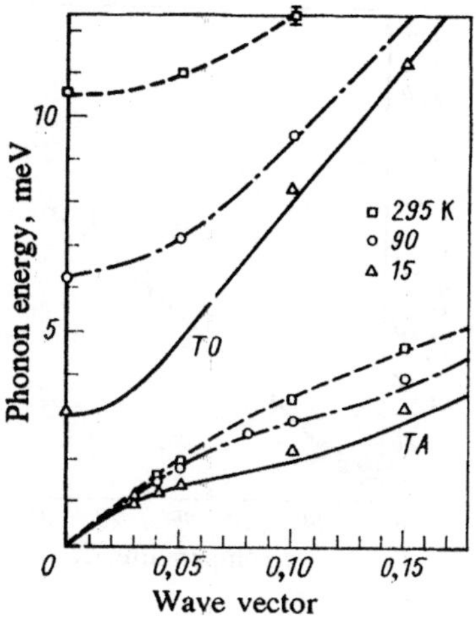

Fig. 43. Acoustical (TA) and optical (TO) transverse phonon branches in the case of $KTaO_3$ [112].

We will not fully expound these results but report only on a few examples illustrating the basic ideas of the soft-mode concept (see [145]).

It has been shown that a lattice instability of the cubic O_h^1 phase relative to the phonon mode $\Gamma - \tau_{10}$ can be observed in a number of crystals ($KTaO_3$, $PbTiO_3$, $BaTiO_3$, etc.). It follows from Figure 41 that this optical mode describes the movement of atoms of the center of the octahedron and of its apices in the directions of the edges of the cube. Condensation of this mode at some temperature T_c implies a structural phase transition which is accompanied by the appearance of a ferroelectric phase; therefore the described soft mode is occasionally termed a ferro-distortion mode.

In $KTaO_3$ crystals, which do not experience a structural transition, a soft long-wave mode representing a transverse phonon in the initial cubic phase is observed. The dispersion curves of the acoustic and optical branches are illustrated in Figure 43 for various temperatures [112]. An abnormal reduction of the frequency of the optical mode with decreasing temperature is clearly visible. (The acoustic mode is smeared by the interaction with the optical phonons.) This mode belongs to the irreducible representation $\Gamma - \tau_{10}$ (see Figure 42, scheme *a*). The temperature dependence of the square of the frequency of the soft mode is illustrated in Figure 44 which shows that the temperature dependence has the classical behavior $\omega^2 \sim (T - T_c)$ in a wide temperature interval. The dashed line on the same figure represents the temperature dependence of the inverse dielectric constant ε^{-1} determined from optical data. The proportionality $\varepsilon^{-1} \sim \omega^2$ holds and reflects the relation between the

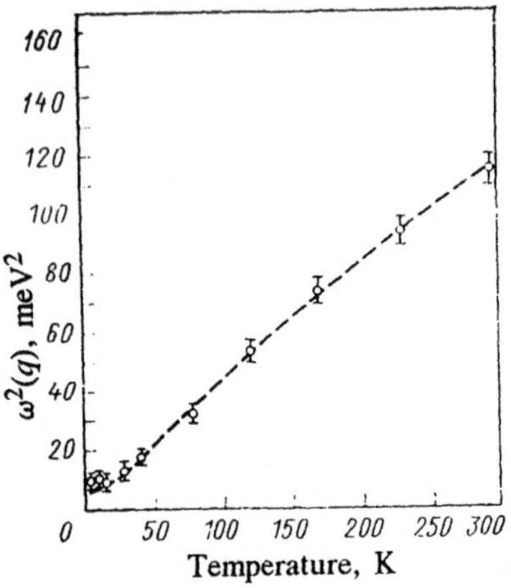

Fig. 44. "Soft" optical mode in $KTaO_3$ [112] [experimental data of the measured frequencies (O) and the inverse dielectric constant (dashed line)].

dielectric properties and the phonon properties of the crystal. This relationship is a special case of the well-known relation [143]

$$\frac{\varepsilon(0)}{\varepsilon(\infty)} = \prod_{\nu} \frac{\omega_{\nu}^2(\mathrm{LO})}{\omega_{\nu}^2(\mathrm{TO})}, \tag{18.3}$$

which relates the dielectric constants in the static limit and in the limit of high frequencies (much greater than all phonon frequencies) with the frequencies of the optical vibrations. It follows from Eq. (18.3) that if one of the transverse optical frequencies tends to zero, the dielectric constant $\varepsilon(0)$ must diverge. This behavior is corroborated by Figure 44.

Similar soft modes were also detected in perovskites undergoing ferrodistortion transitions: $PbTiO_3$ [317], $KNbO_3$ [274], $BaTiO_3$ [192], etc. Neutron experiments have shown that the soft mode is over attenuated in $BaTiO_3$ and $KNbO_3$ and that the attenuation is highly anisotropic; weak attenuation occurs in $PbTiO_3$. A strong relationship between the soft optical mode and the acoustical phonons was also observed.

Antiferro-distortion soft modes. $SrTiO_3$ was most extensively studied among the perovskites experiencing a structural phase transition with $\mathbf{k} \neq 0$. It turned out that the transition at $T_c = 105$ K is related to an instability of the phonon at the boundary of the Brillouin zone (point **R**) [158].

Figure 45 shows the spectra of inelastic neutron scattering at $SrTiO_3$ above and below T_c. Obviously, when the temperature decreases, the frequency of a phonon with **q** at the point **R** decreases in proportion to the

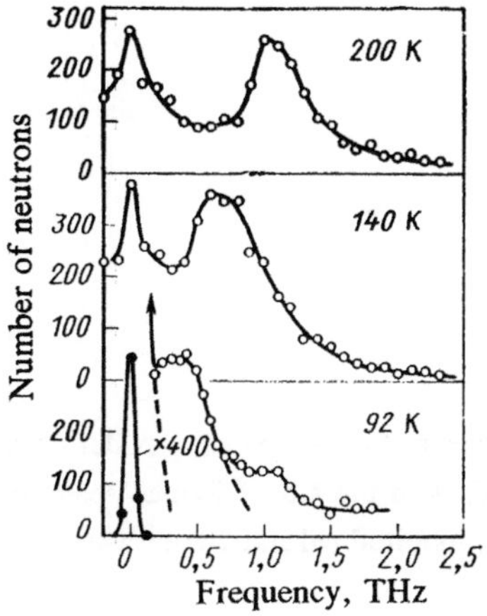

Fig. 45. Spectra of inelastic scattering of neutrons at $SrTiO_3$; the spectra correspond to the absorption of phonons with the wave vector $\mathbf{q} = 2\pi/a(1/2\ 1/2\ 1/2)$; the Bragg superlattice peak below T_c = 110 K is indicated [158].

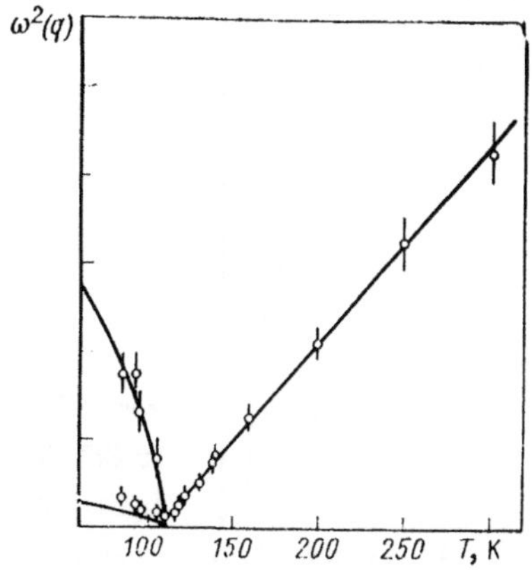

Fig. 46. Temperature dependence of the "soft" mode (at the point **R**) above and below T_c in the case of $SrTiO_3$ [158].

approach to T_c. The soft-mode frequencies derived from this data are illustrated in Figure 46 for points above and below T_c [158]. The triple-degenerate frequency in the initial phase ($T > T_c$) (frequency belonging to the three-dimensional irreducible representation $\mathbf{R} - \tau_8$) is in the dyssymmetric phase split into two frequencies forming a doublet and a singlet. Figure 47 shows for comparison analogous data obtained with $SrTiO_3$ at $T < T_c$ with the aid of Raman scattering of light [180]. The two methods of studying soft modes, namely neutron spectroscopy and optical spectroscopy, give equivalent information.

A similar behavior of the soft modes near a structural transition is also observed in $LaAlO_3$. In both cases, the phonon mode is smeared near the point **R**, and the condensation of the phonon mode leads to a doubling of the cubic cell in all three directions. The structural transitions in $SrTiO_3$ and $LaAlO_3$ differ in regard to the dyssymmetric phases developing at $T < T_c$. In the first case, a tetragonal phase with the space group D_{4h}^{18} develops, whereas in the second case a trigonal phase with the space group D_{3d}^{6} develops. This difference is explained by the fact that one of the triple-degenerate modes belonging to the representation $\mathbf{R} - \tau_8$

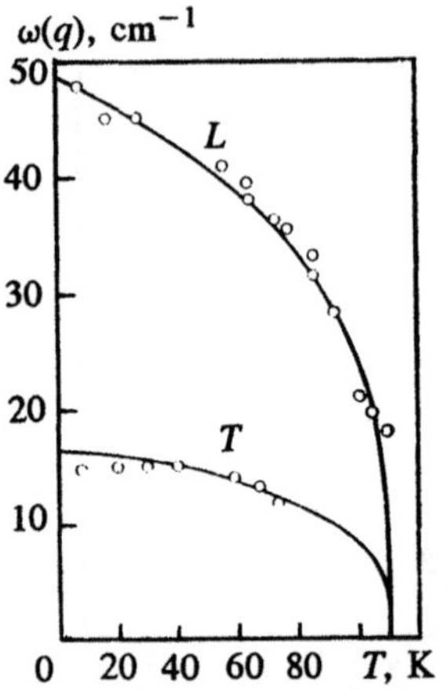

Fig. 47. Frequencies of the two phonon modes observed in $SrTiO_3$ below T_c with the aid of Raman scattering of light (L denotes the longitudinal and T the traverse optical modes [180]).

condenses in $SrTiO_3$ [mixing coefficient of the (00φ) mode], whereas all three modes condense in $LaAlO_3$ with the same weight [mixing coefficient of the $(\varphi\varphi\varphi)$ mode]. As a consequence, different space groups $D_{4h}{}^{18}$ and $D_{3d}{}^{6}$ develop, as could be expected from the group-theoretical scheme [see the last line in (18.2)]. Thus, the phase transition in $SrTiO_3$ can be represented as the result of a rotation of the TiO_6 octahedra around one of the axis of the octahedron over an angle φ, whereas in $LaAlO_3$ a simultaneous bending over the angle φ occurs around all three axes of the octahedron; this is equivalent to a bending around the three-dimensional diagonal of the cube over the same angle

A phase transition of the second kind occurs in both crystals, and the angle φ, which is the parameter of order, changes smoothly with decreasing temperature. Figure 48 illustrates the temperature dependence which was obtained for the angle φ from EPR measurements [266], though this dependence could also be obtained from measurements of the temperature dependence of the neutron reflections of the superlattice.

An example of a structural transition resulting from condensation of the two modes $\mathbf{R} - \tau_8$ with a mixing coefficient $(\varphi\varphi 0)$ was not found among the perovskites. According to the scheme of Eq. (18.2), this would lead to an orthorhombic distortion with the space group $D_{2h}{}^{28}$. Though the pure concepts of symmetry would allow such a phase, it turns out to be thermodynamically unstable when one considers the free energy of the crystal in which the structural distortion described by phonons with the symmetry $\mathbf{R} - \tau_8$ occurs. The results of such an analysis, which was made in Section 6, are depicted in Figure 4 where the stability regions of

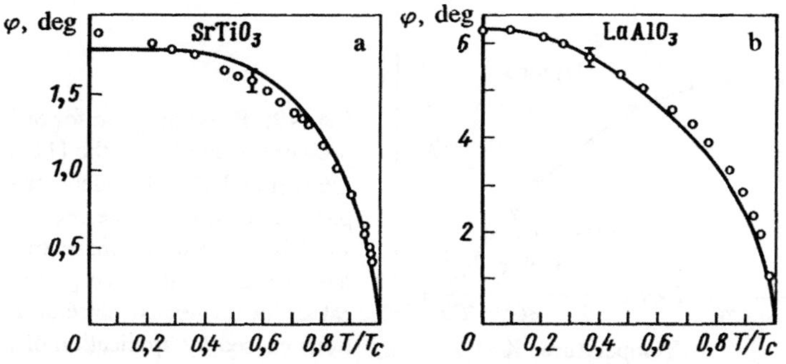

Fig. 48. Temperature dependence of the angle of rotation of the octahedra below T_c in $SrTiO_3$ and $LaAlO_3$ [266] (the solid lines correspond to the theory of [176]).

the tetragonal (00φ) and the trigonal ($\varphi\varphi\varphi$) phases are outlines. The rhombohedral ($\varphi\varphi 0$) phase indeed proved to be unstable [340].

It was determined from the experimental data of [180] that $u_2 \approx u_1$ in $SrTiO_3$, with $u_2 > u_1$. This means that on the (u_2, u_1) plane, points corresponding to the crystal $SrTiO_3$ are near the boundary of the stability region of the tetragonal phase. Consequently, when the external parameters change, e.g., when pressure is applied, a transition to the trigonal phase may occur in this crystal. This was experimentally confirmed: the phase diagram for $SrTiO_3$ is shown in Figure 49 on the pressure-temperature plane. The trigonal phase can be obtained from both the cubic phase (by a transition of the second kind) and the tetragonal phase (transition of the first kind).

These experimental data on the structural phase transitions in perovskites confirm convincingly the concept of a soft mode and can be fully described by the Landau theory of the symmetry of phase transitions of the second kind. However, near T_c, the usual deviations from the Landau theory are observed in the special section 21 which deals with the critical phenomena in structural phase transitions.

19. Other Structural Transitions of the Displacement Type

Nonintrinsic Ferroelectrics of the $Gd_2(MoO_4)_3$ type. Structural phase transitions in two isomorphous compounds, the rare-earth molybdates $Gd_2(MoO_4)_3$ and $Tb_2(MoO_4)_3$, have been thoroughly studied. One

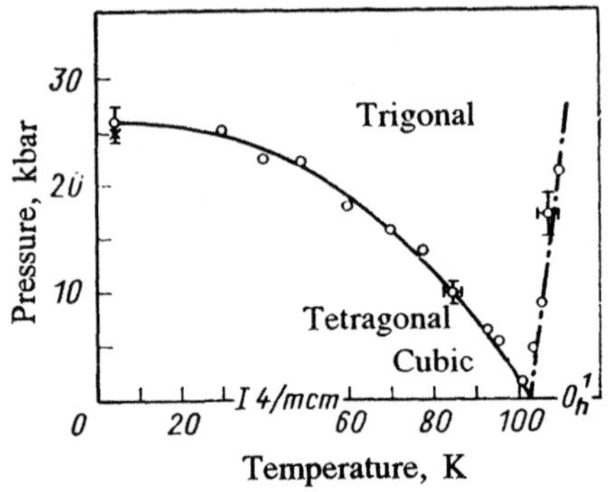

Fig. 49. Phase diagram for $SrTiO_3$ when pressure is applied in the [111] direction [265] (the dashed line denotes the line of phase transitions of the second kind; the solid line relates to phase transitions of the first kind; the space groups are indicated for neglected structural deformations caused by application of pressure).

observes in them a transition from the tetragonal phase to the orthorhombic phase with a change in symmetry: $P\bar{4}2_1m \rightarrow Pba2$.
This transition occurs in the compounds at $T_c \approx 432$ K. The observed change in the symmetry corresponds to an increase in the volume of the unit cell of the orthorhombic phase by a factor of 2. It is easy to show that if this structural transition occurs via the soft-mode mechanism, the phonon mode with the wave vector $\mathbf{q} = 2\pi/a(1/2\ 1/2\ 0)$ must condense at $T \leq T_c$, with the wave vector being situated on the boundary of the Brillouin zone of the tetragonal crystal (point **M**).

It is very difficult to study the phonon spectrum in such a complicated crystal because two structural units, i.e., 34 atoms, are in the unit cell of the initial tetragonal phase and this leads to 102 phonon branches. The problem can be simplified when the MoO_4 group is considered a rigid unit; then the number of phonon branches is reduced to 48, but this still poses a difficult problem for an experimental study.

The authors of [164, 111] carefully studied the phonon spectrum of the $Tb_2(MoO_4)_3$ crystal by inelastic neutron scattering for the purpose of establishing soft modes. Since soft modes were expected in the neighborhood of the point **M**, the low-frequency modes in the various directions of the Brillouin zone, starting from the point **M**, were considered. The dispersion curves which were measured in a wide temperature range above T_c are illustrated in Figure 50 which shows that one of the phonon modes is indeed a soft mode at the point **M**. The typical neutron groups used for obtaining dispersion curves are illustrated in Figure 51 where the evolution from a weakly attenuated phonon to an overattenuated phonon can be recognized. In the case of an overattenuated phonon, i.e., near the temperature of the phase transition, the temperature dependence of the

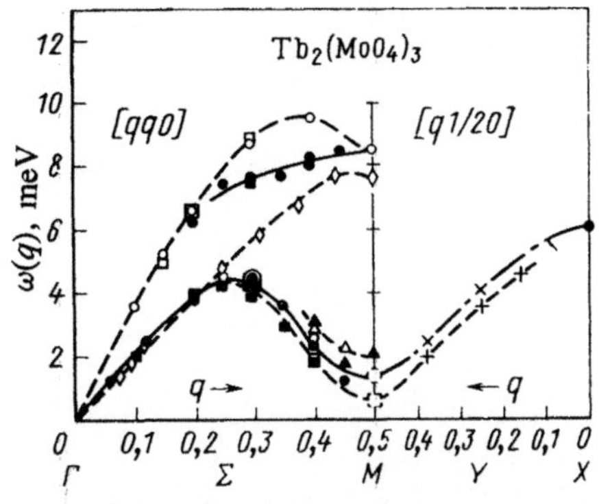

Fig. 50. Some of the dispersion curves for soft phonons in $Tb_2(MoO_4)_3$ [164]. The measurements were made at the temperatures 260°C (○, ●, ×), 184°C (□, ◇, +), and 400°C (▵).

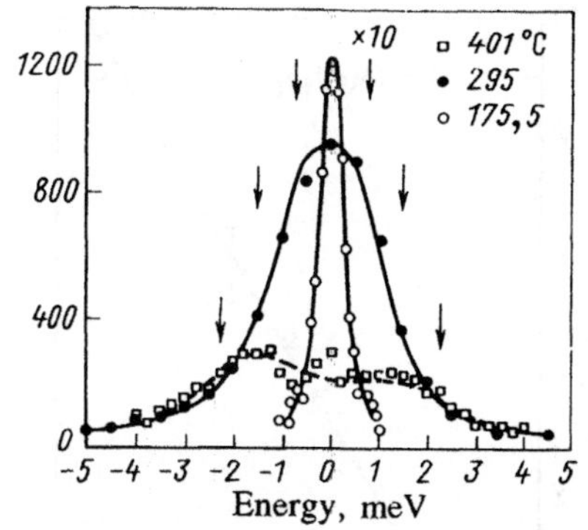

Fig. 51. Spectrum of the neutrons scattered at $Tb_2(MoO_4)_3$ at the point $\mathbf{q} = 2\pi/a(9/2\ 5/2\ 0)$ at various temperatures [111].

frequency of the soft mode was determined from the temperature dependence of the integral intensity of the inelastic peak [111]. The square of the frequency of the soft mode at the point **M** depends linearly upon the temperature in accordance with the result of the theory of an average field (Figure 52).

The authors of [164] considered the special problem of determining the oscillation mode proper for the crystal $Tb_2(MoO_4)_3$ in the symmetric phase as this relates to a soft mode (at the point **M**) whose frequency changes with the temperature as illustrated in Figure 52. The phonon mode will be given when for each atom of the primitive crystal cell the vector of the displacement from the equilibrium position (at a given temperature) can be indicated. In the case of a soft mode we have $\omega(q) << kT_c$ in the neighborhood of the transition point and therefore the intensity (integral over the energies) of the inelastic peak is given by Eq. (6.34). Equation (6.34) can be used to determine the structure amplitudes $\xi(\varkappa)$ from measurements of the $d\sigma/d\Omega$ values for a number of scattering vectors corresponding to scattering at a particular phonon; the general expression (2.22) can be used to determine from the structural amplitudes the unknown atomic displacements in the mode, i.e., the polarization vectors $\mathbf{e}_j^2(\mathbf{q}\,|\,i)$. In the case of scattering at the $Tb_2(MoO_4)_3$ crystal, the intensities of 68 phonon peaks with the indices $(h + 1/2,\ k + 1/2,\ 0)$ were measured; these peaks correspond to scattering at a phonon at the

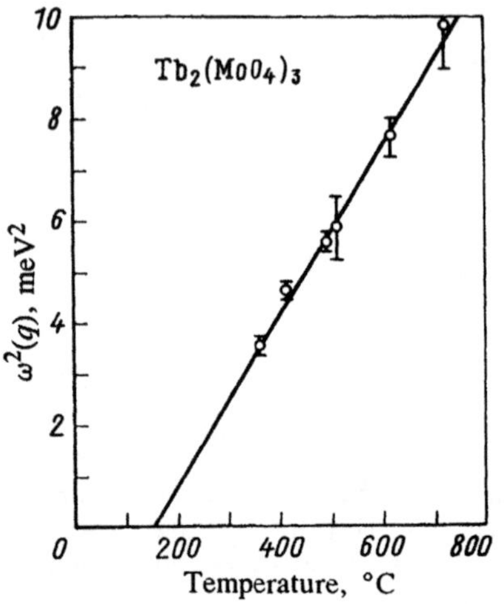

Fig. 52. Temperature dependence of the square of the frequency of the soft phonon at the point **M** in the $Tb_2(MoO_4)_3$ crystal [164].

point **M**. The numerical values of the polarization vectors of the mode could be determined by data evaluation with the least square method and the group theoretical analysis of the phonon modes in this crystal. The relative values of these displacements are temperature-independent and can be compared with the static atomic displacements observed in these crystals below the structural phase transition ($T < T_c$) with x-ray analysis [213].

Figure 53 shows the projection of the unit cell of the rhombic $Tb_2(MnO_4)_3$ crystal upon the (a, b) plane. The crystal is composed of octahedra (formed by MoO_4 groups with molybdenum atoms at the center) whose apices are aligned with the c axis (or opposite to it) or are on the (a, b) plane. The dashed line renders the positions of these octahedra in the rhombic phase (at $T < T_c$). The lattice parameters in this phase are: $a = 1.035$ nm, $b = 1.038$ nm, and $c = 1.066$ nm so that the rhombic distortion of the initial tetragonal crystal is small. The subsequent Figure 54 shows how the unit cell of the rhombic phase is obtained from the unit cell of the initial tetragonal phase. (The lattice parameter is $a_{\text{tetr}} = 0.737$ nm so that $a_{\text{rhomb}} \approx \sqrt{2}a_{\text{tetr}}$). It can be inferred from Figure 53 that the structural transition in this crystal can be described by slight twisting of the octahedra but this pattern of the atomic displacements is obtained only in a first approximation. Apart from the twisting of the octahedra, a displacement of the rare-earth ions and an overall distortion of the tetragonal cell to the rhombic type take place.

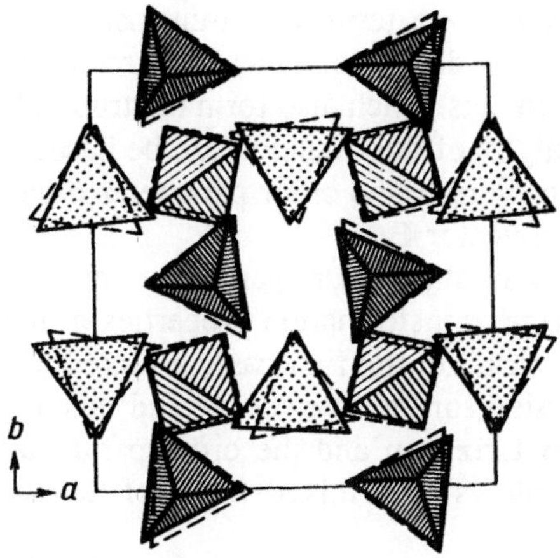

Fig. 53. Projection of the unit cell of the rhombic phase of the $Tb_2(MoO_4)_4$ crystal upon the a, b) plane.

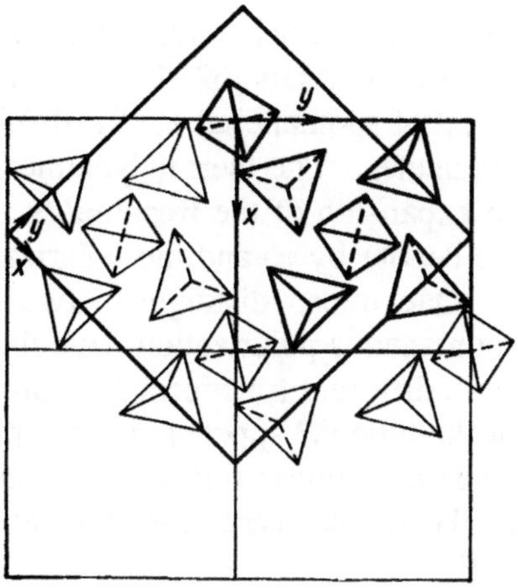

Fig. 54. Position of the unit cell of the $Tb_2(MoO_4)_3$ crystal of the rhombic phase in the initial tetragonal phase (projection upon the base plane).

A comparison of the atomic displacement pattern with the dynamic pattern of atomic displacements in the soft mode [164] shows that these patterns almost fully coincide. A slight discrepancy seems to be associated with the fact that the MoO_4 groups (which in the analysis of the dynamic mode were assumed to be relatively rigid) are to some extent deformed. But the soft mode, which condenses in the rhombic phase (phonon at the point M), is associated with the other (accompanying) modes transforming according to the unique representation of the group *Pba*2 and, hence, the pattern of the static displacements possibly must be somewhat more complicated. One can conclude from the analysis of [164] that the two patterns of atomic displacements, namely static and dynamic, coincide and that the phase transition in $Tb_2(MoO_4)_3$ [as well as in $Gd_2(MoO_4)_3$] occurs by way of "freezing in" the soft phonon mode in the initial crystal. The cited paper is an excellent example of a dynamic structure analysis based on measurements of inelastic neutron scattering at the crystal.

Let us now consider the change which occurs in the electrical properties of rare-earth molybdates in a structural phase transition. These crystals are below T_c ferroelectric crystals with a weak spontaneous moment, but the polarization which develops at $T < T_c$ is not an order parameter. Such ferroelectric materials are termed non-intrinsic ferroelectric materials [62, 104]. Polarization develops in them as a secondary effect of spontaneous atomic displacements which also form the true order parameter. The thermodynamic analysis of this situation can be based on an expansion of the free energy in powers of the order parameter (which we denote by η) and the polarization vector **P**.

One must distinguish two cases: the order parameter and the spontaneous polarization have the same transformation properties or they have different transformation properties. In the first case, a mixed term of the type ηP appears in the expression for the free energy and this term leads to a linear relation between polarization and the order parameter. In the second case, the symmetry allows only mixed terms of the type $\eta^2 P$.

Obviously, in rare-earth molybdates, η and P have different transformation properties because the order parameter transforms in accordance with the representation of the space group $P\bar{4}2_1m$ with a nonvanishing wave vector (point **M**), whereas the spontaneous polarization transforms according to the representation with $\mathbf{k} = 0$. The order parameter belongs to a two-dimensional irreducible representation [164] and therefore a phase transition in these crystals must be described by an expansion of the type

$$F = r\,(\eta_1^2 + \eta_2^2) + u_1\,(\eta_1^2 + \eta_2^2)^2 + \\ + u_2\,(\eta_1\eta_2)^2 + aP^2 + b\eta_1\eta_2 P, \tag{19.1}$$

where $r \sim (T - T_c)$; the quantities u_1, u_2, a, and b are weakly temperature dependent. When this expression is minimized with respect to the variables η_1, η_2, and P, the solution $\eta_1 = \eta_2 = 0$, $P = 0$ is obtained for $T > T_c$ and describes the paraelectric phase; the solution

$$\eta_1 = \pm\,\eta_2, \quad P \sim \eta_1\eta_2 \tag{19.2}$$

for $T < T_c$ corresponds to the ferroelectric phase. Thus, the spontaneous polarization turns out to be proportional to the square of the parameter of order. This is typical for non-intrinsic ferroelectric materials. When the order parameter changes with T like $\eta \sim (T - T_c)^{1/2}$, the spontaneous polarization near T_c changes linearly with temperature: $P \sim (T - T_c)$.

The compounds $Tb_2(MoO_4)_3$ and $Gd_2(MoO_4)_3$ are excellent examples of nonintrinsic ferroelectric materials in which an antiferro-distortional soft mode leads to the appearance of "ferro" properties, i.e., spontaneous polarization. The real situation in these crystals is described by an

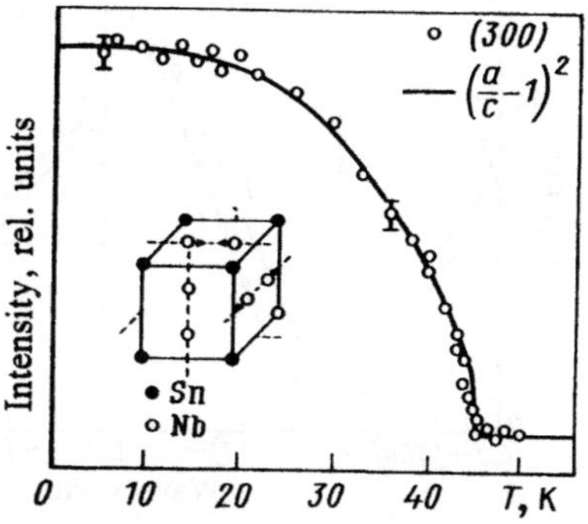

Fig. 55. Comparison of the intensity of the "superlattice" Bragg peak and the parameter of the tetragonal distortion of an Nb_3Sn crystal after the structural transition from the cubic to the tetragonal phase [113].

expansion of even greater complexity than that of Eq. (19.1), in which the spontaneous lattice deformation capable of description by a macroscopical deformation tensor must be brought into account.

Structural transitions in *A*-15 compounds. So far we have cited examples of structural phase transitions in nonmetallic crystals where in all cases the transition is accompanied by the appearance of a ferroelectric or antiferroelectric state. Instability of the initial lattice, compensated for at high temperatures by anharmonicity, is the reason for the structural transitions in the substances considered. Let us now consider distortion-type phase transitions of a completely different physical nature, which in a metal are caused by electron–phonon interaction effects.

Compounds with *A*-15 type structure can serve as an excellent example. In some of them (first of all, in Nb_3Sn and V_3Si) one observes structural phase transition from the cubic to the tetragonal phase. These transitions occur at a low temperature (T_c = 21 K in the case of V_3Si and T_c = 45 K in the case of Nb_3Sn) slightly higher than the transition temperature to the superconductive state. Since the "high-temperature" superconductivity of these compounds is related to a structural instability, a large number of papers have been written on the structural transitions (for their review, see the monograph [16]). We consider in this section only briefly the basic results of theoretical and experimental research on the lattice instability of V_3Si and Nb_3Sn.

The *A*-15 structure is characterized by the cubic space group O_h^3 – *Pm3m* of symmetry and has 8 atoms per unit cell. For example, in the case of Nb_3Sn, the Sn atoms have their positions at the apices and at the center of the cube, whereas the atoms of the transition metal Nb are situated in pairs on the faces (see inset in Figure 55). After the structural transition to the tetragonal phase, the Nb_3Sn crystal has the space group

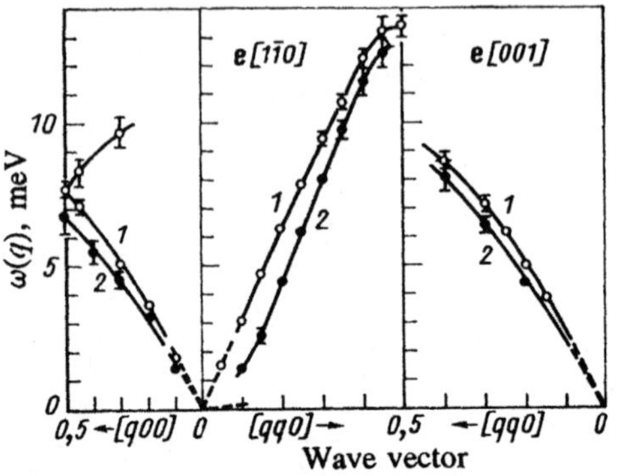

Fig. 56. Dispersion relations for the transverse acoustic phonons in Nb_3Sn at the temperatures 1) 295 K [113] and 2) 46 K.

$D_{4h}{}^9$ – $P4_2/mmc$. The transition to the new structure is obtained by Nb atoms moving along the chains (see Figure 55) and by an overall tetragonal deformation of the cell. In this way, the unit cell is preserved in the phase transition, i.e., the transition is characterized by the wave vector $\mathbf{k} = 0$ of the superlattice.

A comparison of the initial $O_h{}^3$ and final $D_{4h}{}^9$ space groups of the crystal can readily serve for a determination of the symmetry of the soft optical mode whose "freezing in" in the initial crystal results in the structure shown in Figure 55. This is the mode which belongs to the two-dimensional irreducible representation τ_5 of the star $\mathbf{k} = 0$ (Γ_{12} in the other generally accepted nomenclature) [123, 110]. Experimental research which was made with Raman spectroscopy and neutron spectroscopy could not establish a substantial temperature dependence of the frequencies of the optical modes. However, a softening of certain acoustic phonon modes was observed, with the softening starting at a high temperature far before the point of the structural conversion. In V_3Si and Nb_3Sn, the acoustical measurements revealed a complete softening of the shear modulus $C_{11}-C_{12}$ when T_c is approached, with the shear modulus determining the energy of the transverse acoustic phonon with the wave vector parallel to [110] and the polarization [1$\bar{1}$0].

Measurements of the dispersion curves of the phonons in Nb_3Sn [113] have shown that the phonon modes are softened in the various directions of the Brillouin zone; the softening is most pronounced in the transverse acoustic mode with $\mathbf{q} \sim$ [110] (Figure 56). "Freezing in" of the acoustic phonons at the point T_c causes the observed uniform tetragonal deformation of the cubic lattice, the deformation belonging to the irreducible representation τ_5. Thus, changes in the cubic structure of the Nb_3Sn

manifest themselves in a relative displacement of the Nb atom and a tetragonal deformation of the cubic cell. The two modes of distortion (optical and acoustical) belong to the same irreducible representation τ_5. Both the amplitude of the Nb atomic displacements in the chains and the tensor components of the tetragonal deformation must be the order parameter between which a linear relation must exist. This is demonstrated in Figure 55. The experimental points indicate the temperature dependence of the intensity of the (300) Bragg peak (appearing only in the tetragonal phase), the intensity being proportional to the square of the atomic displacements Q. The solid line is proportional to the square of the spontaneous deformation ε. Thus, we have indeed $Q^2 \sim \varepsilon^2$.

A simple phenomenological analysis helps to understand how the softening of the optical mode manifests itself in the associated acoustic mode [113]. We write the free energy of a crystal with the two associated modes in the following form:

$$F = \frac{1}{2} \omega_0 Q^2 + \frac{1}{2} C\varepsilon^2 + gQ\varepsilon. \tag{19.3}$$

The interaction energy can be quadratic because the two modes have the same symmetry. The parameter ω_0 has the meaning of the frequency of the optical mode, and C denotes the modulus of the lattice. By minimizing the free energy with respect to ε, we obtain the relation between the displacements of the atoms and the deformation:

$$Q = -g\varepsilon/C, \tag{19.4}$$

As could be anticipated, this relation is linear.

After substituting this expression in Eq. (19.3), we find that the effective modulus of elasticity is composed of two terms because of the interaction with the optical mode:

$$\tilde{C} = C - (g^2/\omega_0^2). \tag{19.5}$$

When a softening of the frequency **M** of the optical phonon is observed in a symmetric phase, the expression for $\tilde{C}$ can tend to zero earlier than the frequency $\mathbf{M}_0$. Then the phase transition must manifest itself by a spontaneous deformation at the point $T = T_c$, where $\tilde{C}(T) = 0$, though the softening of the lattice modulus originates from a softening of the optical phonons. This situation also exists in Nb_3Sn.

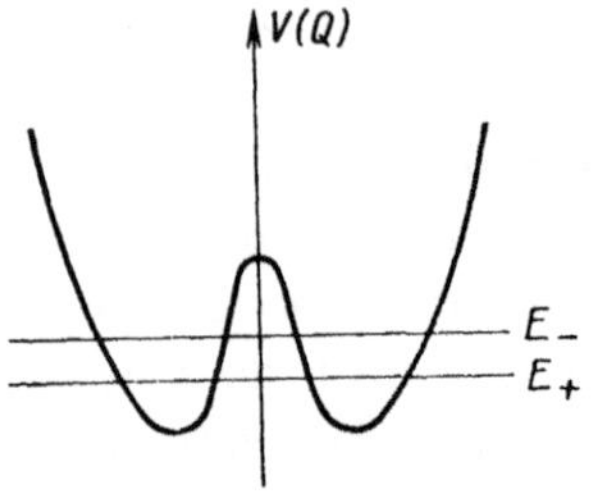

Fig. 57. Single-frequency potential for an atom which undergoes an order–disorder phase transition in a crystal.

In terms of physics, the softening of the optical modes in type A-15 compounds is assumed to result from features of the electron–phonon interaction connected with the quasi-one-dimensional form of the electron states near the Fermi surface [10, 118].

20. Structural Transitions of the Order–Disorder Type

Theoretical Principles. So far we have considered structural phase transitions in which a new (low-temperature) phase arises when certain atoms are shifted from certain symmetric positions. These displacements are assumed to be small and the lattice dynamics is described by harmonic or quasi-harmonic phonons. In some crystals, the majority of which has Seignette or antiferroelectric properties, structural transitions of a completely different type are observed. In them the individual atoms perform finite, discrete movements in the form of jumps between two equivalent quasi-equilibrium positions. These equivalent positions are filled with equal probability at a high temperature, but starting from some temperature $T = T_c$, the positions of the one type are predominantly filled and this disturbs the initial symmetry of the system and corresponds to a phase transition. The difference of the probabilities of filling the two positions plays the role of a parameter of order; the difference gradually increases when the temperature is reduced below T_c so that the order–disorder type phase transition under consideration is a transition of the second kind. We describe below specific cases of such transitions and will now briefly expound the theoretical principles used in the description.

Let us consider an atom which moves in a double potential well with equivalent minima (Figure 57). The ground state is in this well double degenerate but owing to quantum-mechanical tunnelling, the degeneracy is removed and two close energy levels are formed. The lower level corresponds to a symmetric wave function (with maxima in each well), whereas the upper level corresponds to an antisymmetric wave function. When the distance between the levels of the doublet is much smaller than the distance between the ground state and the excited vibrational levels in each of the wells, the description of such a system can be restricted to the two-level approximation assuming that a state is fully defined by indications of particle positions (in the right or the left well). The state of the entire crystal is then fully given when the position of each of the atoms participating in this motion is given in the double potential well.

In the case of a two-level system, the system is conveniently described in terms of the pseudo-spin of magnitude $S = 1/2$, with the state of a particle in the double potential well characterized by the eigenvalues of the spin operator S^z. The Hamiltonian of a system of atoms performing jump-like motions can be stated in the form of the very simple spin Hamiltonian of the following form:

$$\mathcal{H} = -\Omega \sum_l S_l^x - \frac{1}{2} \sum_{ll'} J_{ll'} S_l^z S_{l'}^z. \tag{20.1}$$

The coordinates axes were chosen so that S_l^z, i.e., the component of the pseudo-spin, is diagonal and that its average over the entire ensemble of pseudo-spins renders the difference in the population of the left and right equilibrium positions. At the same time, the operator S_l^x, which characterizes transitions between these positions, and its average value determine the tunnel current. In Eq. (20.1), Ω denotes the matrix element of the transition between two states in a particular double well and $J_{ll'}$ denotes the magnitude of the interaction of the atoms in the various double wells. The Hamiltonian of Eq. (20.1) was derived, for example, in [6]. Let us note that Eq. (20.1) is identical with the Hamiltonian of the Ising model in a transverse field so that the corresponding statistical mechanics is quite well known.

Let us state the results of research made in the molecular field approximation in which the Hamiltonian of Eq. (20.1) is replaced by the effective Hamiltonian of the free pseudospins situated in a molecular field $\mathbf{H}_l$:

$$\mathcal{H} = -\sum_l \mathbf{H}_l \mathbf{S}_l^z, \quad \mathbf{H}_l = \left\{ \Omega, 0, \sum_{l'} J_{ll'} \langle S_{l'}^z \rangle \right\}, \tag{20.2}$$

where the brackets $\langle ... \rangle$ denote averaging over the temperature. The following system of self-consistent equations is immediately obtained for the average value of the pseudospin in the lth cell:

$$\left.\begin{aligned} &\langle S_l^x \rangle = \frac{1}{2}\frac{\Omega}{H_l}\,\text{th}\,\frac{H_l}{2kT_c};\\ &\langle S_l^y \rangle = 0;\\ &\langle S_l^z \rangle = \frac{1}{2}\frac{\sum_{l'} J_{ll'}\langle S_{l'}^z \rangle}{H_l}\,\text{th}\,\frac{H_l}{2kT_c}, \end{aligned}\right\} \tag{20.3}$$

where H_l denotes the absolute value of the vector $\mathbf{H}_l$ of the molecular field:

$$H_l = \sqrt{\Omega^2 + \left[\sum_{l'} J_{ll'}\langle S_{l'}^z \rangle\right]^2}. \tag{20.4}$$

One of the solutions

$$\langle S_l^z \rangle = \langle S_l^y \rangle = 0,\ \langle S_l^x \rangle = \frac{1}{2}\,\text{th}\,\frac{\Omega}{2kT_c} \tag{20.5}$$

of this system exists at any temperature and describes a state in which the dipole moment (being proportional to $\langle S_l^z \rangle$) vanishes (paraelectric phase). Below T_c, which is defined by the equation

$$\frac{2\Omega}{J(0)} = \text{th}\,\frac{\Omega}{2kT_c}, \tag{20.6}$$

there exists a state in which $\langle S_l^z \rangle \neq 0$ (ferroelectric phase). Here $J(0) = \Sigma_{l'} J_{ll'}$ denotes the Fourier component of the quantity $J_{ll'}$ at the wave vector zero.

Equation (20.6) has a solution if $J(0) > 2\Omega$. At a larger value of the parameter ω of tunnelling, a phase transition does not take place and in this way this system is distinguished from the classical system experiencing an order–disorder transition. Equation (20.6) predicts a strong isotope effect. When the mass of the atom increases, the value of the parameter Ω increases considerably and the value of T_c must accordingly increase while the interaction parameter $J(0)$ remains unchanged. Let us note that $\langle S_l^z \rangle$ increases when the temperature drops below T_c, in accordance with the assumptions for the order parameter, whereas $\langle S_l^x \rangle$ does not depend upon the temperature:

$$\langle S_l^x \rangle = \Omega / J(0). \tag{20.7}$$

Now we consider weakly excited states above the thermodynamically homogeneous state which is described by self-consistent field equations (20.3). In the para-electric phase, the equation of motion for the pseudo-spin leads in the approximation of random phases to three excitation branches of the system, with the branches corresponding to vibrations of the instantaneous values of the quantities $\langle S_l^x \rangle$, $\langle S_l^y \rangle$, $\langle S_l^z \rangle$ around their average values. Their frequencies are

$$\omega_1(\mathbf{q}) = 0 \tag{20.8}$$

$$\omega_{2,3}^2(\mathbf{q}) = \Omega \left[\Omega - \frac{1}{2} J(\mathbf{q}) \operatorname{th} \frac{\Omega}{2kT} \right]. \tag{20.9}$$

The eigenstate with the frequency $\omega_1(\mathbf{q})$ represents the longitudinal mode in which the quantity $\langle S_l^x \rangle$ fluctuates, i.e., the molecular field proper. The two other frequencies $\omega_{2,3}(\mathbf{q})$ correspond to transverse excitations and describe the free precession of the pseudospins with respect to the molecular field parallel to the x axis. Their frequencies are strongly temperature-dependent. At very high $T(kT >> \Omega)$, the frequencies are $\omega_{2,3}(\mathbf{q}) \approx \Omega$, but when the temperature decreases, the frequencies decrease and tend to zero at the temperature T_c:

$$2\Omega - J(\mathbf{q}_0) \operatorname{th} \frac{\Omega}{2kT_c} = 0 \tag{20.10}$$

for a wave vector $\mathbf{q}_0$ at which the function $J(\mathbf{q})$ has its maximum. When a ferroelectric phase is formed below T_c, this occurs at $\mathbf{q}_0 = 0$. Thus, the transverse mode which describes the precession of the pseudospins is a soft mode. Around $(T_c, \mathbf{q}_0)$, the quantity of Eq. (20.9) can be expanded in a series in powers of $T - T_c$ and $\mathbf{q} - \mathbf{q}_0$:

$$\omega_{2,3}^2(\mathbf{q}) = r(q_0)(T - T_c) + c(\mathbf{q} - \mathbf{q}_0)^2, \tag{20.11}$$

where $c \sim \Omega^2$ and

$$r\,(\mathbf{q}_0)=\frac{\Omega^2\,J\,(q_0)}{4\,(kT_c)^3}\,k\,\mathrm{ch}^{-2}\,\frac{\Omega}{2kT_c}\,.$$

When $J(\mathbf{q})$ has a maximum at $\mathbf{q} \neq \mathbf{q}_0$, the paraelectric phase becomes unstable as far as the appearance of antiferroelectric order is concerned.

Thus, an order–disorder phase transition can also be interpreted as "freezing in" of a mode with the wave vector $\mathbf{q}_0$ at $T = T_c$. The value $\langle S_l^z\rangle$ describes the statistical independence of the left and right positions of the atoms at $T < T_c$. We have also three branches of frequencies in the ferroelectric phase: $\omega_1(\mathbf{q}) = 0$ and the two other frequencies:

$$\omega_{2,3}^2\,(\mathbf{q})=[J\,(0)\,\langle S^z\rangle]^2+\Omega^2\,(1-J\,(\mathbf{q})/J\,(0)). \qquad (20.12)$$

The latter are soft modes only in the long-wave limit $q \to 0$. In this limit, $\omega_{2,3}(\mathbf{q}) \sim \langle S_l^z\rangle$ is proportional to the spontaneous polarization. This holds also for the antiferroelectric state though the mode with the case of some structure of the ordered state below T_c there appears a soft mode only with $\mathbf{q} = 0$ though the soft mode can be with $q = q_0 \neq 0$ in the phase without order. In this way, phase transitions of the order–disorder type are distinguished from structural transitions of the type of displacements.

We will now derive the cross section of neutron scattering at a soft pseudospin mode. We must start from the very general expression for the double differential cross section of Eq. (1.10) in which the amplitude of nuclear scattering is stated in the form of Eq. (1.13). We separate from that expression that part which is related to atoms performing jump-like motions between the minima of the double potential well. Assume that the coordinates of the two minima of one well are $\mathbf{R}_l \pm \rho_l$, where $\mathbf{R}_l$ denotes the radius vector of the center of the potential well (as usual, the subscript l stands for $l = \{ni\}$, where n denotes the number of the cell and i denotes the number of the atom). The expression for the scattering amplitude [Eq. (1.13)] is conveniently written with the aid of the pseudospins S_l^z which comprises information on the position of a particular atom in the left or the right well:

$$V=\sum_{ni}\frac{1}{2}\,\alpha_i\,\{(1+2S_{ni}^z)\,\delta\,(\mathbf{r}-\mathbf{R}_{ni}+\rho_i)+(1-2S_{ni}^z)\,\delta\,(\mathbf{r}-\mathbf{R}_{ni}-\rho_i)\}. \qquad (20.13)$$

We substitute this equation in Eq. (1.10) and state the cross section of nuclear scattering at the atoms under consideration in the form

$$\frac{d^2\sigma_h}{d\Omega d\varepsilon} = \frac{4k}{k_0} |F_h(\varkappa)|^2 \frac{1}{2\pi\hbar} \int_{-\infty}^{\infty} dt\, e^{-i\varepsilon t} \langle S^z_{\varkappa}(t)\, S^z_{-\varkappa}(0)\rangle, \qquad (20.14)$$

where

$$F_h(\varkappa) = \sum_i b_i\, e^{i\varkappa x_i} \sin \varkappa \rho_i, \qquad (20.15)$$

with $F_h(\varkappa)$ denoting some structure amplitude. The cross section of scattering is expressed through the correlator of the pseudospins and can be stated with the aid of the corresponding Green's function as in Section 14:

$$\frac{d^2\sigma_h}{d\Omega d\varepsilon} = \frac{4k}{k_0} |F_h(\varkappa)|^2 \frac{1}{e^{\beta\varepsilon} - 1} \frac{1}{\pi} \operatorname{Im} G^{zz}(\varkappa, \omega)\Big|_{\omega=-\varepsilon/\hbar}. \qquad (20.16)$$

When the Green's function is calculated on the basis of the Hamiltonian of Eq. (20.1) with proper regard for attenuation, the latter expression leads to the following formula for the cross section of scattering at a soft mode with the frequency $\omega(\mathbf{q})$ and the attenuation $\gamma(\mathbf{q})$:

$$\frac{d^2\sigma_h}{d\Omega d\varepsilon} = 4 |F_h(\mathbf{b})|^2 \frac{kT}{\hbar\pi} \frac{\gamma(\mathbf{q})}{[\omega^2 - \omega^2(\mathbf{q})]^2 + \omega^2\gamma^2(\mathbf{q})}, \qquad (20.17)$$

where the wave vector $\mathbf{q}$ of the mode is related to the scattering vector by the momentum conservation law: $\varkappa = \mathbf{b} \pm \mathbf{q}$.

This expression renders the same energy dependence of the cross section of scattering at soft pseudospin modes as in the case of scattering at soft modes near a structural transition of the displacement type [see Eqs. (6.16) and (6.22)]. When Eq. (20.17) is integrated over the energy, the angular dependence of the cross section of the integral intensity of the peak is obtained:

$$\frac{d\sigma_h}{d\Omega} = 4 |F_h(\mathbf{b})|^2 \frac{kT}{\hbar^2} \frac{1}{\omega^2(\mathbf{q})}, \qquad (20.18)$$

and this corresponds to Eq. (6.34) for phase transitions of the displacement type. Equations (20.18) and (6.34) differ by the factors defining the angular part of the cross section. In the one case, the angular part depends upon the structure amplitude of elastic scattering. In the latter

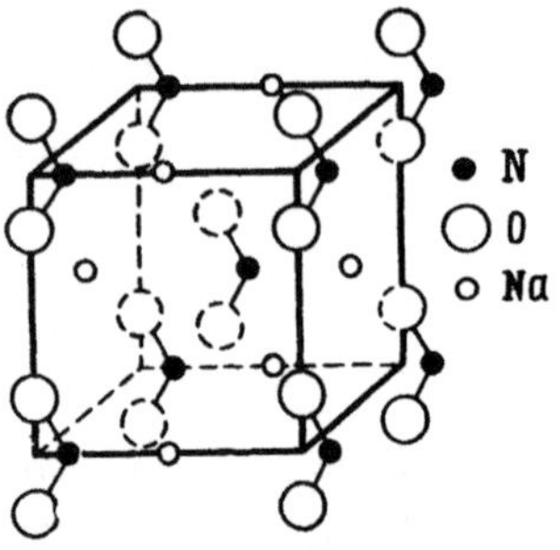

Fig. 58. Crystal structure of $NaNO_2$ in the ferroelectric phase.

case, this is caused by the approximation approach to the problem of the order–disorder phase transition for which the two-level approximation was employed.

We must particularly mention the case in which tunnelling can be disregarded because the parameter Ω is small. Since the frequency of the pseudospin waves is proportional to Ω in the paraelectric phase [see Eq. (20.9)], this means that no pseudospin waves exist in such a system. The system which can be described by the Hamiltonian of Eq. (20.1) without the term accounting for tunnelling is an Ising system, and diffuse modes with a lifetime $\tau \to \infty$ at $T \to T_c$ exist in its phase without order [6]. The Green's function of the longitudinal spin components has an imaginary pole i/τ, and this leads to the following equation for the cross section

$$\frac{d^2\sigma_h}{d\Omega d\varepsilon} \backsim |F_h(\mathbf{b})|^2 \frac{kT}{T-T_c} \frac{1/\tau}{\omega^2+(1/\tau)^2}, \tag{20.19}$$

describing a quasi-elastic peak of width $1/\tau$. The width of the peak tends to zero when T_c is approached; this is a consequence of the critical slowing down of the fluctuations. As far as its frequency and temperature dependences are concerned, Eq. (20.19) is identical with Eq. (6.23) describing the response of a strongly attenuated oscillator though the phenomena which are described by these two expressions are different.

We consider below two examples of order–disorder transitions: in $NaNO_2$ and in hydrogen-containing compounds of the type of KH_2PO_4.

$NaNO_2$. The transition to the ferroelectric phase in the $NaNO_2$ crystal is a classical example of an order–disorder phase transition. The substance has a body-centered orthorhombic lattice and the space group $C_{2v}{}^{20}$ – *Im*2*m*. The crystal becomes ferroelectric at about 163°C. The crystal has in this state the structure shown in Figure 58. It is formed by NO_2 molecules which are oriented toward on side parallel to the *b* axis.

In the paraelectric phase, the molecules are randomly oriented parallel to the *b* direction so that there occurs an order–disorder type transition which can be described with a model potential for the nitrogen atom, the potential having two equivalent minima.

The phonon spectrum of such a crystal has been studied with the aid of inelastic neutron scattering [302]. The primitive cell of the crystal contains 4 atoms and therefore there can be a total of 12 phonon branches; three of them, describing the relative movement in the molecule NO_2 group, have very high frequencies and, as far as structural phase transitions are concerned, only 9 of the phonon branches can be of interest. The dispersion curves of the nine phonon modes have been measured for some symmetric directions of the crystal at various temperatures; substantial changes at temperature changes near the phase transition to the para-electric state could not be detected. Thus, it was established that the structural transition in $NaNO_2$ is unrelated to any soft mode which is usually present in displacement-type transitions. Near T_c, a quasi-elastic peak with an extremely small width ($\sim 10^9$ sec^{-1}) was detected near T_c; this agrees with determinations of the relaxation time obtained from the measurements of the dielectric susceptibility. This fact indicates that a soft (diffusion) mode is present in the system but this mode does not fit into the group-theoretical scheme of classifying the phonon modes of the crystal. The scheme is based on the harmonic approximation when the crystal oscillations are described and the amplitudes of all atoms are assumed to be small. By contrast, in crystals which undergo an order–disorder transition, certain atoms make transition between two equivalent positions separated by a large (finite) distance. This movement is completely disregarded in the group-theoretical approach, and when such a movement exists, an additional mode must correspond to it. this conclusion refers not only to the diffusion mode as in $NaNO_2$ but also to soft pseudospin modes developing in crystals with finite tunnelling.

Ferroelectric materials with hydrogen bonds. Crystals with hydrogen bonds form a large class of ferroelectric materials in which a spontaneous polarization develops as the consequence of a structural phase transition. KH_2PO_4 (abbreviated KDP) is a typical representative. In its paraelectric phase, the crystal has the space group $I\bar{4}2d$ with two structure units per primitive cell. The low-temperature structure (T_c = 123 K) is *Fdd2* orthorhombic with a primitive cell containing the same number of atoms but with the *a* and *b* axes rotated 45° relative to the corresponding

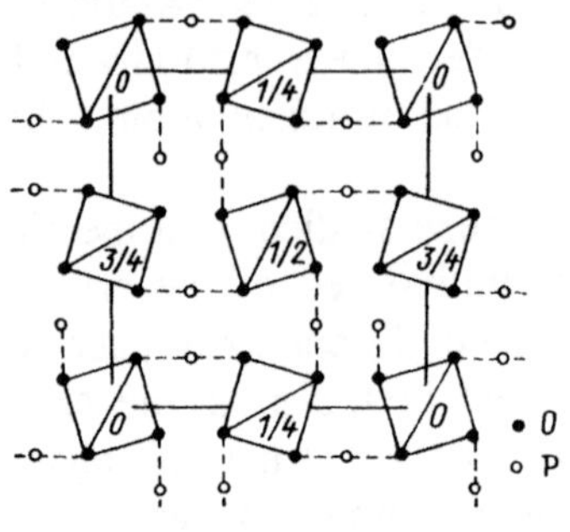

Fig. 59. Projection of the tetragonal KH_2PO_4 structure on the base plane [323] (the numbers denote the positions of the centers of the PO_4 tetrahedra in the direction of the *c* axis. The solid line outlines the unit cell of the body-centered tetragonal crystal).

axes in the initial crystal. Thus, the structural transition is characterized by the wave vector $\mathbf{k} = 0$.

A KDP crystal can be considered composed of rigid PO_4 tetrahedra with a phosphorus atom at the center; the tetrahedra are connected by hydrogen atoms (see Figure 59). On the O—H—O bond, the potential in which the hydrogen atom is situated forms a well with two equivalent minima; therefore the protons assume positions which are asymmetric relative to the center of the bond and tunnel between these positions. In the para-electric phase, the protons fill the two possible positions at random and therefore the mean statistical position of the hydrogen atoms on the O—H—O bond is symmetric as indicated in Figure 59. Below T_c there develops an asymmetry in the position of the protons and this is accompanied by a shifting of the other atoms and, accordingly, the development of a spontaneous dipole moment. The main component of the displacement vector of the hydrogen atom is on the base plane, and a spontaneous moment develops in the perpendicular direction (in the direction of the *c* axis) so that the contribution to it is related to the components of the proton displacement in this direction as well as to the components of the phosphorus and potassium atoms. The displacement of K and P in opposite directions (parallel to the *c* axis) in the ferroelectric phase was experimentally established by x-ray diffraction [181] and neutron diffraction [114].

Cochran [114] used these structure data to construct a dynamic model of the soft mode the condensation of which could render the observed KDP structure below T_c. The displacements assumed for the atoms are illustrated in Figure 60. An experimental verification of this model was made with a three-crystal spectrometer equipped with an KD_2PO_4 single crystal [323] because deuterium has a greater coherent scattering

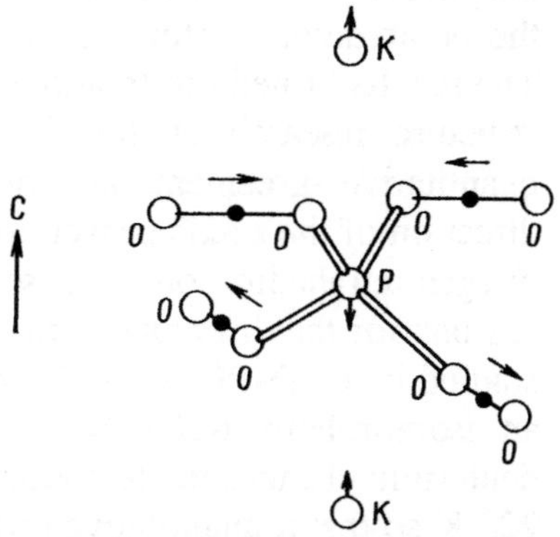

Fig. 60. Model of the soft ferroelectric mode in the KH_2PO_4 crystal (according to Cochran).

amplitude. The technique used in investigations on $Tb_2(MoO_4)_3$ crystals was used to determine the polarization vectors of the soft mode; this technique is based on a comparison of the integral peak of inelastic scattering with the theoretical equation (6.34). In the case of KDP, the problem is complicated by two details: the soft modes turned out to be greatly attenuated, and the peaks of inelastic scattering are superimposed on the basic Bragg peaks (because a structural transition occurs with $\mathbf{k} = 0$ by condensation of the soft phonons with the wave vector $\mathbf{q} = 0$). For this reason, inelastic scattering was not measured at the soft phonon with $q = 0$ but at other phonons with small $\mathbf{q}$ values (corresponding to scattering beyond the Bragg peaks); then an extrapolation to the point Γ was made.

For this purpose one must determine the symmetry of the ferroelectric soft mode and its compatibility with other modes with $\mathbf{q} \neq 0$. Symmetry analysis has shown that the mechanical representation of the KH_2PO_4 crystal in the paraelectric phase with $\mathbf{q} = 0$ (dimensionality 48 × 48) splits into the following irreducible representations of the $I\bar{4}2d$ group at the point **Γ** [101]:

$$d_m^{\mathbf{q}=0} = 4\tau_1 + 5\tau_2 + 7\tau_3 + 6\tau_4 + 13\tau_5$$

(the representations τ_1-τ_4 are one-dimensional, whereas the representation τ_5 is two-dimensional); the representation τ_3 has the ferroelectric properties. The seven sets of basis functions form the ferroelectric optical mode and also the acoustic modes. One of the acoustic modes, which belongs to the representation $\mathbf{\Gamma}$-τ_3, is compatible with the acoustic mode Σ_2. By studying the temperature dependence of this mode at $\mathbf{q} = 0$ near the point $\mathbf{\Gamma}$, the ferroelectric optical mode $\mathbf{\Gamma}$-τ_3 can be derived [323], i.e., the polarization vectors can be determined at all atoms of the crystals. The results turned out to agree with the Cochran model and the results of structure research on the ferroelectric phase of KH_2PO_4, though full quantitative agreement was not obtained. The shift of the D atom in the direction of the z axis proved to be too large and also the distortion of the oxygen tetrahedron on the base plane was excessive. One of the possible reasons for the discrepancy may reside on the fact that the soft mode was studied in a KD_2PO_4 crystal, whereas the structure studies were made on an isomorphous KH_2PO_4 crystal. The replacement of hydrogen by deuterium changes the temperature of the structural transition from 123 to 220 K so that a quantitative difference of the dynamic properties of these crystals can be expected. Nevertheless, the concept of a soft mode in crystals with hydrogen bonds was obviously confirmed in definite form.

Antiferroelectric materials are among the numerous representatives of hydrogen-containing crystals of KH_2PO_4, which have ferroelectric properties. They result from a structural phase transition in which the unit cell of the crystal increases. $(ND)_4D_2PO_4$ is an example: a structural transition with $\mathbf{k} = 2\pi/a(\frac{1}{2}\ \frac{1}{2}\ 0)$ was observed in it and studied with neutron diffraction techniques by the authors of [249].

21. Critical Dynamics in a Structural Phase Transition

The central peak. The numerous examples of structural phase transitions considered in the present chapter show that they involve the soft-mode mechanism. Therefore, in the energy distribution of neutrons scattered above the transition temperature, one observes two peaks corresponding to absorption or emission of a soft phonon; the spacing of the peaks appearing under conditions of an overattenuated phonon.

However, measurements made on $SrTiO_3$ and $KMnF_3$ with spectrometers of the highest resolution quite unexpectedly revealed a more complicated pattern of the dynamics which near T_c manifests itself in a strong singularity of the cross section at low transferred energies [308]. Figure 61 shows the neutron spectrum of two isomorphous perovskites above the temperature of the structural conversion from the cubic to the tetragonal phase (see Section 18). Along with the broad maximum corresponding to scattering at phonons, one observes in both cases a very narrow central peak. This peak appears for the first time at a temperature

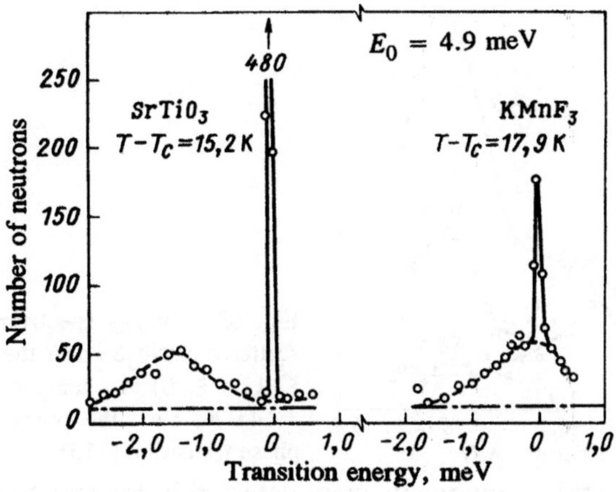

Fig. 61. Energy spectrum of the neutron scattered at $SrTiO_3$ and $KMnF_3$ near the temperature of the structural phase transition [308].

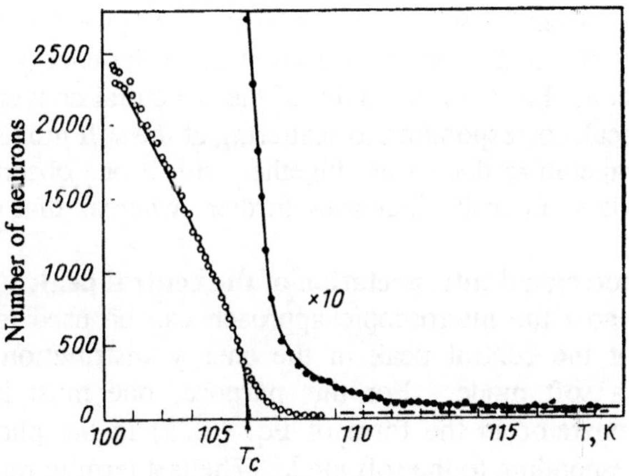

Fig. 62. Intensity of the quasi-elastic neutron peak at the **R** point of the reciprocal lattice as a function of the temperature for $SrTiO_3$.

which exceeds T_c by about 70 K and the intensity of the peak has a critical divergence at the transition point proper (Figure 62); the width becomes smaller than 0.02 meV. In $KMnF_3$, the central component can be recognized on the background of an overattenuated soft phonon which becomes overattenuated already at T_c + 40 K.

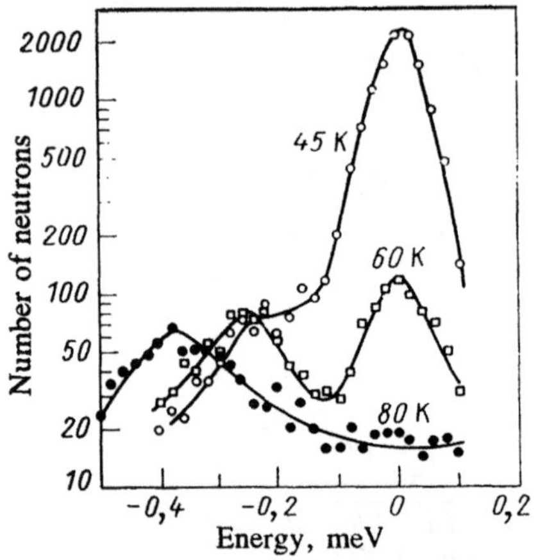

Fig. 63. Energy spectrum of the neutrons scattered at Nb_3Sn near the Bragg peak ($2 - \xi$, $1 + \xi$, 0) at a temperature exceeding $T = 45$ K, i.e., the point of the structural phase transition [113].

Similar central components were detected in the spectrum of scattered neutrons (and of scattered light) in many substances undergoing a structural phase transition which is associated with a soft mode (see, e.g., the review [18]). We cite as an example the Nb_3Sn crystal in which the soft mode is a transverse acoustic phonon. Figure 63 illustrates the dependence of the scattering cross section upon the energy at a constant scattering vector. Far from the point of the structural conversion, a well-pronounced peak corresponding to scattering at the soft mode is observed. When the temperature decreases, together with it one observes a central component whose intensity increases further, whereas that of the lateral peak drops.

Phenomenological interpretation of the central peak. It was shown in Section 6 how the microscopic approach can be used to explain the appearance of the central peak in the energy distribution of neutrons scattered at a soft mode. For this purpose, one must introduce the polarization operator in the form of Eq. (6.25) in the phonon Green's function corresponding to the soft mode. The last term in the polarization operator originates from the interaction of the soft mode with some relaxation mode having the lifetime τ. Since Eq. (6.25) was postulated, the theory outlined above can be assumed a microscopic theory and therefore a truly phenomenological analog of it should be provided.

The soft mode can be described as a harmonic oscillator immersed in a thermal reservoir formed by the totality of all other degrees of freedom. The equation of motion of the oscillator must be stated for the amplitude $\dot{c}(\mathbf{q}, t)$ of the oscillations occurring with the frequency $\omega(\mathbf{q})$ and must comprise the term $\Gamma\dot{c}(\mathbf{q}, t)$ accounting for the friction force [according to

Eq. (7.2), $c(\mathbf{q})$ represents also the fluctuation of the order parameter]. We assume that the particular oscillator interacts with some relaxation mode which is characterized by the amplitude $b(\mathbf{q}, t)$ and the relaxation term γ_b. The dynamics of the oscillator which is in this medium acted upon by an external force $F(t)$ must be described by the following system of equations [18]:

$$\left.\begin{aligned} &\ddot{c}(\mathbf{q}, t) + \Gamma \dot{c}(\mathbf{q}, t) + \Phi_{cc}\, c(\mathbf{q}, t) + \Phi_{cb}\, b(\mathbf{q}, t) = F(t); \\ &\gamma_b \dot{b}(\mathbf{q}, t) + \Phi_{bb}\, b(\mathbf{q}, t) + \Phi_{cb}\, c(\mathbf{q}, t) = 0, \end{aligned}\right\} \tag{21.1}$$

where all coefficients are assumed to be time-independent. The quantities Φ_{cc}, Φ_{bb}, and Φ_{cb} are the second derivatives of the thermodynamic potential in powers of the order parameter.

We will assume that the driving force obeys a harmonic law with the frequency ω: $F(t) = F(\omega)e^{-i\omega t}$; then the solution of Eqs. (21.1) must be stated in the form $c(\mathbf{q}, t) = c(\mathbf{q}, \omega)e^{-i\omega t}$ and $b(\mathbf{q}, t) = b(\mathbf{q}, \omega)e^{-i\omega t}$. By eliminating the quantity $b(\mathbf{q}, \omega)$ from the two equations, we can write the equation for $c(\mathbf{q}, \omega)$ in the form

$$[\Phi_{cc}(\omega) - i\omega\Gamma(\omega) - \omega^2]\, c(\mathbf{q}, \omega) = F(\mathbf{q}, \omega), \tag{21.2}$$

where the frequency-dependent attenuation and the frequency-dependent modulus were introduced:

$$\Gamma(\omega) = \Gamma + \frac{\delta^2 \gamma}{\omega^2 + \gamma^2}, \quad \Phi_{cc}(\omega) = \Phi_{bb} - \frac{\delta^2 \gamma^2}{\omega^2 + \gamma^2}, \tag{21.3}$$

the new constants δ and γ are given by the equations

$$\gamma = \Phi_{bb}/\gamma_b, \quad \delta^2 = \Phi_{cb}^2/\Phi_{bb}. \tag{21.4}$$

The quantity $c(\omega)/F(\omega)$ is proportional to the susceptibility $\chi(\omega)$. The imaginary part of the susceptibility defines the differential scattering cross section and is obtained from Eq. (21.2):

$$\chi(\mathbf{q}, \omega) \backsim \frac{\omega\left[\Gamma + \dfrac{\delta^2 \gamma}{\omega^2 + \gamma^2}\right]}{\left[\omega_\infty^2(\mathbf{q}) - \dfrac{\delta^2 \gamma^2}{\omega^2 + \gamma^2} - \omega^2\right]^2 + \omega^2\left(\Gamma + \dfrac{\delta^2 \gamma}{\omega^2 + \gamma^2}\right)^2}, \tag{21.5}$$

where we introduced the square of the frequency of the soft mode with the aid of the expression

$$\omega_\infty^2(\mathbf{q}) = \Phi_{cc} \sim (T - T_c). \tag{21.6}$$

In this form Eq. (21.5) coincides with Eq. (6.26) which was earlier obtained with the aid of the Green's function and renders the energy dependence of the cross section near the phase transition. From the viewpoint of Eq. (21.6) for $\Phi_{cc}(\omega)$, the quantity $\omega_\infty{}^2(\mathbf{q})$ is the high-frequency limit. Similarly, one can introduce the low-frequency limit $\omega_0{}^2(\mathbf{q})$ which is related to $\omega_\infty{}^2(\mathbf{q})$ by the equation

$$\omega_0^2(\mathbf{q}) = \omega_\infty^2(\mathbf{q}) - \delta^2. \tag{21.7}$$

As has been explained in detail in Section 6, Eq. (21.5), together with the two lateral peaks corresponding to scattering under emission or absorption of soft phonons, can render the central peak. The central peak arises under conditions for which $\gamma << \omega_\infty$. In this case, fast changes in the order parameter $c(\mathbf{q})$ are tuned to slow changes of the amplitude $b(\mathbf{q})$ of the relaxation mode. In the other limit ($\gamma >> \omega_\infty$), there arises a spectral pattern comprising only the lateral lines which at a sufficiently large attenuation Γ of the soft mode becomes a single broad maximum with $\omega = 0$. This situation corresponds to the overattenuated oscillator.

The phenomenological approach [18] based on Eqs. (21.1) is advantageous because the results have a general form: a central peak develops when the order parameter interacts with a slow relaxation mode. The interpretation of the mechanisms of this slow relaxation, i.e., the determination of the meaning of the quantities $b(\mathbf{q})$ in terms of physics, is essentially also the problem of the central peak. The reader can find in [18] a description of the various details of this problem and we switch now immediately to the discussion of the experimental results related to the critical dynamics of structural transitions interpreted with the phenomenological approach.

Figure 64 depicts the results obtained in an evaluation of the spectra of inelastic scattering at $SrTiO_3$ at various temperatures; Eqs. (21.5) were used and it was assumed that the attenuation quantities Γ and γ are temperature-independent quantities. Then the frequency $\omega_\infty(\mathbf{q})$ of the soft

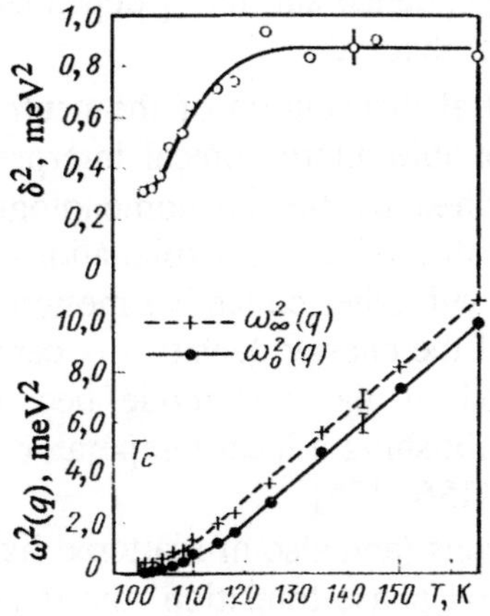

Fig. 64. Temperature dependence of the parameters $\omega_0{}^2$, $\omega_\infty{}^2$, and δ of the soft mode in the case of $SrTiO_3$ [308].

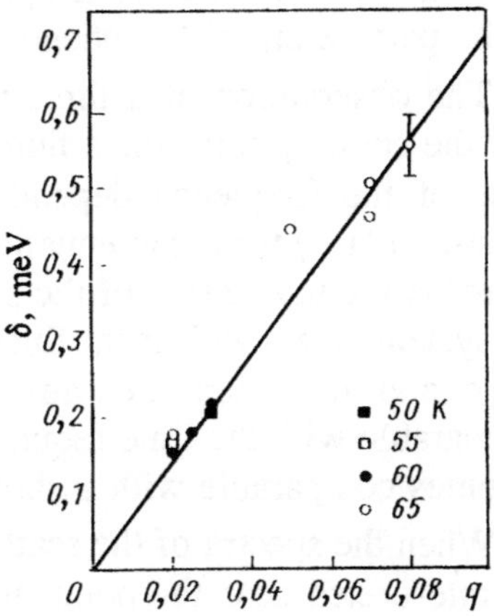

Fig. 65. Dependence of the coupling constant δ of the soft mode and the slow relaxation mode upon the wave vector **q** in Nb_3Sn [113].

mode and the parameter δ of the interaction of the soft mode with the relaxation mode turn out to be temperature-dependent.

It follows from Figure 64 that $\omega_0{}^2(0)$ and $\omega_\infty{}^2(0)$ depend linearly upon the temperature at $T - T_c < 15$ K in accordance with the classical theory, whereas the quantity δ^2 can be assumed constant at $T - T_c > 25$ K. However, when T_c is approached, the latter quantity decreases rapidly to some finite value which is in good agreement with the theoretical calculations of [308].

A similar comparison [113] of the critical scattering at Nb_3Sn with Eqs. (21.5) and (21.7) of the phenomenological theory was made. By contrast to [308] where the neutron spectra of $SrTiO_3$ and $KMnF_3$ were obtained at small **q** (measured from the point **R**), the **q** dependence of the neutron scattering (near the point **Γ**) was especially studied at various temperatures in the case of Nb_3Sn. The ratio of the intensity of the central peak to the intensity of the lateral peaks is proportion to $\delta^2/\omega_0{}^2$ [according to Eq. (6.31)] and is depicted in Figure 65 as a function of **q**. It follows from the data that the quantity δ is weakly temperature-dependent but that there exists a clear linear dependence of δ upon **q**. In order to determine the nature of the slow relaxation mode in substances in which a central peak is observed in the scattering spectrum, it is very important to establish the temperature dependence and the **q** dependence of the parameters of the phenomenological theory.

The observation of a three-peak spectral distribution of the neutrons was the starting point for a number of possible microscopical interpretations of the frequency dependence provided by the phenomenological equation (21.3) for the attenuation of the soft mode. Our explanations are based on the interaction of the soft mode with other collective motions of the system, e.g., with local fluctuations of the energy density. A central mode appears when the vibration period of the soft mode becomes comparable with the time required for establishing a local temperature or becomes comparable with a shorter time [156, 175].

When the spectra of the scattered neutrons (and also of scattered light) are interpreted near the point of a structural phase transition, the importance of the various types of defects (impurities, dislocations, domain boundaries, etc.) must be taken into account. Scattering at defects can near phase transitions can increase strongly upon approaching the transition point [18, 61, 183]. In terms of physics, this effect is based on the fact that when local changes in the order parameter occur near a defect, the size of the region disturbed by the defect increases strongly near T_c because the correlation length ξ increases sharply at $T \rightarrow T_c$. An increase in the size of the region distorted by a defect increases the scattering cross section of a single defect and, hence, increases the intensity of scattering caused by fluctuations of the defect concentration. As has been shown in [191, 304], the intensity maximum of scattering at deflects can exceed the intensity of scattering in the regular system by several orders of magnitude when the defect concentration is about 10^{18} cm^{-3}.

Defects can also lead to the appearance of a central peak. Obviously, a considerable contribution to the intensity of the central peak observed with neutron or light scattering at $SrTiO_3$ is provided by the defects. It was shown [194] that when deuterium is intentionally admixed to an $SrTiO_3$ crystal, the intensity of scattering of light increases near T_c in proportion to the impurity concentration.

The authors of the theoretical study [61] found that defects can substantially change the form of the spectral distribution of the scattering cross section and, in particular, may cause a substantial broadening of the lines corresponding to the soft mode. This is associated with the fact that the region distorted by the defect is a region of increased energy dissipation. Since the size of this region increases at $T \rightarrow T_c$, also the contribution of the defects to the attenuation of the vibrations increases. It is well possible that the considerable line broadening, which is observed in certain crystals near T_c and which correspond to a soft mode, is associated with crystal defects (see [18]). The reader can find a discussion of the various theoretical aspects of this problem in [61].

Critical indices. When we expounded the dynamic investigations of structural phase transitions, we did not yet treat the critical phenomena arising very close to T_c. As we observed earlier, the appearance of a central peak in the fluctuation spectrum and a sharp growth of this peak when T_c is approached are of a particular nature and are not associated with the critical behavior of the system. We described in Section 17 the critical dynamics which arises in the critical state of ferromagnets, i.e., when the wave vector is much greater than the inverse correlation length ($q >> \varkappa_1$). In this region, it is hard to obtain microscopic expressions for the correlators determining the cross section, and therefore the hypothesis of dynamic scaling must be employed. The essence of this hypothesis is that the correlation function is in the critical region a homogeneous function of the momentum $\mathbf{q}$ and the energy ε [Eq. (17.18)], with the characteristic energy $\Omega(\mathbf{q})$ — which defines the width of the spectral distribution — depending upon $\mathbf{q}$ and $\mathbf{k}_1$ in accordance with Eq. (17.20).

Since the behavior of systems at phase transitions of the second kind is of a universal form, we can assume that the hypothesis of dynamic scaling is valid also in structural phase transitions. In this case, which of the quantities we use must be identified with the characteristic energy $\omega(\mathbf{q})$? We observed that the simplest expression for the cross section of scattering at an attenuated soft mode [see Eq. (6.22)] describes two approaching phonon peaks and their merging into a single peak under conditions of an overattenuated soft mode ($\omega^2(\mathbf{q}) < (1/2)\Gamma^2$). In this case, the width of the peak must be identified with $\Omega(\mathbf{q})$. In other words, the hypothesis of dynamic scaling becomes operative when an overattenuated phonon was already formed.

An other situation arises when the phase transition is characterized by the diffusion mode as, e.g., in the case of ferroelectric order–disorder transitions without tunnelling. Then the cross section of scattering is described by Eq. (20.19) and the quantity $1/\tau$ must be identified with $\Omega(\mathbf{q})$. The temperature dependence of the relaxation time τ is characterized by the critical index $\Delta(\tau \nleftarrow (T - T_c)^{-\Delta})$. In the classical theory, $\Delta = 1$; the authors of [195] established in an experiment with $NaNO_2$ that Δ differs from 1.

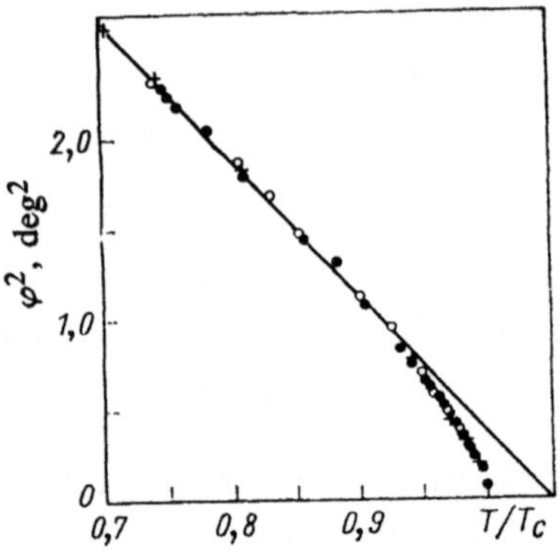

Fig. 66. Temperature dependence of the order parameter in $SrTiO_3$ [264].

Unfortunately, little research has been done on the dynamics of a structural transition in the critical region and to date one cannot assess how the hypothesis of dynamic scaling is fulfilled in these transitions. When compared with magnetic transitions, the data interpretation is difficult because a number of complicating circumstances develop in structural transitions: a central peak appears and the soft mode or the diffusion mode is coupled with the other optical modes. The authors of the reviews [58, 136] have analyzed this problem in greater detail.

In research on static critical phenomena in structural phase transitions, particular difficulties are encountered because the parameter of order is often strongly coupled with both homogeneous and local deformations: local deformations can cause broadening of the phase transition over several degrees and the width of the critical region in ferroelectric and antiferroelectric materials is small. But coupling with a homogeneous deformation can change the critical indices of the system in comparison with the indices in a rigid lattice.

A phase transition of the second kind occurs only in very few ferroelectric materials. The critical behavior was most thoroughly studied in triglycine sulfate (TGS) (see, e.g., [134]). However, since we have not described the structural transition in this substance, we will not discuss its behavior at the critical point. Among the ferroelectric materials of the perovskite group, the solid solution $PbZr_{0.9}Ti_{0.1}O_3$ is characterized by a structural phase transition of the second kind and the value $\beta \approx 0.33$ was obtained for the order parameter [141], with this value differing significantly from the classical value $\beta = 0.5$.

The transition in the antiferroelectric perovskite $SrTiO_3$ is an excellent example of a phase transition of the second kind. Figure 66 illustrates the

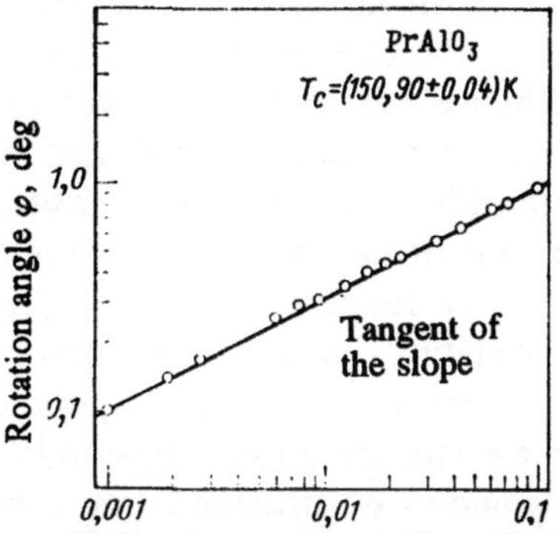

Fig. 67. Temperature dependence of the order parameter in $PrAlO_3$ [120].

temperature dependence of the angle of rotation of the TiO_6 octahedra; this angle is the parameter of order in this transition. A noticeable deviation from the classical law $\varphi \sim \sqrt{T_c - T}$ is observed near T_c. Neutron diffraction measurements of the intensity of the Bragg peaks caused by the superlattice have shown that $\beta = 0.34$ [292]. The value $\beta = 0.33$ was obtained for the other perovskite $LaAlO_3$ [281]. Integral measurements of the diffuse critical scattering enabled the authors of [292] to determine the critical indices of the correlation length:

$$\nu = 0{,}76 \pm 0{,}08 \; (T > T_c), \; \nu' = 0{,}55 \pm 0{,}09 \; (T < T_c).$$

Let us compare these results with the predictions of the theory of fluctuations. The structural transition in $SrTiO_3$ takes place via the three-dimensional irreducible representation of the initial cubic group of the crystal, and therefore the number of components of the order parameter is $n = 3$. For the isotropic n-components model, Eqs. (7.20) were derived for the critical indices; the equations had been obtained in a first approximation involving the ε expansion. From Eq. (7.20) we obtain $\nu = 0.61$ for the index of correlation length in the case $n = 3$. Calculations made in the next order of the ε expansion improved this value to $\nu = 0.69$ [103, 135]. When the cubic anisotropy is taken into account, the critical index of the isotropic vector model is not changed at $n < 4$ (as had been noted in Section 6). Thus, the theoretical ν value for $SrTiO_3$ must be 0.69 and, within the error limits of the measurements, this agrees with the above-indicated experimental value $\nu = 0.76$. According to the theory, the critical index β must be equal to 0.38 in this particular case. The experimental value $\beta = 0.34$ is slightly smaller.

A deviation of the temperature dependence of the order parameter from the classical form was not observed in the other perovskite $PrAlO_3$ (Figure 67). This indicates that either the critical region is extremely narrow or long-range dipole forces play a peculiar role in this phase transition as the phase transition tends to the classical form when the radius of the forces of the interactions between the particles increases.

The fundamental review [134] presents data obtained in research on the static critical indices in structural phase transitions. It should be noted that there are fewer experimental data available than for magnetic materials because, apparently, the research efforts focused on the dynamics of structural transitions; the dynamics turned out to be much more complicated and variable than in the case of magnetic phase transitions. Let us note that the critical scattering near a structural phase transition is often highly anisotropic. Thus, electron paramagnetic resonance measurements in $SrTiO_3$ have shown that the correlation length on the (001) plane is about 40 times greater than between the planes [357]. The layered structure of the fluctuations is a rather common feature of structural transitions in perovskite with a nonvanishing wave vector of the superlattice.

In the last few years, a new class of phenomena has been intensively studied: phase transitions from a noncommensurate structure to a commensurate structure; such phase transitions are observed in a large number of crystals.

In these transitions, the wave vector of the long-period superlattice developing in the crystal at the temperature T_c changes continuously with the temperature and, at some point T_i, changes suddenly to the value corresponding to a new structure which is commensurate with the period of the basic lattice. Near T_i, the structure of the modulated noncommensurate phase can be described by a lattice of solitons each of which is the domain boundary between sections of the commensurate phase.

When T_i is approached from the non-commensurate phase, the dimensions of the domains increase until, finally, at $T = T_i$, the entire crystal volume is no longer filled with the commensurate phase.

Collective excitations of such a system near T_c, the so-called phasons, were detected by inelastic neutron scattering and Raman scattering of light.

Chapter 4

EXPERIMENTAL METHODS OF NEUTRON SPECTROSCOPY OF PHONONS

This chapter concerns mainly the experimental methods of studying the spectra of thermal Bose excitations of crystals, i.e., their phonon and magnon spectra, with the aid of inelastic scattering of neutrons. In principle, the same methods can be used in research on other types of motions in condensed media, e.g., of hydrodynamic motions in liquids or of individual motions (diffusion and vibrations of impurity atoms). However, our considerations will be restricted almost exclusively to phonon excitations.

Exhaustive information on the phonon spectrum of a crystal is included in the set of phonon frequencies $\omega(\mathbf{q})$ and the phonon polarization vectors $\mathbf{e}(\mathbf{q})$ for all resolved values of the phonon wave vectors $\mathbf{q}$ bounded by the first Brillouin zone. In various solid-state problems, one usually employs, depending on the specific features of the problem, two parameters of the phonon spectrum: either the density function $g(\omega)$ of the phonon states or a differential characteristic of the phonon spectrum, namely the dispersion relation $\omega(\mathbf{q})$ for the phonons.

As has been shown in Chapter 1, inelastic neutron scattering can be used to measure the dispersion curves for phonons together with the polarization vectors; these measurements are possible for all directions in the crystal or, what is the same, for the entire $\omega - \mathbf{q}$ space. Of course, also the density function $g(\omega)$ of the phonon states is then determined. But since this work is very laborious, it cannot be done even for one crystal with proper regard for its symmetry properties, i.e., for the irreducible part of the Brillouin zone. Therefore, the experimental methods, which to date have been developed for studying collective excitations in solids with the aid of inelastic neutron scattering in the case of, for example, phonons, are used either for the measurement of $g(\omega)$ or for scanning the $\omega(\mathbf{q})$ space for the purpose of measuring $\omega(\mathbf{q})$, less frequently in combination with $\mathbf{e}(\mathbf{q})$ in selected, usually symmetric directions in a crystal.

In the first volume of the present edition [77], i.e., in the first and fourth chapter, the principles of stationary and pulsed neutron sources and of work with neutron beams were described: the collimation of the neutron beams, their monochromatization, detection, etc. In what follows we briefly expound the distinguishing features of the neutron spectroscopy of phonons; we assume that the reader is familiar with the principles of work with neutron beams.

22. Measurement of the Density Function of Phonon States

Incoherent scattering. It was shown in Section 2 that the cross section of incoherent single-phonon scattering of neutrons can be used to derive the density function $g(\omega)$ of the phonon states. Let us note that in the case of a Bravais lattice with cubic symmetry, the double differential cross section of incoherent scattering is simply proportional to $g(\omega)$ [see Eq. (2.26)]. In practice, one is concerned with systems which scatter neutrons in both coherent and incoherent fashion and, hence, the problem of singling out the incoherent component of the scattering cross section is one of the principal problems in the experimental determination of the function $g(\omega)$.

Among the metals, only vanadium scatters neutrons in an almost completely incoherent manner and crystallizes in a body-centered cubic structure, and the function $g(\omega)$ of vanadium can be rigorously determined in experiments making use of inelastic neutron scattering (see, e.g., [97a, 90a]).

In some cases, the coherent component of the scattering cross section can be effectively extinguished in elements in which at least one isotope with a negative coherent scattering amplitude is present. To this end one must prepare a sample with an especially selected isotope composition. For example, the natural isotope mixture of nickel has a coherent scattering amplitude of $1.03 \cdot 10^{-12}$ cm. There exists the ^{62}Ni isotope with an amplitude of $-0.87 \cdot 10^{-12}$ cm. Consequently, by preparing a sample composed of the natural isotope mixture of nickel and of pure ^{62}Ni, one can almost completely suppress the contribution of the coherent scattering to the double differential cross section. The function $g(\omega)$ of nickel was successfully determined in [97b] with the aid of a neutron spectrometer using the time-of-flight technique and a sample with a modified isotope composition for which the average amplitude of coherent neutron scattering was equal to zero.

Time-of-flight spectrometer. The time-of-flight technique using a mechanical chopper in a beam from stationary neutron sources is very efficient in research on systems with incoherent neutron scattering. The weak dependence of the probability of scattering upon the momentum transferred in a neutron scattering event by such objects makes it possible to use large solid angles for the energy analysis of the scattered radiation, and this means a high aperture ratio of this technique.

To date a large number of neutron time-of-flight spectrometers have been built [205] for research on the dynamics of condensed systems in various ranges of energies transferred in a neutron scattering at a sample.

The basic idea of a spectrometer experiment is to create a monochromatic beam of neutrons incident on the sample under inspection and to measure the neutron energy and the neutron wave vector after inelastic scattering by determining the scattering angle and the time required by the scattered neutrons for travelling a certain distance.

As a result, one obtains information on the collective excitations in a crystal (e.g., phonons), with the excitations generated or annihilated in a neutron scattering event in the sample.

We consider as an example the neutron time-of-flight spectrometer which for many years was used in the I. V. Kurchatov Institute of Atomic Energy [32]. The high aperture ratio and resolution of this spectrometer were obtained by a cryogenic propane source of cold neutrons and a mechanical high-speed chopper of the neutron beam, with the chopper also acting as the monochromator means.

Figure 68 shows the basic scheme of the spectrometer together with the source of the cold neutrons. The position of the chopper relative to the sample under inspection makes it possible to simultaneously record the scattered neutrons in a wide interval of angles so that, in addition to the energy analysis, the dependence of the probability of scattering upon the transferred momentum can be analyzed. A system of detectors of the

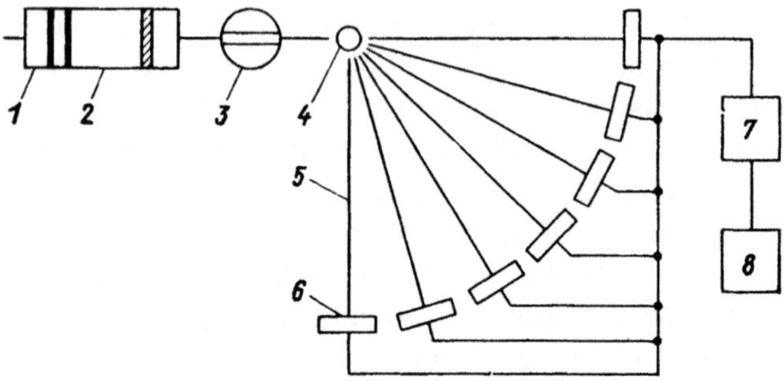

Fig. 68. Basic scheme of a time-of-flight spectrometer with a source of cold neutrons.

scattered neutrons is situated on the vertical plane. A moderator 1 of liquid propane at a temperature of ~ 100 K is the generator of cold neutrons with the wavelength $\lambda \geq 0.40$ nm; the moderator is situated in that part of the cold neutron source which is near the reactor core. The neutrons leaving the source are made monochromatic in a preliminary fashion with the aid of polycrystalline beryllium filter 2 which is cooled to the temperature of liquid nitrogen.

After its collimation, the neuron beam is directed onto a mechanical chopper-monochromator 3 with curved slits to provide additional monochromatization of the neutrons which are incident on sample 4. The energy spectrum of the neutrons scattered at the sample is simultaneously measured under six angles in the interval 15–90°.

Evacuated flight-tubes 5, which end in detectors 6, are used as neutron guides; the detectors signals are applied to a multichannel time analyzer 7. The experimental data are processed in computer 8 with which the spectrometer is connected through a communication channel.

The flight time is measured with this spectrometer in the following fashion: At the moment at which the neutron beam is transmitted, the mechanical chopper applies the start signal to the time analyzer, i.e., the time count begins. Since the neutron flight time from the chopper to the sample is known from the average energy of the monochromatic beam and the chopper–sample distance, the energy of the neutrons scattered at the sample is determined from their flight time on the path of fixed length between the sample and the detector. This time interval is measured with the time analyzer whose memory stores the number of neutron recorded by the detector in the course of a certain time span after starting the analyzer. In this way the fast neutrons appear in the first channels of the analyzer and the slower neutrons appear in subsequent channels.

Therefore, during an experiment one obtains the dependence of the double differential cross section upon the energy transferred in a neutron scattering event at the sample and, after introducing corrections and a mathematical evaluation with Eq. (2.26), one can derive $g(\omega)$.

Strictly speaking, Eq. (2.26) can be used only in the case of spectra of neutrons scattered at an infinitely thin sample when multiphonon scattering can be subtracted and when the measurements are made with an instrument of infinitely high resolution. These requirements are not met in real experiments. Therefore, when $\mathbf{g}(\omega)$ is derived from experimental data, the deformation of the spectra of incident and scattered neutrons in the bulk of the sample, the contribution of multiphonon and multiple single-phonon processes, the influence of the nonmonochromatic primary spectral line of neutrons, and the finite time-of-flight resolution of the spectrometer (determining the total resolution of the setup) must be taken into account. Usually $g(\omega)$ is derived in two stages. First, the experimentally determined spectrum of the inelastically scattered neutrons is reduced to the spectrum of an infinitely thin sample by subtracting multiphonon and multiple processes; after that, from this spectrum obtained with the infinitely thin sample one derives the function $g(\omega)$ with proper regard for the total resolution of the setup.

The factors which were considered in the first stage of the derivation have their influence through the integrals over $g(\omega)$ with weight functions which, in turn, also depend upon $g(\omega)$ but which, in the case of a known phonon spectrum, can be subtracted with any predetermined degree of accuracy. Either the Debye phonon spectrum or the spectrum of the inelastically scattered neutrons proper can be used as a first approximation to $g(\omega)$. In the second stage, the influence of the total resolution (the transfer function) is taken into account; this resolution function changes the form of the weight functions which relate the intensity of the scattered neutrons to the integrals over their product with $g(\omega)$. Therefore, in principle the problem of deriving $g(\omega)$ from experimental data on inelastic neutron scattering is reduced to determining an unknown function from integrals (given with a finite accuracy) over products with known functions; the number of integrals is equal to the number of experimental points in the spectrum of inelastic neutron scattering. This data processing is made with a computer, particularly as the final stage involves the solution of a so-called ill-posed mathematical problem for which certain solution techniques have been developed. This method of deriving information from a spectrum of inelastically scattered neutrons is employed when one wishes to obtain maximum accuracy in a particular experiment. In the case of results obtained with good neutron spectrometers having a total resolution of a few percent, the evaluation of the experimental data is usually restricted to the first stage.

Let us return to the problem of spectrometers. The above-described time-of-flight spectrometer with a cold neutron source (I. V. Kurchatov Institute of Atomic Energy) is operated in a version for obtaining the energy of the neutrons incident on the sample, i.e., in a scattering event, a phonon is absorbed and the neutron is "heated." A low-level background of the experiment and high energy resolution at low energies of up to about 40 meV are the main advantages of a spectrometer working with cold neutrons.

However, this type of spectrometer does hardly allow research on phonons in solids at low temperatures because at low temperatures, the number of phonons is rather small in accordance with the Boltzmann distribution and there is hardly any absorption in a neutron scattering event. Therefore, at low temperatures one must make experiments in which phonons are excited by scattering: a neutron transfers part of its energy to the crystal and "cools." In this case one also uses a time-of-flight spectrometer but the monochromatization of the neutrons incident on the sample by reflection from a single crystal; in this way initial energies of up to 100 meV can be used. The condition that an intermittent is required for the time-of-flight technique is satisfied either by suing a rotating crystal reflecting neutrons with the required wavelength in a specific angular positon or by using a crystal combined with a mechanical beam chopper. Of course, the beryllium filter, which is required in setups working with cold neutrons, is not introduced in spectrometers working with hot neutrons.

The filter function can be successfully fulfilled by a curved neutron guide at the exit of which there are no fast neutrons so that the experimental background is on the minimum level [270]. Depending upon the problems to be solved with the time-of-flight spectrometer, the design of such setups can vary. For example, in order to increase the monochromaticity of the beam of neutrons incident on the sample, systems with a plurality of choppers can be employed. Thus, four choppers are used in one of the cold-neutron spectrometers of the Laue-Langevin Institute (Grenoble, France) [270]. But all these are spectrometers in which the principle of the so-called rectilinear geometry is used, i.e., measurements are made in accordance with the following scheme: monochromatization and chopping of the beam, scattering, energy analysis, and recording. These setups are usually employed with stationary neutron sources.

Spectrometer of inverse geometry. Spectrometers of inverse geometry are to be preferred in work with pulsed neutron sources such as the IBR-30 and IBR-2 reactors of the Joint Institute for Nuclear Research (Dubna) or with linear accelerators in which a pulsating neutron beam is generated via the $\gamma - n$ reaction after stopping electrons in a special target of a heavy element (U, Ta).

In this case, the sample under inspection is irradiated with a pulsed flux of neutron of various energies; after scattering, only neutrons of a certain energy, which is singled out with a monochromator crystal or a threshold filter of beryllium, are detected. The initial energy of the neutrons which are incident on the sample is determined with a time analyzer from the total flight time required for the source-sample and sample-detector paths. Let us note that the latter quantity is known because the neutron energy after scattering is recorded.

The recording of neutron scattering events with an energy spread is the characteristic feature of the spectrometer with the inverse geometry. In the case of a pulsed neutron source with a short flash, it becomes possible to make investigations with a rather high resolution in a wide range of transmitted energies up to 0.5 eV, which is extremely difficult in the majority of the spectrometers working on stationary reactors, particularly when the sample temperature is low.

However, the spectrum of the neutrons which are incident on the sample must be precisely known for the proper determination of the quantities measured in setups of this type.

A spectrometer with inverse geometry has been working for a long time at the Joint Institute for Nuclear Research, especially in research on high-energy excitations in molecular crystals and metal hydrides at low temperature; in addition, a compact spectrometer of inverse geometry was recently built for the linear electron accelerator "Fakel" in the I. V. Kurchatov Institute of Atomic Energy [13c].

We have considered problems related to the experimental derivation of the density function $g(\omega)$ of the phonon states from measurements of the double differential cross section of inelastic neutron scattering.

Incoherent approximation. In the case of samples with coherent scattering, the correct derivation of the function $g(\omega)$ requires averaging over the inelastic structure factor which appears in the cross section of coherent scattering [see Eq. (2.21)]; the averaging must be performed over a surface of equal frequency and the vectors of the reciprocal lattice. This means that the factor must be averaged first over the directions of the wave vector and then over the absolute values of the wave vectors in the irreducible part of the Brillouin zone. In the case of the crystals having a Bravais lattice of some symmetry, an analogous averaging must be performed over the various Brillouin zones. The procedure of averaging the inelastic structure factor is termed transition to the incoherent approximation; a sufficiently correct transition is made on polycrystalline samples when the transmitted momenta exceed the maximum wave vector in the Brillouin zone several times [29, 80a]. This occurs under standard conditions of neutron scattering experiments in the majority of structures when a flight time spectrometer is used at initial energies of ~ 50 meV and the scattered neutrons are analyzed in the angular interval of up to $\sim 150°$.

When systems with approximately identical cross sections of coherent and incoherent scattering are studied, one can likewise use the incoherent approximation to derive the function $g(\omega)$. But the authors of [44a] have shown that completely rigorous separation of the incoherent component of the cross section is possible for Bravais lattices. To this end, measurement must be on a single-crystal sample, and the wave vector transmitted to the crystal in the scattering event is fixed and equal to the vector **b** of the reciprocal lattice. In coherent scattering, the change of the neutron energy is $\varepsilon = \hbar\omega(\mathbf{b}) = \hbar\omega(0) = 0$ because the function $\omega(\mathbf{q})$ is periodic and only acoustic excitation branches in a Bravais lattice. This method is not applicable to lattices with the basis $\omega(0) \neq 0$ because optical vibrations are present.

The proposed method of separating the incoherent component of the scattering cross section [44a] was not employed in experiments as it is difficult to practice this method in time-of-flight spectrometers. The method can be used in crystal spectrometers [132]. However, the availability of such spectrometers and the methods developed for them resulted in favorable conditions for research on the dispersion relations $\omega(\mathbf{q})$ of the phonons, and therefore the experimenters' interest in investigations of $g(\omega)$ has slightly decreased. In principle $\omega(\mathbf{q})$ measurements on a single-crystal sample can be made with a time-of-flight

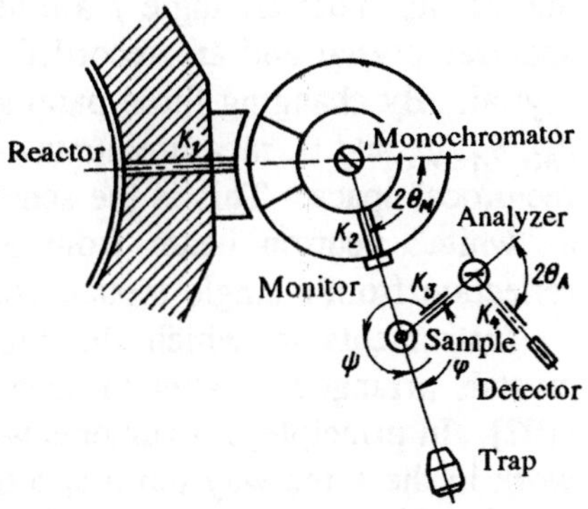

Fig. 69. Scheme of a triaxial crystal spectrometer.

spectrometer too. In accordance with the structure of the cross section of inelastic coherent scattering [see Eq. (2.21)], the spectrum of the scattered neutron represents a set of single-phonon maxima corresponding to the phonons which propagate in various directions of the Brillouin zone of the crystal.

23. Measurement of the Dispersion Curves in the Case of Phonons

The crystal spectrometer. Scanning of the $\omega - \mathbf{q}$ plane along any path is most conveniently made with crystal spectrometers in which monochromatization of the primary neutron beam from the reactor and analysis of the energy of the scattered neutrons are obtained by neutron diffraction at single crystals.

A rapid and relatively uncomplicated change of the experimental resolution of the setup by replacing crystals and collimators is a characteristic feature of crystal spectrometers; similarly, the wavelength λ of the neutrons incident on the sample can be easily modified so that these spectrometers are preferred in experiments in which $\omega(\mathbf{q})$ is measured, i.e., when the parameters of the instrument must be frequently modified in measurements on various sections of the $\omega - \mathbf{q}$ plane.

Figure 69 shows the scheme of a triaxial crystal spectrometer. The neutron beam, which from the reactor is delimited by collimator K_1, is incident on a monochromator crystal. The beam of monochromatic neutrons which were reflected under the angle $2\theta_M$ is via collimator K_2 directed onto a single-crystal sample the orientation of which is characterized by the angle ψ on the plane of scattering. The neutrons scattered at the sample under an angle φ are behind the collimator K_3 incident on the analyzer crystal and are recorded by a detector after reflection from the crystal. By changing three parameters of the four independent variables θ_M, θ_A, φ, and ψ, measurements can be made in any selected direction of reciprocal space. This is the scheme of a triaxial spectrometer in which a "white" neutron beam from a reactor is made monochromatic by reflection from a single monochromator crystal.

Instruments in which the monochromator unit is formed by two crystals arranged parallel to each other ("parallel position") are known [102]. In principle, instruments with one or two monochromator crystals work in the same way but a spectrometer with a double monochromator has certain advantages from the methodological viewpoint. Since the reflectivity of a single crystal is $R \sim \lambda^3$, the monochromatic neutron beam of such an instrument is to a large extent free of admixed higher orders $\lambda/2$, $\lambda/3$, etc. of reflection. Crystals, which are less perfect than a one-crystal monochromatization unit can be used as monochromators. In this case, the experimental resolution is improved in the second reflection. The wavelength of the neutrons is changed without changes in the direction of the monochromatic beam so that the design of the setup is simplified.

However, the additional reflection from the second crystal reduces the neutron flux at the sample by a factor of 2–5; this is important in experiments made on reactors with a neutron flux of $\sim 10^{13}$ neutr/(cm$^2\cdot$sec) because the capabilities of the instrument are limited in this way. Various types of triaxial spectrometers are successfully working in various research centers (see, e.g., [270]); in the USSR, a fully automated spectrometer with a one-crystal monochromator is in the I. V. Kurchatov Institute of Atomic Energy [19]; spectrometers with double monochromators are used in the A. F. Ioffe Physicotechnical Institute and in the Institute of Metal Physics of the Ural Science Center [13b].

Methods of scanning the reciprocal space. The experimental methods of making measurements with triaxial spectrometers were described in detail in [162]. The practice of research work which during many years was made with triaxial spectrometers has shown that the technique of the constant transmitted wave vector ($\varkappa$ = const) and the technique of the constant transmitted energy (ε = const) are the most efficient measurement techniques. In the first technique ($\varkappa$ = const), the wave vector $\varkappa = \mathbf{k} - \mathbf{k}_0$ transmitted to the crystal lattice in a scattering event is recorded (Figure 70).

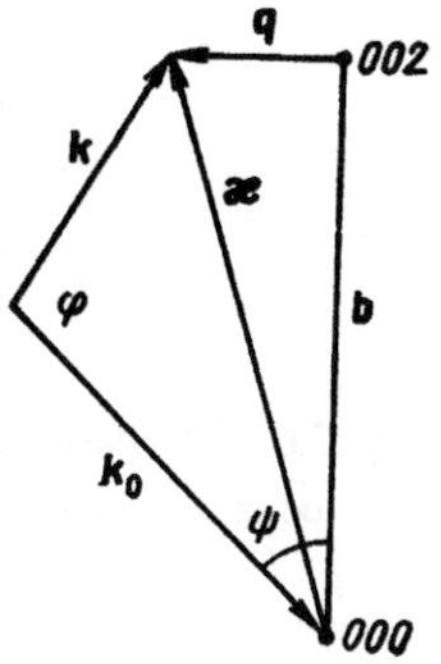

Fig. 70. Scheme of the $\omega(\mathbf{q})$ measurement in the $\varkappa$ = const method.

This means that the wave vector of, for instance, the phonon, $\mathbf{q} = \varkappa - \mathbf{b}$, is fixed. Consequently, in order to determine the dispersion law $\omega(\mathbf{q})$ of the phonons in any direction of symmetry of the Brillouin zone, the phonon frequencies ω corresponding to the previously selected $\mathbf{q}$ value in this direction must be measured. One proceeds as follows: The spectrometer elements are set into positions characterized by the angles θ_M, φ, ψ, and θ_A so that the wave vector conservation law is satisfied for the given $\mathbf{q}$ value (see Figure 70).

The detector count, i.e., the number of neutrons scattered at the sample, is recorded in this configuration. After that, the configuration of the instrument is changed by modifying the spectrometer angles but this is done in a manner such that the condition $\varkappa$ = const is again satisfied. This can be effected in two ways: 1) by fixing the wavelength of the neutrons incident on the sample, i.e., the angle θ_M, and by changing the angles φ, ψ, and θ_A (i.e., the length of the vector $\mathbf{k}$); 2) by fixing the wavelength of the scattered neutrons (i.e., the angle θ_A) and by modifying the angles θ_M, φ, and ψ. After setting the elements of the instrument to the predetermined positions, the number of scattered neutrons is measured again with the detector. By multiple repetitions of these operations, one obtains the dependence of the detector count N upon the energy ε transmitted to the crystal lattice in a scattering event, i.e., $N(\varepsilon)$.

In order to facilitate the processing of the experimental data, the angular positions are in each resetting of the instrument changed so that the energy ε changes with the step width $\Delta\varepsilon$.

The $N(\varepsilon)$ curve also provides information on the collective excitations because a sharp maximum must appear on the curve as the energy conservation and wave vector conservation laws are simultaneously

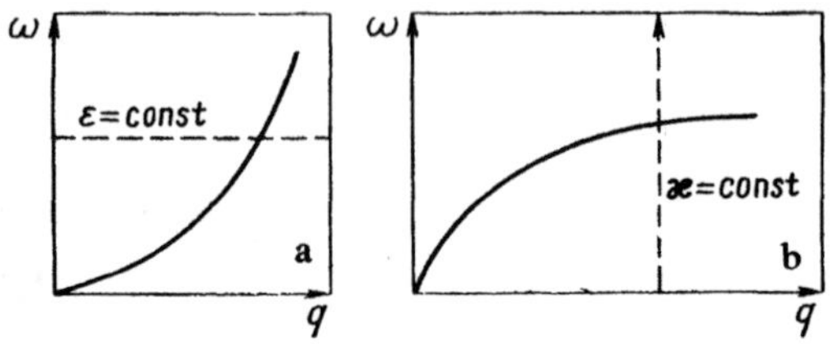

Fig. 71. Comparison of the $\omega(\mathbf{q})$ measurements in the ε = const method (a) and in the $\varkappa$ = const method (b) (the directions of scanning are indicated by an arrow.

satisfied (see Section 2). Since the quantity **q** is fixed in accordance with the definition of the $\varkappa$ = const method, the phonon energy ε corresponding to excitation with the wave vector **q** can be determined from the position of the maximum. In this way, the energy characteristic of the phonon is determined.

Therefore after making a set of measurements with various **q**, the phonon dispersion law $\varepsilon(\mathbf{q})$ or $\omega(\mathbf{q})$ can be obtained in one of the directions within the Brillouin zone of the crystal examined.

It follows from the name that the method of the constant transmitted energy (ε = const) is characterized by a fixed energy ε which is transmitted to the crystal in a scattering event. The wave vectors $\mathbf{k}_0$ and **k** are fixed and the angles φ and ψ are changed so that the end of the vector **k** slides in a selected direction of the vector **q**. In the experiment one measures the curve $N(\mathbf{q})$ on which the maximum is observed while the energy and wave vector conservation laws are simultaneously satisfied in neutron scattering at the crystal. The **q** value corresponding to the selected ε value is determined from the positon of this maximum. Accordingly, this method also makes it possible to determine the dispersion law of the collective excitations in the crystal.

The question of when the various experimental methods are to be preferred relates to the form of the dispersion curves and the resolution function of the setup. When the curvature of the function $\omega(\mathbf{q})$ is large (this is characteristic of magnetic excitations in a crystal), the ε = const method is to be preferred. One then obtains the shortest passage of the finite resolution function of the instrument through the curve $\omega(\mathbf{q})$ (Figure 71a). Vice versa, when the slopes of the dispersion curves is small, the $\varkappa$ = const method must be used in the measurements for recording narrow single-phonon maxima (see Figure 71b).

As has been mentioned above, the accuracy of the results which can be obtained by inelastic neutron scattering experiments is given by the aperture ratio and the resolution of the spectrometer with which the measurements are made and, quite obviously, also by the intensity of the neutron source on which the spectrometer works. With a given neutron source and a particular problem of physics, one tries to obtain with the spectrometer the energy resolution required for the problem and, since an increase in resolution implies a decrease in aperture ratio, the greatest possible aperture ratio at a given resolution. As a rule, the resolution of the spectrometer is the important function. The quantities which are really measured in experiments are a convolution of the resolution function of the instrument with the true law of scattering. Therefore taking into account the resolution is one of the basic problems in the evaluation of experimental data. In time-of-flight spectrometers, the resolution function depends mainly upon the length of the neutron pulse, the error in the flight path lengths, and the width of the time analyzer channels; in crystal spectrometers, the resolution function depends mainly upon the horizontal and vertical angular divergence of the neutron beam, the mosaic structure of the monochromator crystals, the sample and the analyzer, and the Bragg angles used. We stressed above that, depending upon the "curvature" of the dispersion curve, various methods of scanning the $\omega - \mathbf{q}$ plane must be used for obtaining highest accuracy of the results. It follows from general considerations that the recommendations which were given in Chapter 3 and 4 of the first volume [77] for optimizing diffraction experiments apply also to experiments involving inelastic neutron scattering and crystal spectrometers. The questions which are related to the resolution function of neutron spectrometers have been considered in rather great detail in many papers [205, 339] and will not be discussed here.

Let us only note that the spectrometers under consideration facilitate research on the collective excitations when the energy resolution amounts to a few percent in the energy range of fractions to dozens of a milli-electron-volt. These limits make it possible to obtain exhaustive information on the phonon subsystem of crystals with the required precision. Besides that, it becomes possible to undertake research work on fine details of the phonon spectrum such as the group velocity of the phonons or their lifetime so that neutron-based research on the dynamics of the crystal lattice of solids could be advanced to new frontiers.

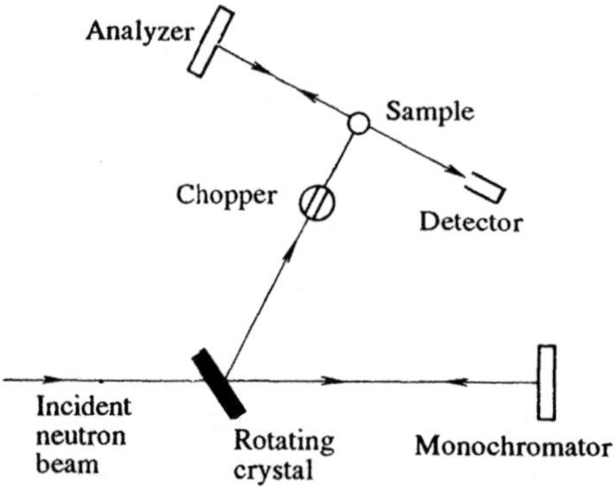

Fig. 72. Scheme of the backscattering spectrometer.

Other spectrometers. Recently there have been built spectrometers which in solids facilitate excitations of extremely low energy while the energy resolution is extremely high (10^{-4}–10^{-5} meV). The backscattering spectrometer of [163] is one of these spectrometers. The principle of obtaining the high resolution with this spectrometer originates from the conventional Bragg condition $2d \sin\theta = \lambda$ of diffraction, which means that $\Delta\lambda/\lambda = \Delta\theta \cot\theta$ is given by the angular divergence of the neutron beam. Consequently, if $\theta \approx 90°$, the nonmonochromaticity $\Delta\lambda/\lambda$ of the neutron beam is almost zero.

The basic scheme of a backscattering spectrometer is shown in Figure 72. The beam is made monochromatic by a crystal under the Bragg angle $\theta_M = 90°$; after that, the neutron beam is rotated by an additional crystal and directed onto the sample. After scattering, the neutron beam is analyzed by a crystal which, like the monochromator, is mounted under the angle $\theta_A = 90°$. The purpose of the chopper is to allow the recording of neutrons which, before being incident on the detector, travelled from the sample to the analyzer and from the analyzer to the detector. The energy of the neutrons incident on the sample is varied within wide limits by gradually moving the monochromator in the direction o the beam or, more precisely, by the Doppler effect developing in the reflection of the neutrons from the moving crystal.

The backscattering spectrometer works successfully in the Laue–Langevin Institute (Grenoble) [270] and facilitates studies of excitations in solids at transmitted energies $\varepsilon \leq 30$ meV with an energy resolutions $\delta\varepsilon \simeq 1$ μeV. The parameters of the setup make it possible to study superfine splitting of levels, tunnelling modes in crystals, diffusion processes, low-frequency excitations and structural phase transitions, etc.

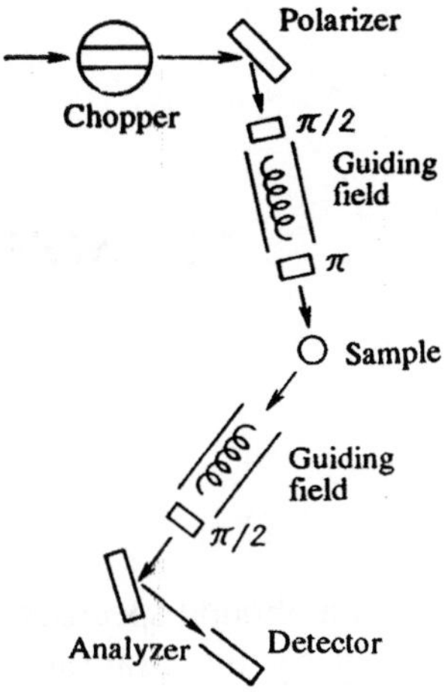

Fig. 73. Scheme of the neutron spin-echo spectrometer.

A setup making use of the spin-echo effect (Figure 73) is an other type of spectrometer with high resolution ($\delta\varepsilon \simeq 10^{-4}$ meV) [250]. Behind the chopper, which effects partial monochromatization, the neutron beam is polarized with a polarizer. The polarized neutrons enter a guiding field in which the neutron spin performs a precession with the Larmor frequency.

At the distance l from the polarizer there occurs a neutron phase shift because the neutrons have different velocities, but if a neutron-spin rotating device is placed at that point, the neutron beam is again in phase at the distance l ("spin echo"). When a sample is placed on the path of the neutrons, the beam will not be precisely in phase because inelastic neutron scattering takes place; the energy transferred to the sample in the scattering even can be determined from measurements of the degree of beam depolarization with the analyzer. By contrast to the backscattering spectrometer, one actually measures in this experiment the Fourier transform of the scattering function and therefore information can be obtained only with additional mathematical data processing.

The experiments and data evaluation show that, in principle, an energy resolution $\delta\varepsilon \simeq 10^{-6}$ meV can be obtained with the spin-echo spectrometer. After reviewing setups with such a unique energy resolution, we must point out that similar instruments working with ultra-cold neutron beams can be built. A gravitational ultracold spectrometer is planned [363]; with it an energy resolution $\delta\varepsilon \simeq 10^{-7}$ meV should be obtained.

Chapter 5

PHONON SPECTRA OF CRYSTALS

The phonon spectra of crystals with an ordered structure can be treated mainly by the example of metals. Depending upon the details of the electron structure, metals can be divided into several groups which we will consider separately, selecting the alkali, polyvalent, nontransition noble and the transition metals. We will not consider rare-earth metals because their systematic investigation is still in the initial stage.

24. Phonon Spectra of Non-Transition Metals

The dynamics of non-transition metals and the pseudopotential. The discussion of the experimental data is conveniently preceded by a short exposition of the theoretical concepts concerning the non-transition metals. At the present time, the greatest progress in the development of a rigorous microscopic theory of metals has been achieved for non-transition metals. This was possible because of the success of the theory of the pseudopotential the principle of which has been expounded in the literature (see, e.g., [94, 95]). The most consistent description of the dynamics of the non-transition metal on the basis of the pseudopotential approach and the methods of multibody theory was obtained in the work of Kagan and Brovman, summarized in the review of [9]. We explain the general scheme which can be applied to simple *sp* metals for describing the phonon spectrum.

The dynamic properties of a metal lattice are given by the potential energy which depends upon the coordinates of the ions and which, in accordance with the adiabatic approximation, can be stated as a sum:

$$E = E_c + E_i + E_{ei}, \tag{24.1}$$

where E_c denotes the energy of the exchange interaction of the ionic skeletons; E_i denotes the energy of the direct Coulomb interaction of the ions which are immersed in a homogeneous, compensating negative background; and E_{ei} denotes the energy of the indirect interaction of the ions via the conduction electrons.

Since in the majority of simple metals, the wave functions of the ionic skeletons are practically nonoverlapping, the first component is usually disregarded. The calculation of E_i is, in principle, without difficulties and is made in the usual fashion with the Ewald and Kellerman technique. The main difficulties of the theory are related to the determination of the last term which is often called the energy of the electron band structure.

In order to solve this problem, one introduces the concept of a relatively weak pseudopotential which describes the interaction of the conduction electrons with an individual ion, and the energy of the electron interaction is defined as a series of perturbation theory in powers of this potential. When one uses the local form of the potential and the perturbation theory is restricted to the second order, one obtains a relatively simple expression for E_{ei} in a regular crystal:

$$E_{ei} = -\frac{\Omega}{2} \sum_{\mathbf{b} \neq 0} |S(\mathbf{b})|^2 |V_{\mathbf{b}}|^2 \frac{\Pi(\mathbf{b})}{\varepsilon(\mathbf{b})}, \tag{24.2}$$

where ω denotes the crystal volume; the summation is extended over the vectors **b** of the reciprocal lattice. This expression is simple because the various factors in it are related to the properties of the crystal lattice, the individual ion, or the Fermi liquid of electrons. The first factor is the structure factor of the crystal:

$$S(\mathbf{q}) = \frac{1}{\sigma} \sum_j \exp(-i\mathbf{q}\,\mathbf{x}_j), \tag{24.3}$$

where $\mathbf{x}_j$ denotes vectors which define the position of σ atoms in the unit cell. The second factor $V_{\mathbf{b}}$ denotes the matrix element of the electron–ion interaction and characterizes the scattering properties of the ion. The third term contains only the characteristics of the electron subsystem, namely the polarization operator $\Pi(\mathbf{q})$ and the static dielectric permeability of the electron gas:

$$\varepsilon(\mathbf{q}) = 1 + \frac{4\pi e^2}{q^2 \Omega_0} \Pi(\mathbf{q}), \tag{24.4}$$

where Ω_0 denotes the volume of the unit cell.

In the simplest approximation of random phases, the polarization operator is expressed by the Lindhard function:

$$\Pi_0(\mathbf{q}) = \frac{3n_0}{2E_F}\left\{1/2 + \frac{1}{4}\,\frac{[1-(q/2k_F)^2]}{q/k_F}\ln\left|\frac{1+q/2k_F}{1-q/2k_F}\right|\right\}, \tag{24.5}$$

where n_0 denotes the density of the electron gas; and E_F denotes the Fermi energy. When exchange effects and correlations in the electron gas are taken into account, the expression for the polarization operator assumes the form

$$\Pi(\mathbf{q}) = \frac{\Pi_0(\mathbf{q})}{1-(4\pi e^2/q^2)\,G(\mathbf{q})\,\Pi_0(\mathbf{q})}, \tag{24.6}$$

where $G(\mathbf{q})$ plays the role of the effective electron–electron interaction. A unique solution of this problem has not been obtained to date and the various versions of the theory lead to a number of different $G(\mathbf{q})$ values while, however, the form of Eq. (24.6) for the polarization operator is maintained. A complete review of the problem can be found [23].

The two last factors in Eq. (24.6) are usually combined by introducing some characteristic function:

$$F(\mathbf{q}) = \frac{\Omega_0 q^2}{4\pi Z^2 e^2}\cdot\frac{|V_b|^2\,\Pi(\mathbf{q})}{\varepsilon(\mathbf{q})}, \tag{24.7}$$

termed also the momentum-energy function which indicates the change in the energy of the electron gas in the interaction of a single component of the perturbation potential with the wave vector $\mathbf{q}$. Having calculated once the function $F(\mathbf{q})$, the energy of the system then can be found for any configuration of the ions.

When the energy of the metal is known, its phonon spectrum can be calculated in the usual way after compiling the force matrix and thereafter the dynamic matrix and solving the secular equation

$$\det|D_{ij}^{\alpha\beta}(\mathbf{q}) - \omega^2 \delta_{\alpha\beta} \delta_{ij}| = 0, \tag{24.8}$$

where $D_{ij}{}^{\alpha\beta}(\mathbf{q})$ denotes the dynamic matrix the elements of which has the subscripts of the Cartesian coordinates α, β and of the atoms i, j in the unit cell. The dynamic matrix, as well as the potential energy of Eq. (24.1), are split into separate terms. The following expression is obtained for the electron–ion contribution in the approximation in which Eq. (24.2) was stated for the dynamic matrix:

$$D_{ij}^{\alpha\beta}(\mathbf{q}) = -\frac{\omega_0^2}{\sigma}\left\{\sum_{\mathbf{k}} \frac{(\mathbf{q}+\mathbf{k})^{\alpha}\,(\mathbf{q}+\mathbf{k})^{\beta}}{|\mathbf{q}+\mathbf{k}|^2}\,\psi(\mathbf{q}+\mathbf{k})\,e^{i\mathbf{k}\,(\mathbf{x}_i-\mathbf{x}_j)} - \right.$$
$$\left. -\,\delta_{ij}\sum_{\mathbf{k}\neq 0}\frac{k^{\alpha}k^{\beta}}{k^2}\,\psi(\mathbf{k})\right\}. \tag{24.9}$$

where $\omega_0 = (4\pi Z^2 e^2/M\Omega_0)^{1/2}$ denotes the plasma frequency of the ions; M and Ze denote the mass and the charge of the ions; and σ denotes the number of atoms in a unit cell.

For the purpose of calculating the phonon spectrum with Eqs. (24.8) and (24.9), the form of the electron–ion potential must be given. The problem of selecting the appropriate form of the potential is a substantial part of the theory. The Abarenkov–Heine potential is one of the most frequently used potentials. It has the following form in the real space:

$$U(r) = -\sum_{l} A_l(r)\,\mathcal{P}_l \quad \text{for } r < r_0;$$
$$U(r) = -ze/r \quad \text{for } r > r_0; \tag{24.10}$$

where $\mathcal{P}_l$ denotes the polarization operator selecting the corresponding component with the orbital quantum number l from the wave function upon which this operator acts. The parameter A_l and r_0 of the potential, i.e., the depth and the radius of the potential well, are selected so that this potential properly describes the spectroscopical data on the levels of the free ion. In this case, the calculation of the properties of the metal can be considered derived from "first principles." But such a calculation is not easy to perform, particularly since the potential of Eq. (24.10) is a nonlocal operator. This nonlocal form results from the dependence of the potential well upon the orbital number and the energy. However, in the case of simple metals such as the alkali metals, Mg, Al, etc., this dependence can be disregarded and a simpler local form of the potential can be used; for this potential, the form factor can be stated as

$$V(\mathbf{q}) = -\frac{4\pi e^2 Z}{q^2 \Omega_0}\left[(1+V_0)\cos qr_0 - V_0 \frac{\sin qr_0}{qr_0}\right]\exp\left[-\xi\left(\frac{q}{2k_F}\right)^4\right]. \quad (24.11)$$

The exponential factor at the parameter ξ was introduced for excluding "nonphysical" vibrations of the potential at large q; the value of ξ usually amounts to a few hundredths. When such a simple form of the potential is used, the parameters V_0 and r_0 are no longer selected in accordance with the electron properties of the ion but by trial and error of the metal characteristics to be calculated: equilibrium volume at the pressure zero, moduli of elasticity, frequencies of the phonons, etc.

The alkali metals. Monovalent elements of the first group of the periodic system, namely Li, Na, K, Rb, and Cs, are the simplest representatives of non-transition metal. In them the conduction electrons behave almost like free electrons and their Fermi surface is very close to a sphere. The postulates of pseudopotential theory are applicable to a very large extent. Therefore, the alkali metals were among the first metals which were the objects of detailed experimental and theoretical research work.

At the present time, neutron data on the dispersion curves of phonons are known in great detail for all elements, except for cesium. The high chemical activity and the low Debye temperature of these metals considerably complicate work on them, especially as far as the preparation of single crystals is concerned. All alkali metals crystallize under normal conditions in a body-centered cubic lattice, and the phonon spectrum, which contains only the acoustic modes, can have the simple form shown in Figure 74 for potassium. The spectra of all the metals are similar. When the frequency of the phonons is expressed in fractions of the plasma frequency ω_p of the ions, differences in the masses of the atoms are excluded and the difference in the interatomic distances is approximately taken in to account. On this scale the $\omega(\mathbf{q})$ for the five elements must almost coincide, and differ by 10-15% at the edges of a zone. However, a small but qualitative difference shows up in the Li spectrum: the longitudinal and transverse modes intersect at $q = 0.5q_{\max}$ in the [100] direction, because the group velocity $d\omega/d\mathbf{q}$ of the longitudinal phonons

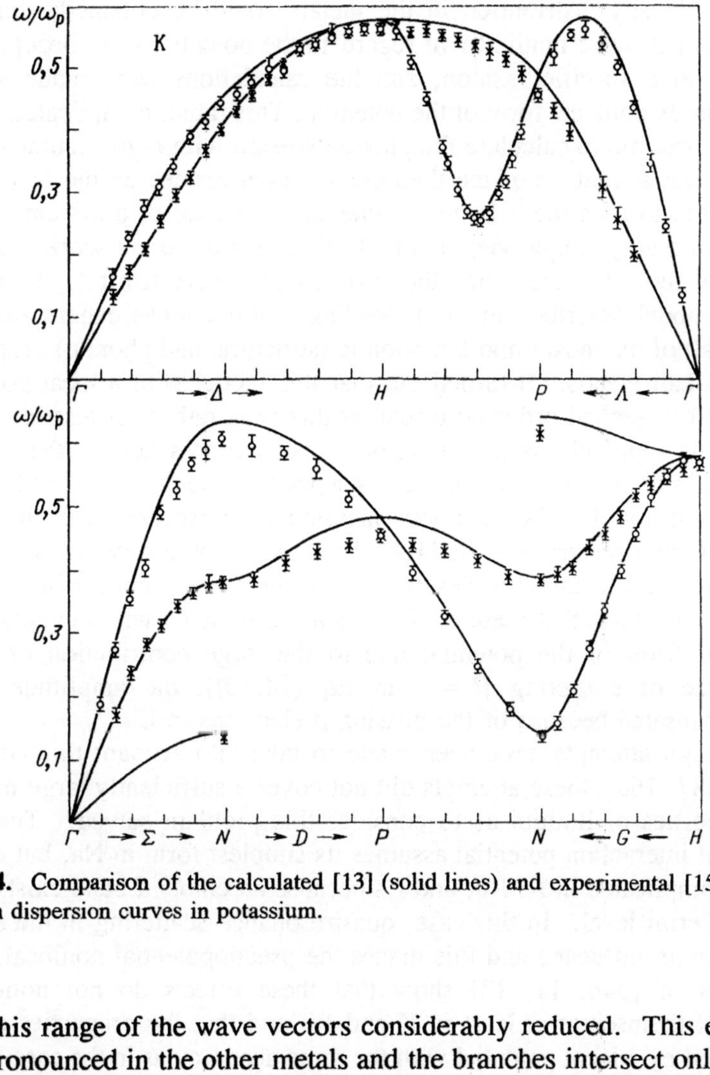

Fig. 74. Comparison of the calculated [13] (solid lines) and experimental [159] (dots) phonon dispersion curves in potassium.

is in this range of the wave vectors considerably reduced. This effect is less pronounced in the other metals and the branches intersect only at the boundaries of the zone where the two frequencies are degenerate because of symmetry relations.

Since the dispersion curves $\omega(\mathbf{q})$ for the alkali metals are simple, attempts to describe them without trial parameters are of interest. One of the latest examples of this approach are the calculations made for a few simple metals [287, 160, 338]. There the potential of the electron–ion interaction was devised so that the most accurate description of the electron density distribution in the metals could be obtained; shielding effects which were nonlinear in regard to the potential were brought into account in a specific fashion, and the calculations were made with a significantly nonlocal form of the potential. This rather complicated model made it possible to calculate the phonon spectra with high accuracy. But the greatest advantage of the theoretical model resides on the fact that it is possible to describe a large number of properties of a system form a unique and very simple viewpoint. In this regard, recent work (see [11, 346] and the references cited therein) has been very fruitful. Using the simple model described in the preceding section, those papers analyzed almost all of the most important atomic (structural and phonon) properties of the alkali metals. It turned out that this model with a local potential restricted to second-order perturbation theory is rather successful.

In this approach, as well as in other calculation schemes, the greatest difficulties are encountered in the description of the properties of lithium. In the majority of cases, the above-mentioned intersection of the branches of the phonon spectrum in the [100] direction is not observed. All atomic properties of Li cannot be described with the same accuracy in a unique scheme. It seems that these difficulties are associated with the noticeably nonlocal form of the potential due to the large contribution of the p amplitude of scattering [$l = 1$ in Eq. (24.10)], the amplitude being uncompensated because of the missing p electrons in Li.

Though attempts have been made to take into account this nonlocal form [287, 160], these attempts did not cover a sufficiently large number of properties of lithium as to consider this problem solved. The electron–ion interaction potential assumes its simplest form in Na, but can be more complicated in K, Rb, and Cs because an empty d band exists close to the Fermi level. In this case, quasiresonance scattering in the d state can have an influence and this makes the pseudopotential nonlocal. The analyses of [346, 11, 13] show that these effects do not noticeably manifest themselves at least in K and Rb and that the properties can be described by a local potential with the same accuracy as in the case of Na. Figure 74 shows how this calculation reproduces $\omega(\mathbf{q})$ in the case of K. The dispersion curves are obtained with the same accuracy (of a few percent) also for the other metals. The state density $g(\omega)$ which was obtained by a similar calculation [347] for the phonons in K is illustrated in Figure 75. Though direct measurements of $g(\omega)$ with the aid of inelastic neutron scattering were not made in the case of the alkali metals, the high accuracy of the calculated $\omega(\mathbf{q})$ points to a good accuracy of the theoreti-

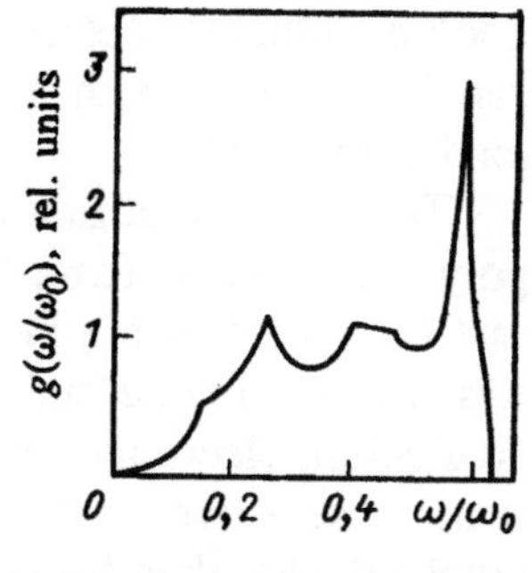

Fig. 75. The function $g(\omega/\omega)$ for potassium (calculations of [347]).

cally determined state density. Of course, the $g(\omega)$ curves of all alkali metals have the same form which is characteristic of body-centered cubic lattices.

Owing to their low Debye temperature and the relatively simple interatomic interaction, the alkali metals are considered suitable objects for research on anharmonicity effects in lattice dynamics. Therefore, the anharmonic properties of these elements – except for Cs – were studied more or less completely and precisely with the neutron technique. The temperature-dependent shifts $\Delta\omega(\mathbf{q})$ of the frequencies and the width Γ of the single-phonon reflections were measured in dependence upon the phonon attenuation. The two quantities increase with the temperature and have an order of magnitude of 5-10% of the phonon frequency near the melting point. The authors of [345] made an almost exhaustive theoretical analysis of the anharmonicity in these metals. They used the above-described model, which had been tested on a large number of harmonic properties, and succeeded in describing, within the error limits, the available experimental data on $\Delta\omega(\mathbf{q})$ and Γ and in analyzing their dependencies upon the wave vector of the phonon and the temperature. This analysis revealed interesting properties of the anharmonic dynamics of the alkali metals. An abnormal temperature dependence, which is characteristic of the soft mode, was detected for the Σ_3 branch. It was shown that not only the harmonic but also the anharmonic properties of all the alkali metals are similar. The relative magnitudes of $\Delta\omega(\mathbf{q})/\omega(\mathbf{q})$ and $\Gamma(\mathbf{q})/\omega(\mathbf{q})$ fluctuate noticeably within the limits of an entire Brillouin zone and, particularly interestingly, remain relatively small up to the melting point proper.

Thus, our understanding of the alkali metals is now very advanced as a consequence of research on the dynamics of the alkali metals with the aid of inelastic neutron scattering and a careful theoretical analysis. A good quantitative description of the atomic properties of the alkali metals can be obtained with a very simple version of the theory with a local model potential.

The polyvalent non-transition metal. The principal difference of the polyvalent metals from the monovalent alkali metals resides on the fact that the radius of the Fermi sphere in them is larger than the size of the first Brillouin zone and that energy gaps appear in the spectrum of the conduction electrons. This complicates the electron shielding and makes it more sensitive to the specific features of the electron–ion interaction potential. We therefore observe a considerable variety of properties in the polyvalent metals. Thus, divalent Mg and trivalent Al seem to be practically the same simple metals as the soft mode. But there exists a large group of elements Zn, Cd, Sn, Pb, etc.) which are much more complicated in regard to their structure and dynamics. Though, in principle, the pseudo potential theory can undoubtedly be applied to these metals, a quantitative description of their properties and, above all, of their phonon spectra in many cases cannot be obtained with the same simple approximations and with the same degree of accuracy as in the case of the alkali metals. Certain limitations of the theory are encountered. However, one can assume that no major modifications are required to overcome these limitations in this case and that the computation scheme can be improved along the earlier established lines. These corrections encompass more flexible, specific forms of the potential, the incorporation of higher orders in perturbation theory and of band effects, the search for more appropriate ways of describing the electron–electron interaction, etc.

The information which was obtained on the phonon spectra with the aid of inelastic neutron scattering is a substantial part of the experimental data on which the theoretical work is based. To date experimental data were obtained for almost the entire group of metals considered, from Be to Pb. Only the earth-alkali elements Ca, Sr, and Ba are an exception: they were bypassed by the experimenters, apparently because of certain difficulties encountered in the preparation of samples. But these difficulties are not unsurpassable and one can assume that this gap will be filled in the near future. First let us consider Al and Mg, i.e., simple metals with face-centered cubic and hexagonal close-packed structures, respectively; after that, we will consider metals of greater complexity on the example of elements of the IIB group.

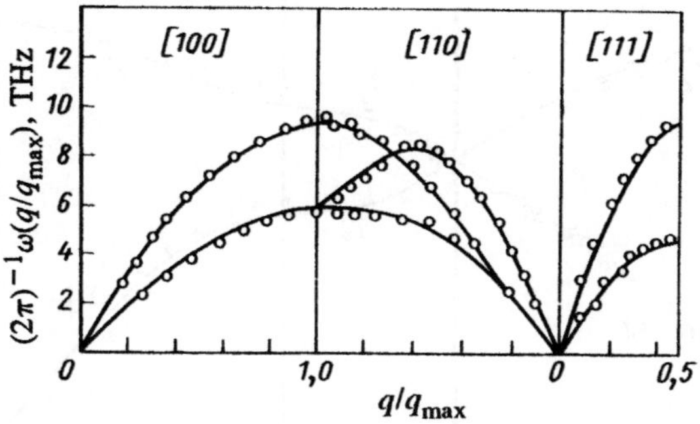

Fig. 76. Comparison of the calculated (solid lines, [9]) and experimental (dots, [330]) dispersion curves for the phonons in aluminum.

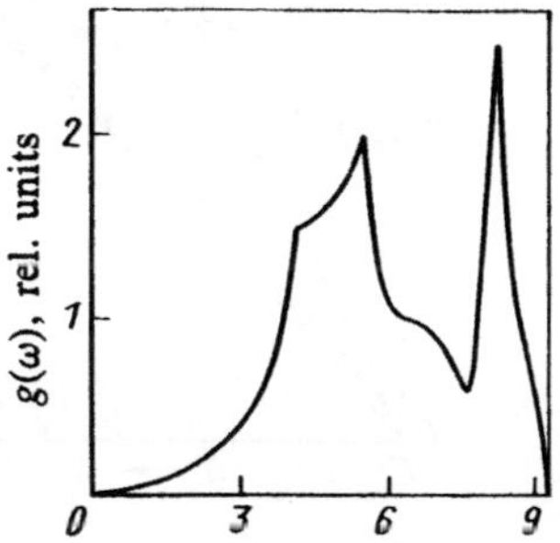

Fig. 77. The function $g(\omega)$ for aluminum [328].

In 1958, the authors of [138] made the first measurements of the phonon dispersion curves on aluminum with the aid of inelastic neutron scattering. Subsequently, $\omega(\mathbf{q})$ was measured several times in Al, and highly detailed results were published in [330] (see Figure 76). The spectrum has a simple structure: there are only acoustic branches with a form characteristic of face-centered cubic lattices. The dispersion curves are smooth and are without noticeable anomalies. However, in a more detailed investigation in which the derivative $d\omega d\mathbf{q}$ is analyzed, one observes that the curves have a well-developed fine structure caused by electron-phonon interaction effects (see Chapter 6), whereas this fine structure is only weakly pronounced in Al, as in the alkali metals, and not visible in Figure 76. In addition to $\omega(\mathbf{q})$, the authors of [328] measured a large number of phonon frequencies in symmetric directions at other

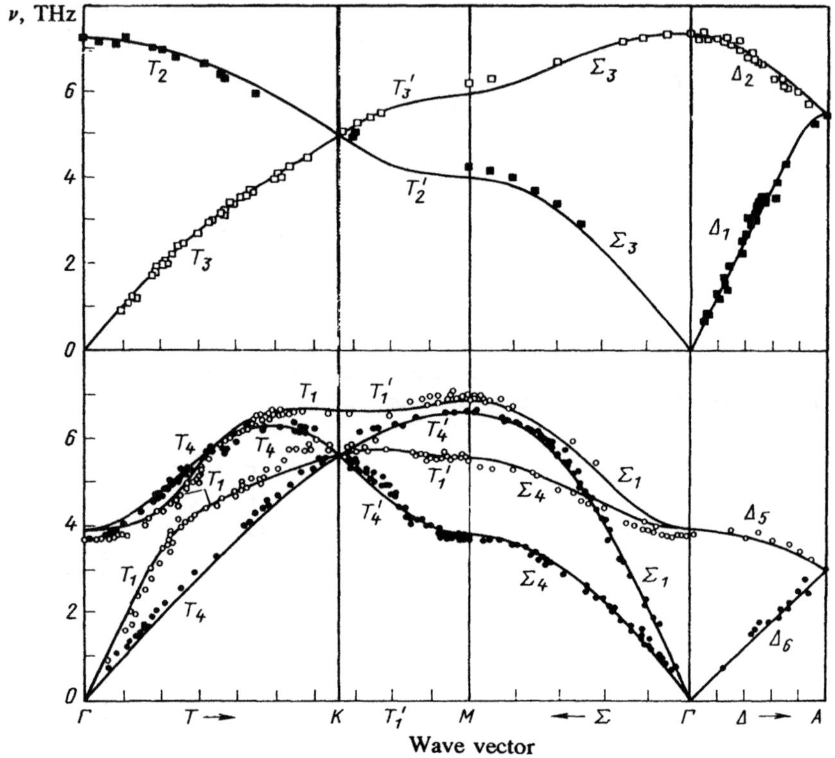

Fig. 78. Comparison of the calculated (solid lines, [9]) and experimental (dots, [284]) dispersion curves for the phonons in Mg.

points of the Brillouin zone. Using interpolation, the authors could derive the frequencies in the entire volume of the zone and calculate the density of the phonon states. The resulting curve is shown in Figure 77 and has also the form characteristic of face-centered cubic lattices with clearly separated regions of transverse and longitudinal vibrations.

Magnesium is a representative of the third group of frequently encountered metal structures: the hexagonal close-packed lattices. The dispersion curves of magnesium were measured [284] with the time-of-flight technique. Though this method is not very convenient for studying $\omega(\mathbf{q})$ in selected directions, the full set of dispersion curves could be measured with relatively high accuracy. The Mg spectrum shown in Figure 78 differs significantly from the spectra of cubic structures: it is

TABLE 19. Properties of Divalent Hexagonal Metals

Metal	Be	Mg	Zn	Cd	Hg
Ratio *c/a* of the axes	1,568	1,624	1,856	1,886	—
Ion-plasma frequency ω_0, THz	49,22	17,78	13,34	8,54	6,10

more complicated due to the appearance of optical branches. But this complication is caused mainly by structural features, and, as noted above, the form of the interatomic interactions of Mg does not significantly differ from the case of Al. The phonon spectra of Al and Mg have been successfully calculated by many researchers using greatly dissimilar methods, among them the pseudopotential method. Though the possibility of describing all atomic properties with a simple, unique scheme was not checked for these metals, $\omega(\mathbf{q})$ was obtained with high accuracy in calculations of this type (see Figure 76). The relative simplicity of the properties of Al and Mg obviously can be explained by the simplicity of the electron–ion interaction potential in which only the s component of scattering is important in this particular case. In order to observe the changes which complications in the form of the potential resulting from complications in the ion structure induce in the phonon spectrum, we will compare Mg with the other elements of the IIB group, namely Be, Zn, Cd, and Hg.

Divalent hexagonal metals. Table 19 lists some information on the lattice properties of the group of metals under consideration, including mercury, though under normal conditions it crystallizes in a trigonal structure.

Not only $\omega(\mathbf{q})$ but also the state density $g(\omega)$ was measured with neutrons for these metals. Direct measurements of $g(\omega)$ are missing only for Hg and the state density was calculated with the Born–von Karman model on the basis of measured $\omega(\mathbf{q})$ vales. The $g(\omega)$ curves are conveniently used for comparing the dynamic properties of the metals under consideration, with the frequency being expressed in fractions of the ion plasma frequency ω_0, as in Figure 79. One realizes that here the similarity observed in the spectra of the alkali metals is missing.

A significant broadening of the band of "reduced" phonon frequencies occurs in proportion to the complication of the ion shell system. One

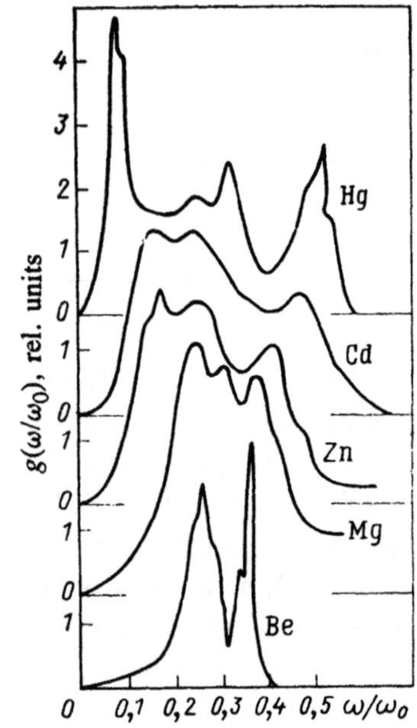

Fig. 79. Reduced density functions $g(\omega/\omega_0)$ of the phonon states.

observes not only a shift of the high-frequency limit of the spectrum but, even more clearly, a shift of the low-frequency maximum toward the coordinate origin. A certain correlation of the shift of this maximum with the appearance of filled *d* shells (in Zn and Cd) and an *f* shell (in Hg) is noted. These changes in the spectrum of the normal vibrations are associated with certain differences in the structural characteristics of these metals.

We infer from Table 19 that the ratio c/a of the axes in hexagonal close-packed lattices increases with increasing atomic number and deviates considerably from the ideal value of 1.633 in the case of Zn and Cd. A large deviation from the ideal close packing is also characteristic of trigonal mercury. The presence of directions in which the interatomic distances are increased and the bonds weakened in the lattice also means an increase in the density of the low-frequency states. An analysis can be made in greater detail on the level of the dispersion curves. We consider for this purpose in greater detail the phonon spectrum of Cd in which c/a exhibits the greatest deviation from the ideal value.

The investigation of Cd with the aid of the neutron technique is complicated because the absorption cross section of slow neutrons is very

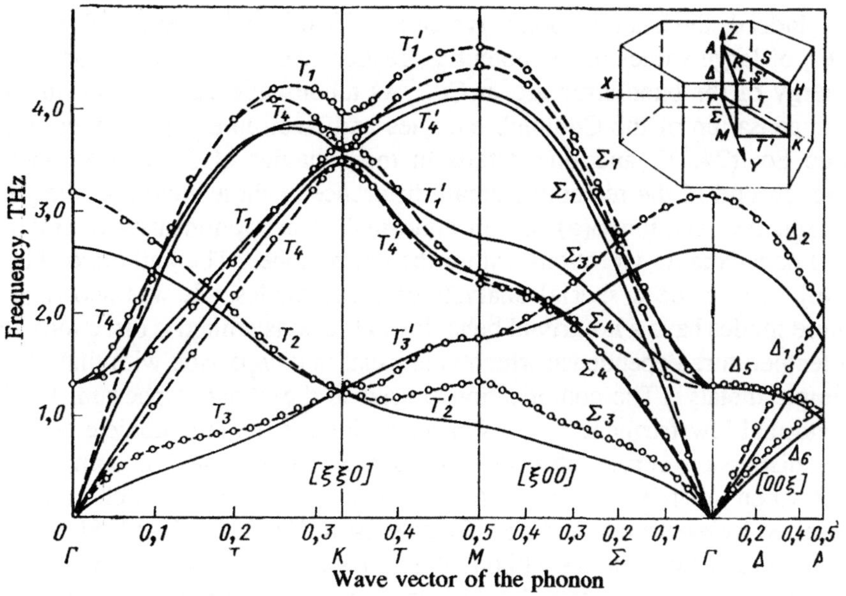

Fig. 80. Dispersion curves of the phonons in Cd (O refers to the experiments of [99, 140]; solid curves refer to the calculation of [165]; and dashed curves refer to smoothed experimental curves).

large in the natural mixture of Cd isotopes. Therefore the phonon spectrum was measured on a sample prepared form the separated ^{110}Cd isotope with weak neutron absorption. Precise $\omega(\mathbf{q})$ measurements were made on this sample at 300 K [99] and 80 K [165] in symmetric directions, and the group velocity $d\omega/d\mathbf{q}$ [140] was measured for the low-frequency branches of the spectrum which were also studied in nonsymmetric directions in the Brillouin zone. Since also g was measured for Cd with the aid of neutrons, Cd can be considered one of the metals in which the dynamic properties have been experimentally studied in great detail. The $\omega(\mathbf{q})$ curve of cadmium is illustrated in Figure 80. One infers from a comparison with the spectrum of Mg (Figure 78) that the frequency of the vibrations which are polarized in the direction of the c axis (branches Σ_3, T_3, $\Delta_{1,2}$) is greatly reduced in Cd; this is a logical consequence of the increased ratio c/a of the axes. These low-frequency modes are of particular interest because the shielding of the vibrations by conduction electrons is most pronounced in them.

Indeed, the **b** lattice point, which is shifted into the region of small q, i.e., of large values of the potential, renders the main contribution to the energy of the band structure in Cd. As a consequence, a strong mutual compensation of the Coulomb energies and electron energies takes place [see Eq. (24.1)], and fine details in the behavior of E_{ei} appear on the background of the remaining small difference of these energies. For this reason the function $\omega(\mathbf{q})$ has an abnormal form: nonlinearity at small **q** and considerable vibrations within the entire zone. The authors of [165, 140] have made a special analysis of these modes. It was shown that these modes have a nontrivial behavior in the nonsymmetric directions and a temperature dependence which is unusual in comparison with that of the simple metals. The complication caused in the phonon spectrum by the abnormal low-frequency modes or the noticeable nonmonotonic effects on the dispersion curves observed in Cd are characteristic of complicated polyvalent metals. Similar low-frequency vibrations, which must be affected by electron shielding effects, exist in Hg [216], Ga [356], and Sn [126] which have lattices without close packing. There are no such modes in Pb with its face-centered cubic structure but $\omega(\mathbf{q})$ is highly nonmonotonic in almost all branches of the spectrum [328].

The possibilities of the pseudopotential theory. The experimental data which were obtained with neutron spectroscopy for the phonon spectra of simple metals allow attempts to answer the question of how the pseudopotential theory explains, and quantitatively describes, the details of the dynamics of such metals. Let us consider Cd as an example. Figure 80 illustrates the results of a $\omega(\mathbf{q})$ calculation made with the scheme used for the alkali metals, Al, and Mg with the potential of Eq. (24.10). A stable spectrum is obtained in the entire range of wave vectors and the accuracy of reproducing the frequencies at the boundary of the Brillouin zone is, on the average, ~ 10% for all spectral branches. The model includes a priori information on the structure of the crystal, i.e., a close-packed hexagonal lattice with the real ratio $c/a = 1.89$ of the axes was used. When the equilibrium value of c/a is determined with this scheme, the value turns out to be close to the ideal value. Of course, when $\omega(\mathbf{q})$ is calculated with such equilibrium values of the structure parameters, the phonon spectrum will be very far form its real form; the positions of the branches are close to the case of Mg. This simple model turns out to be inadequate also for describing the fine structure of the dispersion curves observed in Cd, Zn, Pb, and other metals. They are treated in this model as Kohn anomalies whose positions and magnitudes are determined on the basis of a spherical form of the Fermi surface. But in practice the position and particularly the amplitude of these anomalies are in many cases quite different. As noted before, there exist a few directions in which the calculations can be improved. Let us consider them in greater detail.

One of the basic improvements implies that one goes beyond the second approximation of perturbation theory. This problem was solved when a multiparticle theory of non-transition metals was developed [9]. When third-order and higher-order terms of the potential are taken into account, unpaired interactions between the ions occur in the metal, the interactions being caused by multiple scattering of conduction electrons at certain ions.

One can expect from general considerations that these noncentral forces, which resemble covalent forces, will be particularly important in anisotropic crystals such as Zn, Cd, and Sn. It was observed that these forces can be directly detected in a close-packed hexagonal lattice. As a matter of fact, a paired interaction leads to axial symmetry of the force matrix for each pair of ions. As a consequence, the dynamic matrix can have a symmetry higher than predicted by the space group of the lattice. This feature may lead to additional symmetry in the phonon spectrum at certain points of the Brillouin zone, yet this symmetry is reduced when unpaired forces exist. Thus, the qualitative restructuring of the spectrum and of its scale indicate that there exist unpaired forces and their quantitative characteristic appears. The form of the dispersion law of the phonons in close-packed hexagonal metals near the point K of symmetry of the reciprocal lattice is an example of such a situation. General group analysis shows that the phonon frequencies of the two branches T_1 and the two branches T_4 at the point K are given by the elements of the dynamic matrix in the following way:

$$\begin{aligned}
\omega_1^2 &= \operatorname{Re} D_{11}^{xx} + \operatorname{Im} D_{11}^{xy};\\
\omega_2^2 &= \operatorname{Re} D_{11}^{xx} - \operatorname{Im} D_{11}^{xy} + 2\operatorname{Re} D_{12}^{xy};\\
\omega_3^2 &= \operatorname{Re} D_{11}^{xx} + \operatorname{Im} D_{11}^{xy};\\
\omega_4^2 &= \operatorname{Re} D_{11}^{xx} - \operatorname{Im} D_{11}^{xy} - 2\operatorname{Re} D_{12}^{xx}.
\end{aligned} \tag{24.12}$$

When there is no unpaired interaction, $\operatorname{Im} D_{11}{}^{xy} = 0$ and in this case the frequencies ω_2 and ω_4 are located at the same distance from the degenerate frequency $\omega_1 = \omega_3$. When unpaired forces are present, the degenerate

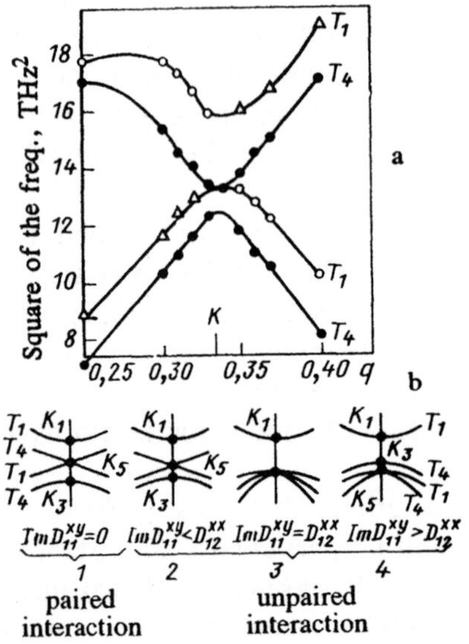

Fig. 81. Appearance of unpaired forces at the point K in the phonon spectrum of Cd: *a*) experimental data for the squares of the frequencies near the point K; *b*) theoretical predictions on various contributions of an unpaired interaction.

frequency is shifted relative to the center $(\omega_2^2 + \omega_3^2)/2$ by 2Im D_{11}^{xy}. Depending upon the scale of the unpaired forces, various approximations of the curves T_1 and T_4 at the point K can be made (Figure 81*b*). Figure 81*a* shows the results of a special investigation of the splitting of the branches at the point K (see [165] for Cd). A clear asymmetry of type 2 indicates that unpaired interaction occurs. A situation of type 4 is encountered in beryllium, and this means that this interaction contributes even more to the dynamic matrix [327]. The well-known disturbance of the Cauchy relations [9] is another proof of the existence of unpaired forces in metals. This means the appearance of additional nonKohn anomalies in $\omega(\mathbf{q})$.

The phonon spectra which were calculated in second-order and third-order perturbation theory differ only in the details mentioned above but their general form and the level of the frequencies are basically unchanged. This refers not only to the simple alkali metals [13] but also to metals of greater complexity, of the type of Cd [165]. When higher orders are taken into account, the ratio c/a is only slightly modified in close-packed hexagonal metals [117].

We note that these negative conclusions relate to calculations based on model potentials with trial parameters. When higher orders are taken into account, the parameters of the potential are redefined, and the two processes often compensate each other. When "first principle" potentials without trial parameters are used, multifrequency interactions must be taken into account for reasons of principle, and this must be substantially affect the calculations.

The use of electron–ion interaction potentials in a form most appropriate for the problem is another important way of increasing the accuracy of the calculations. Our version of the local potential of Eq. (24.11), the form factor $V(\mathbf{q})$ of which depends only upon the change in the electron momentum, is extremely simple. In the general case, the potential is nonlocal and depends upon the energy so that its form factor $V(\mathbf{k}, \mathbf{k} + \mathbf{q}, E)$ has a much larger number of arguments. In the Abarenkov–Heine model of Eq. (24.11), this develops because the depth $A_l(E)$ of the potential well depends upon the energy and the orbital quantum number. In the case of Zn and Cd there exists a reason in terms of physics for this dependence to be noticeable. The reason is that the filled *d* shells of these metals are close to the bottom of the conduction band and even overlap it partially. Under these conditions, the *d* component of the wave function must undergo stronger resonance scattering at the energy E_d.

In recent work these effects were taken into account in calculations of the various atomic properties of Zn and Cd. In one case [128], a resonance term $A_2(E) \sim 1/(E - E_d)$ was introduced in the *d* component of the potential; in other work [257, 139], a potential of $s-d$ state hybridization was introduced to describe noble metals and transition metals. It follows from this work that the hybridization proper and the resonance interactions in Zn and Cd do not play a noticeable role because they are localized at an energy E_d significantly differing from E_f, whereas mainly electrons near the Fermi surface participate in the shielding processes. But this effect influences $A_2(E)$ at $E = E_F$ by increasing it relative to the value obtained in the usual fashion by extrapolation of the atomic terms. This makes the potential in Zn and Cd highly nonlocal.

We note in this connection the work of [313] where an attempt was made to put the elements into a systematic order according to the degree to which their potentials are nonlocal. Depending on the criteria introduced by the authors, almost half of the non-transition metals belong to the category of "nonlocal" metals, among them Li, Be, Mg, Ga, In, Sn, and Hg in addition to Zn and Cd. Though this systematic order is highly arbitrary, it shows once more that any calculation of the atomic properties of these metals, which intends to be accurate, must take into account the nonlocal form of the potential. As far as Zn and Cd are concerned, a c/a value close to the real value [128] was obtained with a nonlocal potential but it was not possible to overcome in this case the main difficulty, namely to describe simultaneously the static and dynamic characteristics with high accuracy.

It was established in a detailed investigation of the dispersion curves of Cd [140] that the simple calculation scheme under consideration does not properly describe the fine structure of the phonon spectra and therefore must be improved, i.e., the deviations of the real Fermi surface from a sphere must be brought into account. It is a well-known fact [9] that when the total energy of a metal is determined, the deviation from the spherical form must be taken into account only in terms of the fourth order and higher orders in the potential. But when the fine structure in $\omega(\mathbf{q})$ is analyzed, one must from the very beginning consider the real surface and bring into account the coherent redistribution of the electron spectrum near the Bragg surfaces. This problem will be considered in greater detail in the chapter on electron–phonon interactions, and we stress here only its importance for the analysis of the spectra of metals such as Zn, Cd, Pb, etc. in which the anomalies in $\omega(\mathbf{q})$ are very pronounced.

Finally, let us note the electron–electron interaction problem on which great efforts were concentrated in the last few years (see, e.g., the review of [23]). Nevertheless, so far a fully satisfactory solution has not been obtained. The problem is to properly describe the exchange and correlation effects in the electron gas during its shielding by the ion potential. One therefore tries to find an expression for the dielectric function $\varepsilon(\mathbf{q})$ of a homogeneous interacting electron gas and to solve the problem of shielding the potential with the aid of this expression.

The correction to the local field $G(\mathbf{q})$ (correction taking into account the exchange and the correlation) differs noticeably in the various papers at $q = 2k_F$ and by the asymptotic value at large $\mathbf{q}$. The question is: to what extent is this difference in the phonon spectra also expressed in the other atomic properties of metals, and which version of $G(\mathbf{q})$ are more realistic; the problem was considered in calculations relating to alkali metal [13], simple non-transition metals [188], and non-transition metals of greater complexity [321]. But the conclusions drawn in work of this type are inconsistent and do not allow a definite adoption of one of the theories expounded.

In addition to the problems listed, the correct calculation of the properties of non-transition metals makes it necessary to take into account a number of less significant factors such as the change of the effective ion charge due to a correlation "hole," changes in the effective mass of the electron, the spin–orbit interaction which is important in heavy elements of the type of Pb, etc. [95]. All this points to the complexity of describing processes which occur in the electron–ion plasma of a metal during vibrations of the crystal lattice.

In this connection, one must remark on becoming familiar with the theoretical and computational papers of this field. Usually the attention of the authors is focussed mainly on the analysis of one of the theoretical aspects listed, and the results of the calculations are compared with a small number of experimental facts (with the phonon spectra, in the majority of cases). It turns out that no definite conclusion can be drawn in this approach. No wonder that of the dozens of published papers, only a few of the authors could not obtain good agreement with experiments! Obviously, in order to obtain real progress, simultaneous self-consistent considerations of all physical factors determining the properties of a metal must be made, and the greatest possible amount of experimental information on the atomic properties of the system must be used for verifying the usefulness of some model; this has been done in, for instance, the work of [11, 13] on alkali metals. The experimental data, in turn, must be sufficiently unambiguous and must not allow many alternative interpretations. The above-considered explanation of multiparticle forces by splitting of the branches at the point K of the phonon spectrum of close-packed hexagonal metals is an appropriate example.

Noble metals. Owing to their electron structure, copper, silver, and gold have intermediate positions between the simple metals and the transition metals. The narrow d zones which in Zn and Cd are near the bottom of the conduction band are in those metals approximately in the middle of the conduction band but they do not yet come to the Fermi level as in the transition metals. Under these conditions, the almost free s–p states are mixed with substantially trapped electrons and a rather complicated structure of the electron spectrum is the consequence [94]. But the states E_F are not very much affected and therefore many properties of the noble metals, including their phonon spectra, are similar to the properties observed in the simple non-transition metals.

The dispersion curves of the phonons have now been measured for all the three elements. They were studied particularly carefully in copper which holds the record of experimental papers describing the measurement of the phonon spectra with the neutron technique (there exists ten papers form which we note the last [227, 272]). Despite the relatively large neutron absorption cross section, reliable $\omega(\mathbf{q})$ results were obtained also

TABLE 20. Reduced Frequencies at the Boundary of a Brillouin zone in the case of Cu, Ag, and Au

Metal	Cu	Ag	Au
Ion-plasma frequency ω_0, THz	7,655	4,857	3,609
Longitudinal phonon in the [100] direction, ω_L/ω_0	0,94	1,03	1,30
Transverse phonon in the [100] direction, ω_T/ω_0	0,66	0,72	0,77
Transverse phonon in the [111] direction, ω_T/ω_0	0,44	0,45	0,51

for Ag [217, 166] and Au [238]. All the spectra have the simple form which is characteristic of face-centered cubic crystals and resemble the aluminum spectrum illustrated in Figure 76. The degree of similarity of the dynamic properties of the three noble metals can be assessed from the frequencies at the edge of the zone; the frequencies are listed in Table 20 in fractions of the ion-plasma frequency. As in the series of the alkali metals, the spectra are similar though one observes that the plasma frequencies do not fully express the existing differences.

In the case of Cu, as well as in the case of Al, the frequencies were measured in nonsymmetric directions and the function $g(\omega)$ was constructed [272]; the temperature-dependent effects [227] were studied in detail and the fine structure of the dispersion curves was analyzed [271]. The anharmonic frequency shift and the width of the single-phonon reflections were of the same order of magnitude as in Al and in the alkali metals. The Kohn anomalies are very weakly pronounced on the dispersion curves and can be established only by a very careful analysis. A significant anomaly, which becomes stronger with increasing temperature and whose nature could not yet be established [227], was observed on the longitudinal branch in the [110] direction at $q \simeq 0.2q_{max}$. Only the dispersion curves in the principal directions of high symmetry are known for Ag and Au.

The properties of the nobel metals can be described with the pseudopotential method only when a certain complication of the computation scheme, which had been used on the simple metals, is introduced. Above all, one must take into account the direct exchange interaction between the d shells of neighboring ions [ther term E_c in Eq. (24.1)]; this interaction is significant because the d shells are much larger than in the alkali metals. In the majority of earlier papers relating to calculations of the phonon spectra, this *d–d* interaction was added to the electron–ion interaction calculated as in non-transition metals with parametric model potentials in local form [96, 269].

The authors of [259, 344] later developed a more comprehensive treatment of metals with filled *d* shell which overlap in the band of the valence electrons. A model of resonance interactions, proposed earlier in the pseudopotential theory of [94, 95], is the basis. The model takes into account the *s–d* state hybridization which makes the potential of the electron–ion interaction highly nonlocal and dependent upon the energy. The difficulties which arise when the energies of a metal with such a potential are calculated were overcome in various ways by the authors of [96, 269]. The method developed in [94] proved to be most practical: a model potential of the Abarenkov–Heine type of Eq. (24.10) with a resonance part for $l = 2$ was used. When the problem of shielding of such a potential is treated, the characteristic function assumes the form of a sum of several terms in place of the simple form of Eq. (24.7):

$$F(\mathbf{q}) = \frac{\Omega_0 q^2}{4\pi e^2 Z_{\text{eff}}^2} [\Phi_{ss}(\mathbf{q}) + \Phi_{sd}(\mathbf{q}) + \Phi_{dd}(\mathbf{q}) + I(\mathbf{q})]. \tag{24.13}$$

where Φ_{ss} corresponds to the contribution resulting from the nonresonance part of the potential, Φ_{dd} corresponds to the contribution resulting from the resonance part, and Φ_{sd} accounts for the hybridization effect. The fourth term contains only the form of the electron–electron interaction with a correction for the local field $G(\mathbf{q})$. The term $\Phi_{dd}(\mathbf{q})$ includes in this version the exchange interaction between the *d* shells, and in this case it is not necessary to separately calculate this interaction as in the other schemes. The calculation of the phonon spectra, of the structures, and of the moduli of elasticity rendered good results in the case of Cu and Ag. In the case of Au, the $\omega(\mathbf{q})$ values were too large, and the authors of [238] explained this by a significant influence of the spin–orbit interaction in this heavy metal.

Let us note in conclusion of this section on the noble metals that the application of perturbation theory is in this case based on the appearance of the small parameter $2\Gamma_d/(E_d - E_F)$, where Γ_d denotes the width of the *d* band. Therefore the method is not applicable to the transition metals in which the *d* band intersects E_F and is not completely filled.

TABLE 21. Transition Metals in the Groups of the Periodic System

III	IV	V	VI	VII	VIII
Sc	Ti	V	Cr	Mn*	Fe Co Ni
Y	Zr	Nb	Mo	Tc	Ru* Rh* Pd
La	Hf	Ta	W	Re	Os* Ir* Pt

25. Transition Metals

The transition elements form the most numerous group among the metals. Table 21 shows the corresponding part of the periodic system from which the classification of these metals with respect to their valency can be inferred. The phonon spectra of a large fraction of the metals listed have been measured to date with the inelastic neutron scattering methods and the number of elements disregarded so far (denoted by an asterisk in the table) decreases rapidly. The results of these neutron studies have shown that the phonon spectra of the transition metals differ considerable from the spectra observed in simple *sp* metals. The complexity of the energy spectrum of the *d* electrons near the Fermi surface implies that the electron–phonon interaction is in these metals usually very strong, and this fact is reflected in their phonon spectra. Before we describe the features of these spectra, let us note that the dynamic properties of the crystal lattice, as well as most of the other physical properties, are very similar in the elements of one group and differ considerably from the elements with different valences.

Transition metals with a hexagonal close-packed lattice. The metals of Groups 3 and 4 have a hexagonal close-packed lattice with an almost ideal dense packing and therefore the general form of $\omega(\mathbf{q})$ in them resembles the spectrum illustrated in Figure 78 for magnesium. The electron–phonon interaction manifests itself in scandium and yttrium in the form of weak Kohn anomalies, but in Ti, Zr, and, apparently, also in Hf, this interaction causes a strong temperature dependence of the frequencies [322]. The unusual form of this dependence is particularly noticeable in the longitudinal phonons which propagate along the hexagonal axis: near the edge of the zone, their frequency decreases with increasing temperature. Figure 82 shows the behavior of these modes in Ti and Zr and in

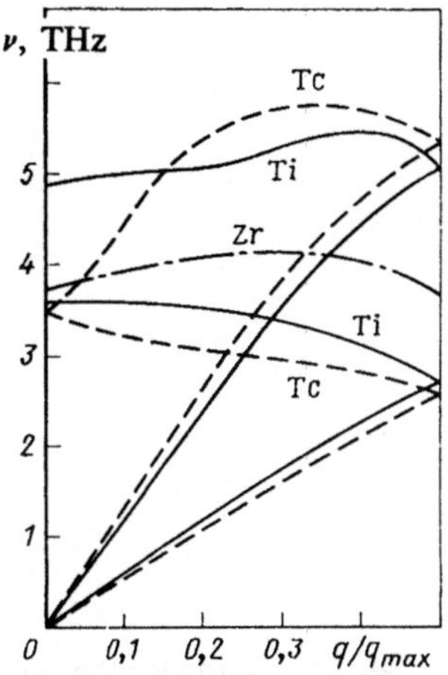

Fig. 82. Dispersion curves for the phonons in Ti, Zr, and Tc.

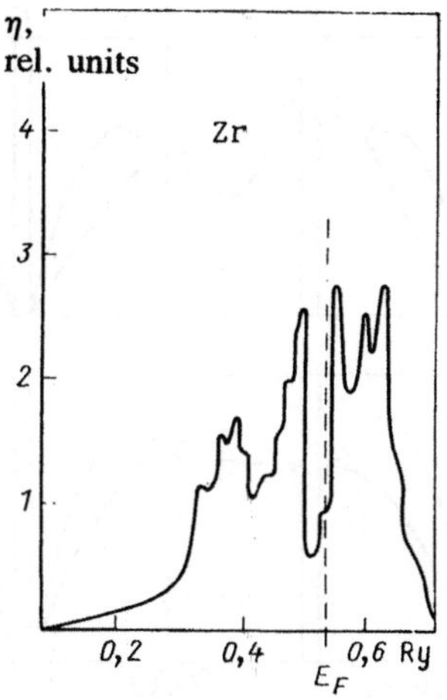

Fig. 83. Curves depicting the density of the electron states in Zr.

a hexagonal metal of Group 7, namely technetium. In Tc, which has a high transition temperature to superconductivity (8 K), the softening of the phonons is greatest, and this means that electron–phonon interaction is involved. A direct calculation of the electron spectrum of these metals makes it possible to qualitatively understand the reasons for the observed effect. Such a calculation shows that the Fermi level of these metals is near a strong maximum of the state density (Figure 83) and shifts with changing temperature toward the maximum; in this way the shielding effect of the electrons upon the interaction between the ions is enhanced and, hence, the frequency of the phonons is reduced.

Let us emphasize that the strong vibrations of the state density near E_F, which are associated with narrow d bands, are in general characteristic of the majority of transition metals and are the reason for many irregularities in their phonon spectra. This holds to the greatest extent for the metals of Groups 5 and 6 whose properties are particularly strongly affected by the electron–phonon interaction.

The transition metals with a cubic lattice. Figure 84 illustrates $\omega(q)$ for Nb and Mo [282]. When these curves are compared with the spectra of the body-centered cubic lattices of the alkali metals, an important complication of the spectrum's structure is noted in the case of the

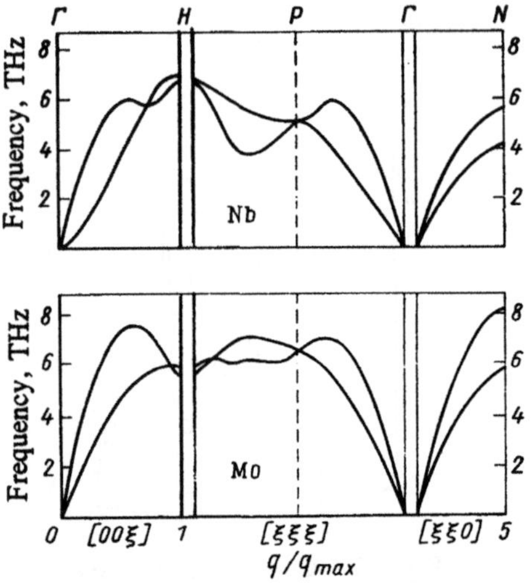

Fig. 84. Dispersion for the phonons in Nb and Mo.

transition metals. The most important feature manifests itself in the longitudinal branch in the [001] direction in Nb: at $q \simeq 0.8q_{max}$, the frequency decreases sharply and the longitudinal branch drops below the transverse branch. It is noteworthy that this anomaly is typical also for the niobium carbides and is correlated with the transition temperature to the superconductive state in these systems. The spectra of Mo, Cr, and W differ noticeably from those of the metals of the niobium group in regard to the structure of the dispersion curves and the frequencies.

The most noteworthy feature of these metals is observed near the point H. A number of explanations of the anomalies observed in these metals have been proposed on the basis of an analysis of their electron structure. They can be explained with a model of strong binding [357a] and with the band theory taking into account resonance processes associated near the E_F with a saddle point originating in the electron spectrum from s–d hybridization [140a]. Besides this, for describing the phonons in these metals, many phenomenological models have been devised; a review of these models can be found in one of the latest papers of this type [177a].

Vanadium has a special place in research on phonons with the aid of neutrons. Belonging to Group 5, the physical properties, including the dynamic properties, of vanadium make it resemble Nb and Ta. But since neutron scattering at the vanadium nucleus is almost completely incoherent, the information obtained by neutrons on the phonon spectrum relates only to the density of state. The function $g(\omega)$ of V was measured a number of times and is at the present time very precisely known. It is typical that the strong anomalies discussed above in relation to the dispersion curves of the metals of Group 5 are observed with rather high accuracy in measurements make also of the function $g(\omega)$ [90a].

The elements of the iron group, and specifically Ni, Pd, and Pt, which close the entire series of transition metals in regard to the electron band structure, resemble the noble metals. The basic differences in the phonon spectra which for these metals were measured with high accuracy result from the more pronounced anomalies which are treated as manifestations of the Kohn effect.

As indicated above, the pseudopotential theory which was developed for *s–p* metals is not applicable to the transition metals. The difficulties which arise in the description of their properties with "first principles" originate from the fact that the *d* electrons, which play the decisive role in these metals, cannot be represented in a model of almost free electrons which is applicable to the *s–p* electrons. But they form rather wide energy bands so that not even the approximation of strong binding can be employed.

The concept of a pseudopotential with strong resonance interaction, i.e., a model which can be used also for the noble metals, is, in principle, applicable to the transition metals. But by contrast to the metals with filled *d* shells in transition elements, the resonance energy is near the Fermi level. The resulting divergence does not allow a proper calculation of the energy of the electron gas with the aid of perturbation theory.

Calculations of the electron spectrum can be made without using perturbation theory and have been made with greater success for the transition metals. Starting from the specific form of the band structure, the observed features of $\omega(\mathbf{q})$ discussed above can be described [357a]. But it is practically impossible to calculate the phonon spectrum as a whole in this way without using perturbation theory. The calculations which were made in [110a, 227] have shown that treatment of the transition metals in analogy to the noble metals can provide acceptable results in certain cases but does not solve the problem in principle.

In recent work [13a, 369] a more consistent description of the transition metals has been developed and can be employed for devising appropriate methods of calculating the phonon spectra on the basis of a pseudopotential.

26. Nonmetal Crystals

Covalent crystals. Covalent bonds exist in the crystals of metals belonging basically to Groups 4, 5, and 6b of the periodic system. As far as the electron-related properties are concerned, one must consider also dielectrics (such as boron, diamond) and semiconductors (Si, Ge, Se, Te), as well as semimetals (Bi, As, Sb). These differences, which stem from the low concentrations of free electrons in the conduction band, are not very important in regard to the atomic properties which depend upon the total energy of the crystal. The majority of the external electrons are in the valence band which, by contrast to the metals, is completely filled in all these materials. In regard to structure, the covalent crystals differ significantly from the metal lattices with close packing: multiatomic structures of lower symmetry are characteristic.

These properties of the structure and of the electron configuration of covalent crystals are also the reason for certain of their dynamic properties to deviate from the properties of metals. Their phonon spectra comprise high optical modes (usually with low dispersion) separated from the acoustic band by an energy gap. As a free Fermi limit does not exist in the electron spectrum of covalent crystals, Kohn anomalies cannot occur in the dispersion curves of the phonons but other manifestations of the electron shielding of the lattice vibrations are possible. The phonon spectra of the group of elements under consideration and of many semiconductor compounds formed with these elements, e.g., GaAs, InSb, etc., were intensively studied during the entire time in which the neutron technique has been used and by now have been rather exhaustively studied in experiments. Germanium was studied particularly thoroughly because it is important for applications, large and perfect crystals are available, and the nuclear properties are suitable for research. Germanium was one of the first objects of research with neutrons: in 1958, the authors of [133a] published the results of first measurements of $\omega(\mathbf{q})$ in two symmetric directions. The authors of [271] obtained highly accurate, complete data for Ge. Figure 85 shows the form of $\omega(\mathbf{q})$ in symmetric directions.

Silicon, which in many aspects is a homolog to germanium, has a phonon spectrum resembling that shown in Figure 85. Two other representatives of Group 4b, diamond and grey tin, have the same structure as Ge and Si and are accordingly similar as far as the form of the phonon spectrum is concerned. However, a special investigation of

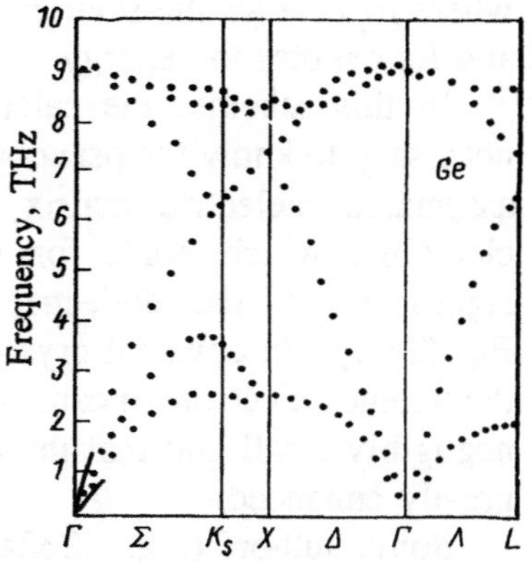

Fig. 85. Dispersion curves for the phonons in Ge.

the similarity of the spectra of the four elements reveals that certain systematic differences reflecting changes in the electron structure do exist [308a]. In the transition from diamond to tin, the gap between their valence bands and the conduction bands changes from 5.4 eV to 0 and the degree of liberation of the valence electrons increases accordingly, implying a modification of the shielding effect of the electrons. The microscopic theory of the covalent bond is at the present time much farther from completion than the theory of the metal bond. A general formulation of the problem of determining the phonon spectrum of semiconductors, as well as of dielectrics and metals, was still given at the end of the sixties [308a, 280a]. The formulation of the problem is based on describing the shielding effect of the valence electrons through the dielectric function of the system.

That part of the dynamic matrix which corresponds to the electron–ion interaction can be stated in the following form (in the case of a uni-atomic lattice):

$$D_e^{\alpha\beta}(\mathbf{q}) = \frac{1}{M\Omega_0} \sum_{\mathbf{bb}'} (\mathbf{q}+\mathbf{b})^\alpha (\mathbf{q}+\mathbf{b}')^\beta V_{\mathbf{q}+\mathbf{b}} V_{-\mathbf{q}-\mathbf{b}'} \chi(\mathbf{q}+\mathbf{b}, -\mathbf{q}-\mathbf{b}'). \quad (26.1)$$

where χ denotes the static response function of the electron density; the function can be expressed through the wave functions of the total Hamiltonian $|n\rangle$ of a static periodic lattice:

$$\chi(\mathbf{q}, \mathbf{q}') = \sum_n \frac{\langle 0|\rho_\mathbf{q}^+|n\rangle \langle n|\rho_{-\mathbf{q}'}|0\rangle}{E_0 - E_n}, \quad (26.2)$$

where $\rho_{\mathbf{q}}$ denotes the Fourier component of the electron density operator, and E_n denotes the energy.

In this scheme, the calculations of the phonon spectrum makes it necessary to know the pseudopotential $V_{\mathbf{q}}$ and the function $\chi_{\mathbf{b},\mathbf{b}'}(\mathbf{q})$ or the associated dielectric matrix, $\varepsilon_{\mathbf{bb}'}(\mathbf{q})$. In the model of almost free electrons, which holds for *s–p* metals, only the diagonal terms are significant in the dielectric matrix, and Eq. (26.1) coincides with Eq. (24.9). In covalent crystals, the local form of the wave functions of the valence electrons means that the nondiagonal elements are no longer negligibly small and that the difficulties in the calculation of $\varepsilon_{\mathbf{bb}'}(\mathbf{q})$ are greatly enhanced.

Some authors (e.g., [232a]) have calculated the full dielectric matrix on the basis of the electron spectrum calculated with the pseudopotential method. Thus, the frequencies of the phonons at certain points of a Brillouin zone were calculated with the "first principles." But this approach to the calculation of the entire phonon spectrum is in practice not useful because it is very laborious. Therefore, in order to bring into account the nondiagonal elements in regard to shielding, various model representations must be employed.

In the case of covalent crystals, the model of charges at the bonds has the best foundation in terms of physics. The model accounts for the fact that an excess density of valence electrons in the intermediate region between the ions is characteristic of a covalent bond. Therefore, the dynamic matrix part defining the contribution of the nondiagonal shielding matrix elements of the pseudopotential is modelled by a potential of the negative charges located halfway between neighboring ions. Beginning with the first paper by Martin [247a], this model, this model has been used several times for calculating the phonons in Ge, Si, and other semiconductors. In some cases (e.g., [1a]), a general formalism of dielectric shielding is used but in the majority of calculations, one introduces many simplifications which render a reduction to the usual phenomenological model with a small number of trial parameters.

But even in the case of complicated structures such as tellurium, one can do with five parameters [157a], whereas the simple trial-and-error determination of its spectrum with the Born–von Karman model requires 62 parameters [280a].

The shell model is an other widely employed model. In it, long-range forces, which are important in covalent crystals, are electrostatic forces and result from the relative displacement of the atomic cores and the

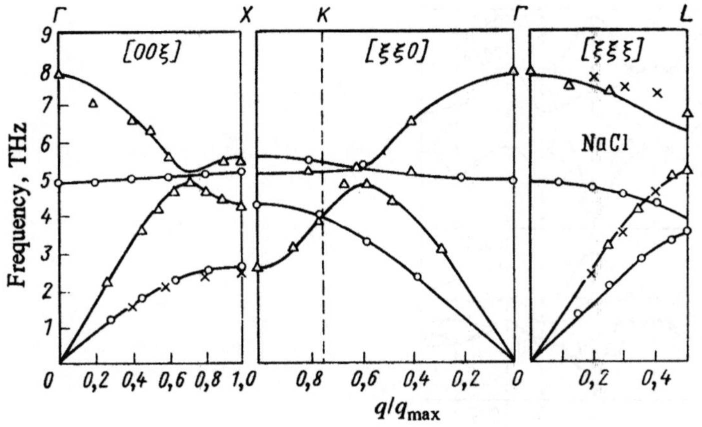

Fig. 86. Dispersion curves of the phonons in NaCl.

spherical shell of the outer electrons. This interaction is usually added to the short-range forces which extend over one or two nearest coordination spheres. This model, in which up to tens of free parameters remain, was widely used for comparatively successful calculations of the phonon spectra of both the semiconductors of Group 4 and many compounds of the Groups 3–5.

This model received a quantum-mechanical foundation shortly after its introduction. It was shown that the model is the lowest dipole approximation of a more general theory describing the lattice dynamics with the aid of a number of electric multipoles [91b]. Though the shell model was successfully used in the case of semiconductors, one must note that it is more adequate for ion crystals.

We terminate our remarks on the model calculations of the dynamics of covalent crystals and refer the interested reader to the review of [308b].

In conclusion, we mention recent work [349a and 349b] in which a theory of dielectric shielding was expounded; this theory can help in the development of rather precise methods of calculating the properties of covalent crystals from first principles.

Ion crystals. This third type of crystal bond is characteristic of simple regular crystals. The five metals (Li, Na, K, Rb, and Cs) and the four halogens (F, Cl, Br, and I) form twenty types of ionic crystals almost all of which to date have been studied with the aid of neutrons to obtain the phonon dispersion curves in symmetric directions. Also anharmonicity

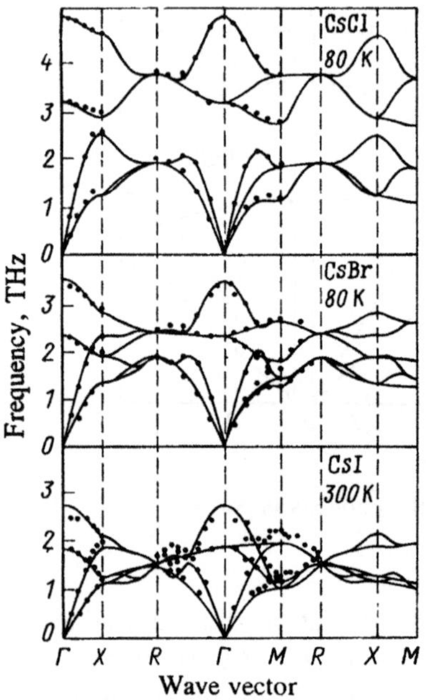

Fig. 87. Dispersion curves of the phonons in CsCl, CsBr, and CsI.

effects, the frequency of the phonons at asymmetric points of the zone, and polarization effects have been studied in some of the crystals. Since for materials of this type exhaustive information on their optical modes had been obtained with infrared and Raman spectroscopy, one can safely say that the dynamics of the ionic crystals have been rather fully studied in experiments.

All alkali halides crystallize in very simple diatomic structures with cubic symmetry of the type of NaCl or CsCl; in the majority of cases the first type is found. Figure 86 shows the phonon spectrum of the NaCl crystal. By contrast to the uni-atomic hexagonal face-centered lattices, a band of optical frequencies with a relatively complicated dispersion law is observed in addition to the acoustic modes. The width of the energy gap between optical and acoustic vibrations is the most characteristic distinguishing feature in the behavior of $\omega(\mathbf{q})$ in the various compounds with such a structure. When the difference in the atomic masses of the metal and the halide increases, the gap becomes wider. A similar relationship is also found in the spectra of crystals with CsCl structure (see Figure 87).

In addition to the alkali halide compounds, there exists a large number of crystals with ionic bonding or mixed ionic–covalent bonding; these crystals are formed from elements of other groups: copper and silver halide compounds and the halides of other metals, metal oxides, etc. To date numerous interesting data have been obtained for them with neutron methods. But their analysis goes beyond the scope of the present book, while the data concerning alkali halide compounds obviously suffice for confirming the available theoretical models of the dynamics of ionic crystals.

Compounds with an ion bond are traditionally considered one of the simplest types of crystal bodies. Experimental and theoretical studies of them have a long history. Until neutron data on the phonon spectra became available, it was assumed that the properties of ionic crystals can be reliably described with a simple model of point charges. The first calculations of the phonon spectrum of NaCl were made with this model by Kellermann as early as in 1940 [219a]. There the ions were considered rigid, nondeformable, nonpolarization spherical particles which interact through their electrostatic long-range potential and a short-range potential (overlapping of the electron shells). The development of neutron research revealed the limitations of this model and led to new, more sophisticated models which to date have increased in number.

When the deformability and polarizability of the ions in lattice vibrations were included, the old model was improved along two lines. One of these relates to a model of dipole deformations and, later on, quadruple deformations (the work of [5a]) may serve as an example of the latter); the second line involved a shell model in the first versions of which the cloud of the outer electrons could shift only relative to the center of the ion.

Afterwards, the two approaches were generalized by introducing various forms of shell deformations into the shell model. We note the three most popular models: 1) the "breathing" shells; 2) the deformable shells; and 3) a shell model with three-particle forces. The authors of [187a, 187b] analyzed the accuracy of these approximations for describing the phonon spectra and several other atomic properties of alkali halide crystals and could show that these models, particularly the model with the third version of the deformations, describe with high accuracy the properties of both the static and the vibrating ion lattice. But it must be emphasized that this was accomplished by introducing a rather large number (eight) of free parameters.

Obviously, a reduction of the number of trial parameters leads to a rigorous microscopic theory. But, as in the case of the covalent bond, this is still problematic for ionic crystals. The difficulties which were noted in the theory of dielectric shielding are even more aggravated because the states of the valence electrons in ionic crystals are more localized. Other alternative approaches considered in a review [118b] are possible but the development of a first-principle theory of the ion bond is a matter of the future.

Chapter 6

ELECTRON–PHONON INTERACTION AND ANOMALIES OF THE PHONON SPECTRA OF METALS

As has been described in the preceding chapter, by now a large volume of experimental information on the phonon spectra of metals has been obtained with the aid of inelastic neutron scattering (see also [162]). This has greatly stimulated new interest in the theoretical description of their static dynamic properties mainly on the basis of the pseudopotential approximation [94, 95].

However, even in the case of the simple non-transition metals, the pseudopotential methods which to date have been developed proved to be inadequate for a detailed description of the phonon spectrum. Only in the case of alkali metals was the pseudopotential approach allowed to describe, with a single model and high accuracy, a large number of properties of the alkali metals ranging from lithium to cesium, the properties including the phonon spectrum, the thermodynamic and elastic characteristics, and their dependencies upon temperature and pressure [9, 13]. Such a successful description of the properties could not yet be obtained for polyvalent metals.

The ideas about the static and dynamic properties of the metals were improved mainly by analyzing the specific importance of electrons for the development of the phonon spectrum [9]. Qualitatively new information required for improving the conclusions referring to metals can be obtained from research on the electron–phonon interaction which must manifest itself in a fine structure of the phonon spectrum of metals and must determine both their local properties and anomalies of the phonon dispersion curves. Until recently it was assumed that the slight deformation which appears in the dispersion curves of phonons in metals owing to the effect of an electron subsystem and to poor performance of systems measuring inelastic neutron scattering with neutron sources of inadequate intensity make it impossible to expect any significant progress in this field. It is therefore extremely important to come to a new experimental frontier in research on inelastic neutron scattering, namely to neutron spectrometry with high energy resolution and high phonon–wave vector resolution; these instruments are now built in a number of research centers around the world. This became possible as a consequence of considerable improvements of both sources and setups for inelastic scattering. Precision experiments made with crystal spectrometers of neutrons now make it possible to refine the dispersion curves of phonons and also provide a unique possibility of investigating anomalies in the group velocity of the phonons and to study the attenuation of collective excitations and a number of other effects associated with electron–phonon interaction. The information obtained in these experiments is of principal importance for research on the peculiar role of the conduction electrons not only in the development of the phonon spectra of solids but also in phenomena such as superconductivity and electron transitions. We consider in the present chapter the results of research in which the fine structure of phonon spectrum in metals was studied with the aid of inelastic neutron scattering.

27. Kohn Anomalies on the Phonon Dispersion Curves

Up to date, features which on the dispersion curves of phonons in metals are associated with the Kohn effect [222] have been most thoroughly studied from the theoretical and experimental viewpoints; the effect originates from a sharp boundary in the Fermi distribution of electrons in metal. The essence of the effect is as follows. Let us consider the expression obtained by Lindhard [133] for the polarization operator:

$$\Pi(\mathbf{q}, \omega) = \frac{-1}{(2\pi)^3} \int \frac{(n_{\mathbf{\kappa}+\mathbf{q}} - n_{\mathbf{\kappa}})\, d^3\mathbf{\kappa}}{E_{\mathbf{\kappa}+\mathbf{q}} - E_{\mathbf{\kappa}} + \hbar\omega}, \tag{27.1}$$

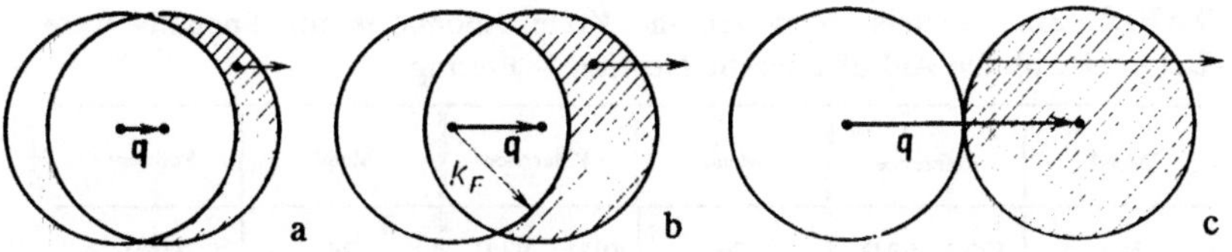

Fig. 88. Scheme illustrating the Kohn effect (k_F denotes the radius of the Fermi surface; and **q** denotes the phonon momentum).

where $n_{\mathbf{k}}$ denotes the Fermi distribution function of the electrons; and $E_{\mathbf{k}}$ denotes the dispersion law. In the case of free electrons, a spherical Fermi surface of radius $\mathbf{k}_F$, and the temperature zero, we have

$$\operatorname{Im}\Pi(\mathbf{q}, 0) = 0;$$

$$\operatorname{Re}\Pi(\mathbf{q}, 0) = \frac{m\kappa_F}{2\pi^2\hbar^2}\left[\frac{\kappa_F}{q}\left(1 - \frac{q^2}{4\kappa_F^2}\ln\left|\frac{q+2\kappa_F}{q-2\kappa_F}\right|\right) + 1\right]; \qquad (27.2)$$

where m denotes the effective mass of the electrons; the frequency ω is assumed to be equal to zero because the energy of the phonons is much lower than the electron energy $\hbar^2\mathbf{k}^2/2m$. For $q = 2\mathbf{k}_F$, the logarithmic term in Eq. (27.2) has a singularity associated with the term $n_{\mathbf{k}+\mathbf{q}} - n_{\mathbf{k}}$ appearing under the integral sign in Eq. (27.1). This means that we must consider only those **k** values for which a state with the wave function $\psi_{\mathbf{k}+\mathbf{q}}$ is filled and for which the state $\psi_{\mathbf{k}}$ is empty or vice versa. For small q values, such states belong to two regions on the Fermi surface (Figure 88*a*). When q increases, these regions become broader (Figure 88*b*), and finally, at $q = 2\mathbf{k}_F$, the entire sphere contributes to $\Pi(\mathbf{q}, \omega)$ (Figure 88*c*). After that, $\Pi(\mathbf{q}, \omega)$ as a function of **q** changes, and though the polarization operator remains a continuous function of q, $d\Pi(\mathbf{q}, \omega)/d\mathbf{q}$ diverges logarithmically at $q = 2\mathbf{k}_F$. In this way the Kohn effect develops in the phonon spectrum of a crystal lattice.

Some component of the lattice potential corresponds to a phonon with the wave vector **q**. The electrons are redistributed and thereby shield this field; as a consequence, the interion interaction becomes dependent upon $\Pi(\mathbf{q}, \mathrm{m})$ or the dielectric function $\varepsilon(\mathbf{q}, \omega)$. The phonon frequency also depends upon $\varepsilon(\mathbf{q}, \omega)$ and the logarithmic singularity in the dielectric function is reflected in the phonon frequency. When the condition $q = 2\mathbf{k}_F$ is satisfied, which with proper regard for the periodicity of the of the lattice can be stated as

$$|\mathbf{q} + \mathbf{b}| = 2\kappa_F, \qquad (27.3)$$

TABLE 22. Metals in Which the Kohn Anomalies for Phonons were Detected with the Aid of Inelastic Neutron Scattering

Metal	Reference	Metal	Reference	Metal	Reference
Mg	[207, 284]	Cu	[271, 273]	Pd	[169]
Al	[329, 360]	Zn	]107, 206]	Cd	[140]
Cr	[312]	Nb	[268, 311]	Pt	[169]
Ni	[169]	Mo	[311, 364]	Pb	[133, 328]

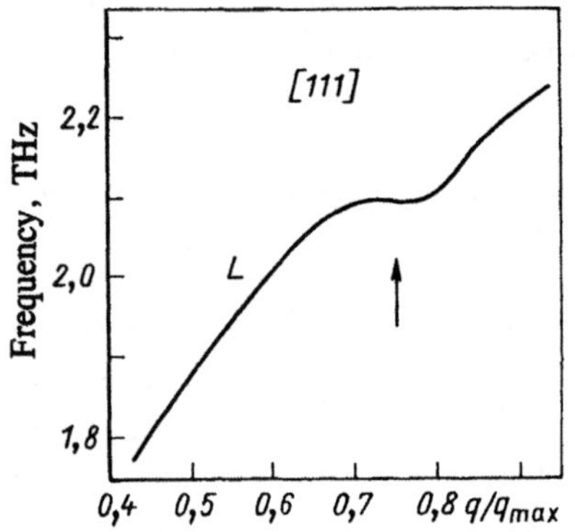

Fig. 89. The Kohn effect on the longitudinal phonon branch of Pb in the [111] direction.

where **b** denotes the reciprocal lattice vector, there appears an inflection on the dispersion curve for the phonon and causes a strong anomaly in the group velocity of the phonons.

The Kohn effect was for the first time experimentally observed in the spectrum of the lattice vibrations of lead [133] (Figure 89). Subsequently, research work was undertaken to detect the Kohn features in the dispersion curves of the metals. A list of the elements studied and the corresponding references are compiled in Table 22. The results of these investigations helped to reliably establish that Kohn features exist in the phonon spectrum and that the anomalies manifest themselves in various ways in different elements.

The singularities are rather strong in some metals [e.g., in Mg, Al, Zn, Nb, Cd, and Pb, clearly nonmonotonic dispersion curves $\omega(\mathbf{q})$ were experimentally recorded], whereas the singularities are less pronounced in other metals. For example, for localizing the singularities in the case of Cu, the measured dispersion curves were compared with the curves derived in the Born–von Karman model and the apparently "undisturbed" conduction electrons [273]. In this analysis, the anomalies are "enhanced"

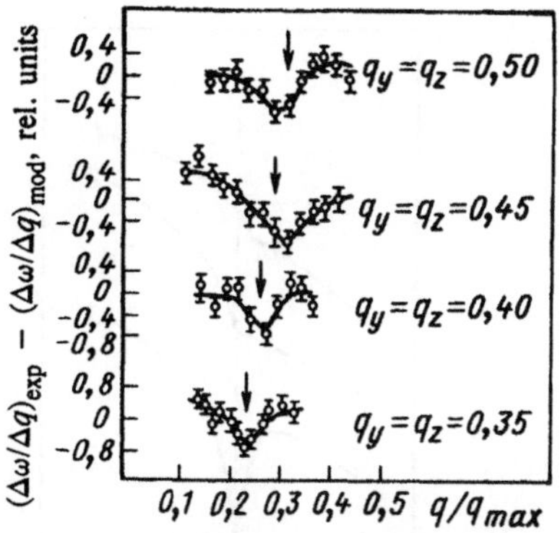

Fig. 90. Detection of the Kohn effect in Cu.

as they appear as a difference $(\Delta\omega/\Delta q)_{\rm exp} - (\Delta\omega/\Delta q)_{\rm mod}$ (Figure 90). Similar investigations were made on individual branches of the vibrations in Pt, Pd, and Ni [169] for which rather weak irregularities in the group velocity of the phonons had been recorded at points of possible Kohn anomalies.

The various forms in which the Kohn anomaly manifests itself were of interest for both experimenters and theoreticians studying the fine structure of dispersion curves of phonons and the characteristics of the Kohn anomalies proper. It was noted in [363] that when in the case of increasing $\mathbf{q}$, the sum $\mathbf{q} + \mathbf{b}$ passes through the Fermi surface from the inside, the phonon frequency increases sharply, provided that condition (27.3) is satisfied, and vice versa. In other words, anomalies both "upwards" and "downwards" are possible in the group velocity of the phonons, and this is entirely caused by geometrical factors. For example, in alkali metals, the Fermi surface as a whole is contained in the first Brillouin zone and, hence, only "downward" Kohn singularities can appear in these metals.

An important characteristic of the Kohn singularities is their magnitude which in the majority of cases is defined as the frequency discontinuity relative to the "undisturbed" $\omega(\mathbf{q})$ dependence in the detuning region Δq. The topology of the Fermi surface has the greatest influence upon the magnitude of the singularities. It was shown [2] that the singularities are significantly increased for a certain class of surfaces. For example, in the case of a cylindrical Fermi surface with the radius $\mathbf{k}_F$, the real part of the polarization operator, which determines the singularity, is

$$\operatorname{Re}\Pi(\mathbf{q},\omega) = \frac{m\kappa_F}{\pi^2\hbar^2}\operatorname{Re}\left[1 - \left(1 - \frac{q^2}{4\kappa_F^2}\right)^{1/2}\right], \qquad (27.4)$$

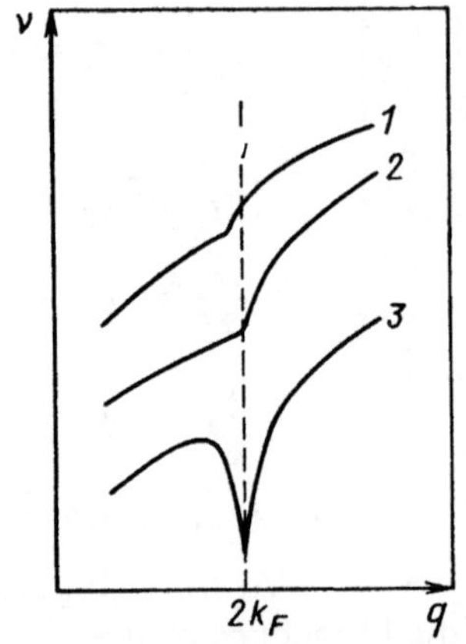

Fig. 91. Appearance of the Kohn effect on the phonon dispersion curve in dependence upon the topology of the Fermi surface in the case of 1) spherical, 2) cylindrical, and 3) plane surface.

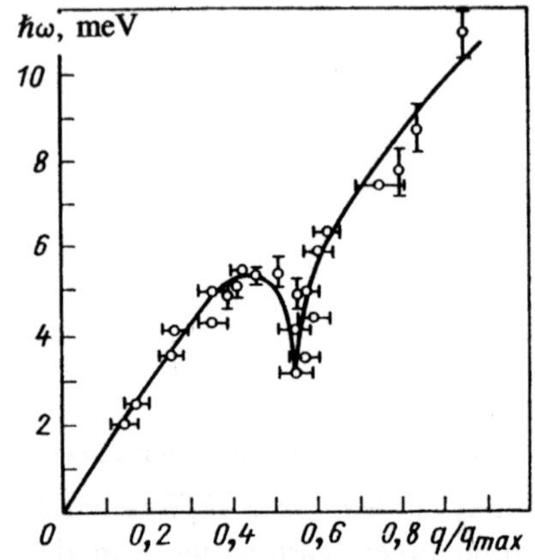

Fig. 92. The Kohn anomaly on the dispersion curve of phonons in $K_2Pt(CN)_4Br_{0.3} \cdot 3H_2O$.

and a root-type anomaly differing from the logarithmic anomaly in the case of the spherical Fermi surface develops in the phonon spectrum at $q = 2\mathbf{k}_F$ (see [94]). The form of the singularity is enhanced in the transition to surfaces close to cylindrical surfaces (due to an increased phase volume) because the phase volume accounting for the generation of electron/hole pairs with a fixed total momentum increases. Enhancement will occur whenever the Fermi surface contains a noticeable region with a radius of curvature substantially greater than $\mathbf{k}_F$. The maximum increase in the phase volume occurs in the case of Fermi surfaces containing plane or almost plane sections. Then the expression for the polarization operator has the form [2]:

$$\Pi(\mathbf{q}, \omega) = \frac{m}{\kappa_F (2\pi)^3} \int d^2\kappa \ln\left(\frac{4\kappa_F}{q - 2\kappa_F}\right)^2, \tag{27.5}$$

and the group velocity behaves like $1/(q - 2\mathbf{k}_F)$ in the region of the singularity, which corresponds to a very sharp anomaly in the phonon spectrum. Figure 91 illustrates the possible situations which are associated with the appearance of the Kohn effect in systems of different Fermi surface topologies. The results of [2] were experimentally confirmed in research on compounds having quasi-one-dimensional conductivity. $K_2Pt(CN)_4Br_{0.3} \cdot 3H_2O$ (KCP) is representative of such compounds. The quasi-one-dimensional character of the conductivity indicates that plane sections are present on the Fermi surface of KCP; in accordance with [2], this implies the appearance of a "powerful" Kohn anomaly in the phonon spectrum. Indeed, in an experiment in research on the fine structure of the longitudinal vibration branch of KCP [290], a giant Kohn anomaly was observed (Figure 92). Later on, a powerful Kohn singularity [318] was observed in the compound TTF–TCNQ which also has unusual quasi-one-dimensional electron properties. Combined studies of quasione-dimensional systems (structural instability, phase transitions, chemical properties, etc.) are now an independent line of work, but research on the fine structure of the vibration spectra – as in model objects – is of principal importance for a proper understanding of the electron–phonon interaction in metals.

Problems related to the importance of the electrons for the development of a fine structure in phonon spectra of metals were most consistently treated by the authors of [9]. In the multielectron theory developed in that work for non-transition metals, the dynamic matrix of the lattice vibrations is represented as a series in powers of the potential V_h/E_F of the electron–ion interaction:

$$D_{\alpha\beta}(\mathbf{q}) = D_{\alpha\beta}^{(i)}(\mathbf{q}) + D_{\alpha\beta}^{(2)}(\mathbf{q}) + D_{\alpha\beta}^{(3)}(\mathbf{q}) + \ldots \tag{27.6}$$

Here and below, a single-atom lattice is considered; $D_{\alpha\beta}^{(i)}(\mathbf{q})$ denotes the dynamic matrix of the vibrations of the ion lattice in the field of a homogeneous noninteracting electron gas; and $D_{\alpha\beta}^{(n)}(\mathbf{q})$ denotes the correction resulting from nth order perturbation theory. The contributions resulting from $D_{\alpha\beta}^{(2)}(\mathbf{q})$ and $D_{\alpha\beta}^{(3)}(\mathbf{q})$ to the square of the phonon frequency are expressed in the form

$$[\omega_j^{(2)}(\mathbf{q})]^2 = 2\frac{\omega_0^2}{M}\left\{\sum_b [\zeta_j^2(\mathbf{q},\mathbf{b})\, V_{\mathbf{q}+\mathbf{b}}^2\, \Gamma^{(2)}(\mathbf{q}+\mathbf{b}, -\mathbf{q}-\mathbf{b})] - \right. \\ \left. - \sum_{\mathbf{b}\neq 0} [\ldots]_{\mathbf{q}=0}\right\}; \tag{27.7}$$

$$[\omega_J^{(3)}(\mathbf{q})]^2 = 6\,\frac{\omega_0^2}{M}\left\{\sum_{\mathbf{b}_1\,\mathbf{b}_2} [\zeta_J(\mathbf{q},\mathbf{b}_1)\,\zeta_J(\mathbf{q},\mathbf{b}_2)\,V_{\mathbf{q}+\mathbf{b}_1}V_{-\mathbf{q}-\mathbf{b}_2}V_{\mathbf{b}_2-\mathbf{b}_1}\times\right.$$

$$\left.\times\,\Gamma^{(3)}(\mathbf{q}+\mathbf{b}_1,\,-\mathbf{q}-\mathbf{b}_2,\,\mathbf{b}_2-\mathbf{b}_1)] - \sum_{\substack{\mathbf{b}_1\neq 0\;\mathbf{b}_2\neq 0\\ \mathbf{b}_1\neq\mathbf{b}_2}} [\ldots]_{\mathbf{q}=0}\right\}, \qquad (27.8)$$

where $\zeta_J(\mathbf{q}, \mathbf{b}) = (\mathbf{q} + \mathbf{b})\mathbf{e}_J(\mathbf{q})$.

The so-called multiterminal networks $\Gamma^{(2)}$ and $\Gamma^{(3)}$ describe the properties of the homogeneous electron gas at a given density and, hence, affect the phonon spectrum of the metals. More particularly, for certain sets of wave vectors these multiterminal networks become singularities because of the nonanalyticity caused by the sharp boundary of the Fermi population in phase space. In order to analyze specific properties of metals, explicit analytic expressions are required for the multiterminal networks but when the fine structure of phonon spectra is described, only the multiterminal singularities are of great interest. The authors of [18] have developed a method which makes it possible to determine the singularities of the multiterminal networks without their explicit analytic calculation and have analyzed in detail the singularities developing in the form of the spectrum. When the dynamic problem was treated and only one second-order term of the electron–ion interaction $D_{\alpha\beta}{}^{(2)}(\mathbf{q})$ was included in the considerations, the singular part of the two-terminal network is expressed by the equation

$$I^{(2)}(\mathbf{q}+\mathbf{b},\,-\mathbf{q}-\mathbf{b}) \sim \left(\kappa_F^2 - \frac{q^2}{4}\right)\ln\left|\kappa_F^2 - \frac{q^2}{4}\right|, \qquad (27.9)$$

and a logarithmic singularity of the Kohn type appears in the vibration spectrum at $q = 2\mathbf{k}_F$. It is significant that in the theory developed by the authors of [9], the anomalies associated with the appearance of the Kohn effect are automatically obtained in the second-order perturbation theory or, which is the same, in the approximation of a paired interaction between ions. The term $D_{\alpha\beta}{}^{(3)}(\mathbf{q})$ in Eq. (27.6) actually means that the so-called unpaired interion interaction through the conduction electrons was brought into account. In this case, the electrons can be additionally scattered at a third ion, creating in this way an indirect interaction between ion pairs; this interaction assumes the form of a three-particle interaction and makes it possible to predict qualitatively new singularities in the vibration spectrum of the metal. The singular part of the three-terminal network $\Gamma^{(3)}$ is as follows:

$$I^{(3)}(\mathbf{q}+\mathbf{b}_1, -\mathbf{q}-\mathbf{b}_2, \mathbf{b}_3) \sim \sqrt{\kappa_F^2 - q_R^2}\,, \quad q_R < \kappa_F, \tag{27.10}$$

where q_R denotes the radius of the circle circumscribing the triangle with the sides $|\mathbf{q} + \mathbf{b}_1|$, $|\mathbf{q} + \mathbf{b}_2|$ and $|\mathbf{b}_3| = |\mathbf{b}_2 - \mathbf{b}_1|$. When the wave vector $\mathbf{q}$ has a value on the dispersion curve, there develops a root-type singularity when an acute-angle triangle with the sides indicated is inscribed into the circle of radius $\mathbf{k}_F$.

In Eq. (27.8) for $[\omega_J^{(3)}(\mathbf{q})]^2$ there is the additional parameter $V_{\mathbf{b}}/E_F$ which is small compared with $[\omega_J^{(2)}(\mathbf{q})]^2$. But the more pronounced root-type singularity of $\Gamma^{(3)}$ as compared with the logarithmic singularity of $\Gamma^{(2)}$ (Kohn singularity) makes us hope that a three-particle singularity will appear in experiments. Nevertheless, the additional factor $V_{\mathbf{b}}/E_F$ which appears every time at increasing n (number of the multiterminal network) seems to pose difficulties to the determination of singularities with $n > 3$. The "hierarchy" of singularities reflecting the topology of the Fermi surface on the dispersion curves reflects very clearly the features of the electron–phonon interaction in metals.

A group velocity $d\omega/d\mathbf{q}$ tending to infinity corresponds to singularities in the vibration spectrum. They therefore must manifest themselves most distinctly on the curves depicting the dependence of the group velocity of the phonons upon the wave vector. First detailed measurements of the group velocity of the phonons in directions of high symmetry of the Brillouin zone of aluminum were made in a unique experiment with high energy resolution and wave vector resolution [360]. The data show that in the transition to detailed investigations of the phonon spectrum. The rather smooth phonon dispersion curves of aluminum seem to reveal the developed fine structure (Figure 93) which reflects the specific aspects of the electron–ion interaction in this metal. The presence of clearly expressed anomalies in $d\omega/d\mathbf{q}$, which are associated with the Kohn effect, have confirmed the prevailing ideas about the Fermi surface of aluminum.

But the work of [360] is valuable mainly because it could show that qualitatively new information can be obtained from detailed measurements of the phonon dispersion curves; this opens new possibilities for research on the electron–phonon interaction and the improvement of the modern theory of metals. In this connection, the authors of [140] carefully studied the fine structure in the spectra of the non-transition metals of the second group of the periodic system, which have a hexagonal lattice with two atoms per unit cell. Among the metals considered (Be, Mg, Zn, Cd), Cd

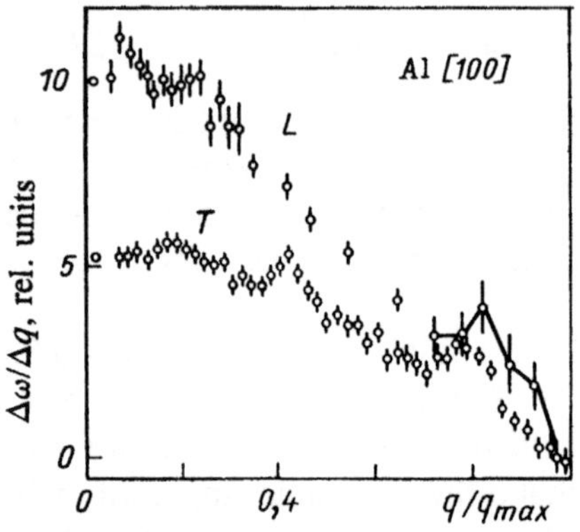

Fig. 93. Group velocity of the longitudinal and transverse phonons of the [100] direction in Al.

is of greatest interest because it has a very complicated structure of the ion and crystallizes in the lattice with the maximum anisotropy $c/a = 1.89$, whereas $c/a = 1.63$ holds for an ideal close-packed hexagonal structure. Such a strong structural anistropy causes substantial compensation of the direct Coulomb contribution and the indirect (through the conduction electrons) contribution to the dynamic matrix and manifests itself in low-frequency branches of the vibration spectrum of cadmium [94, 165]. Such branches allow studies of their fine structure since the effects associated with the electron–phonon interaction must be clearly recognizable on the background of the compensated ion and electron contributions to the vibration spectrum of the lattice.

Measurements which were made on the ^{110}Cd isotope with its weak neutron absorption [140] helped to derive the dependences of the phonon group velocity $d\omega/d\mathbf{q}$ upon the wave vector in the principal directions of high symmetry in a Brillouin zone. A sharp drop of $d\omega/d\mathbf{q}$ at increasing distances from the center of the Brillouin zone is the characteristic feature of the dependencies obtained (Figure 94); this substantially differentiates Cd from metals such as Cu, Al, etc. The group velocity is also nontrivial up to the boundary of the Brillouin zone, vibrates strongly, and shows the existence of pronounced anomalies which manifest themselves on the dispersion curves $\omega(\mathbf{q})$ proper. The identification of the singularities, which was made with the concepts concerning the real Fermi surface of cadmium in a scheme of broadened bands, confirmed the assumption that closed electron surfaces exist in a third band above the (101) faces or, as they are also termed, "butterflies" are missing in real situations. The absence of "butterflies" on the Fermi surface of cadmium means that a surface of free electron states is missing in the [101] direction and that the

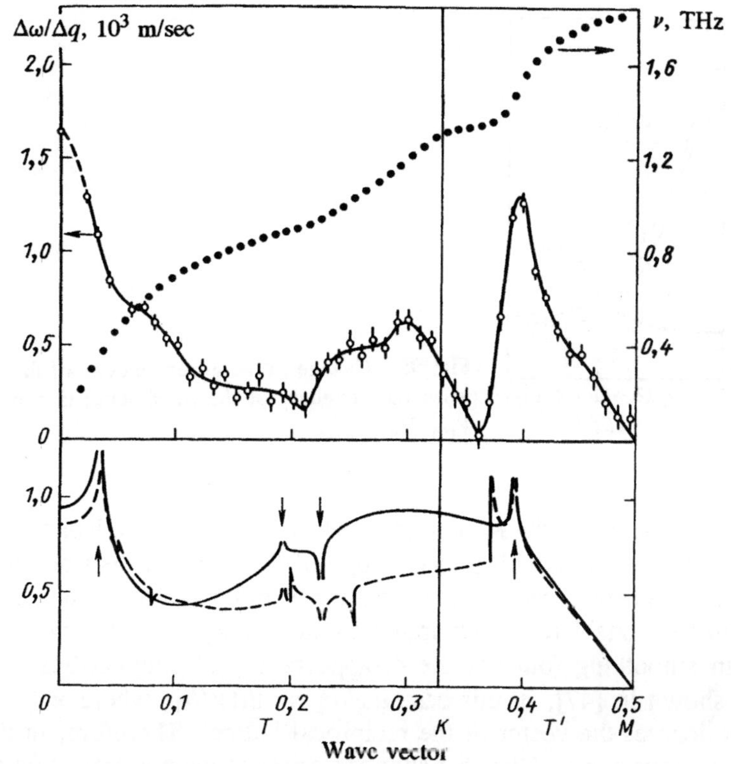

Fig. 94. Group velocity of phonons in Cd for the T_3 branch (●): experimental data (○) and calculation with the second-order (solid curves) and third-order (dashed curves) perturbation theory of electron–ion interaction [140]; the position of the Kohn singularities are denoted by vertical arrows.

(101) planes, which confine a second Brillouin zone, are the boundaries of the filled states.

When the size of the spherical segment cut off by the (101) face decreases, the positions of all singularities associated with the vector **b** (101) of Eq. (27.3) must be shifted toward $\mathbf{q} = 0$. A comparatively small (about 4%) deviation of the boundary of the filled electron states in the [101] direction from the spherical form makes part of the Fermi surface disappear and, consequently, causes the disappearance of an entire group of Kohn anomalies.

The shifting of the boundary of the valence electron distribution on the (101) face of the Brillouin zone must imply a frequency discontinuity at

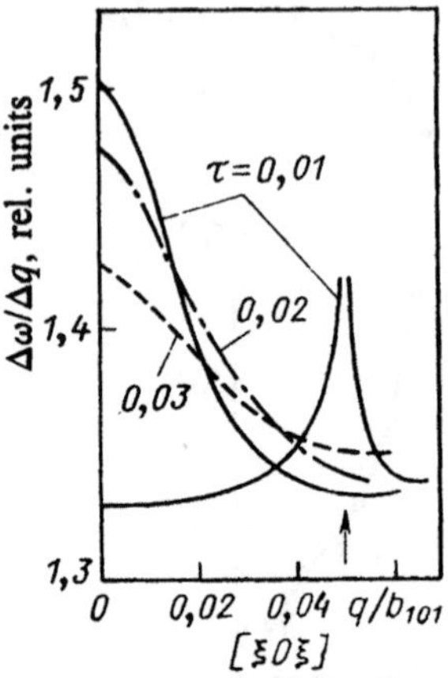

Fig. 95. Change in the group velocity of the phonons when the boundary of the filled states is moved ($\tau = V_{101}/2E_{\text{bound}}^0$).

$\mathbf{q} = 0$ in analogy to the Kohn effect. However, while the Kohn singularities were associated with a sharp edge of the electron distribution in the continuous energy spectrum, the regions of filled and free states are in the case under consideration separated by an energy gap; this may cause a certain smoothing (but not the disappearance) of singularities. As has been shown in [47], in this case $d\omega/d\mathbf{q} \simeq \ln|\varkappa/b|$, where $\varkappa = (V_b/b)$, and **b** denotes the vector of the reciprocal lattice. Therefore, in the case of small gaps $\varkappa << 1$ [such a situation arises in the non-transition metals, e.g., we have $(101)\varkappa = 0.02$ in cadmium for the (101) faces], in the $d\omega/d\mathbf{q}$ function there may appear singularities corresponding to the boundary of the band filled with electron states. It cannot be ruled out that the influence of band effects upon the group velocity in the long-wave region ($\mathbf{q} \to 0$) accounts for the lack of success in describing the experimental values of the moduli of elasticity of cadmium by perturbation theory and a spherical Fermi surface [99].

The authors of [47] have exhaustively treated the behavior of the group velocity of the phonons at an anomaly when an energy gap appears at the boundary of the filled states. Figure 95 shows that the shift of the boundary of the filled states toward the edge of the band leads to the disappearance of the usual Kohn anomaly which appears at a finite value of the wave vector and increases the group velocity at $\mathbf{q} \to 0$. The effect has a significant magnitude when the width of the gap varies within the limits which are characteristic of non-transition metals. The experimental [140] and theoretical investigations of the singularities in cadmium at $\mathbf{q} \to$

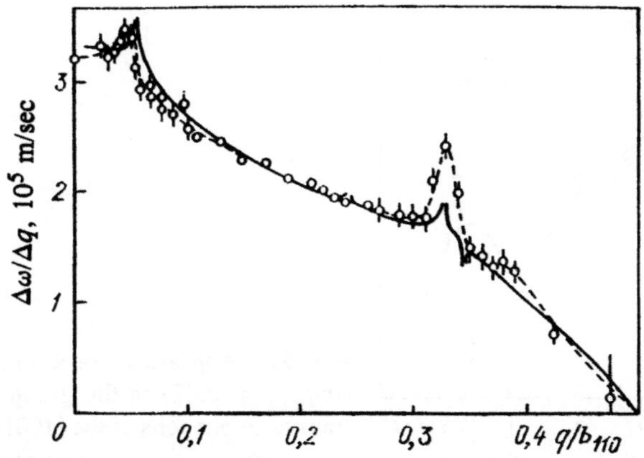

Fig. 96. Group velocity of the phonons in Mg; comparison of the calculated (solid curve) and experimental (dots) data.

0 revealed for the first time a new type of Kohn anomalies in the acoustic range. In magnesium, in which the Fermi surface comprises "butterflies," the long-wave part of the phonon spectrum shows the usual Kohn anomaly related to the vector **b** (101) of the reciprocal lattice by virtue of condition (27.3) [140b] (Figure 96). In this connection, an experimental investigation of the change of the electron spectrum of cadmium and magnesium alloys is of interest; there the Kohn anomalies in the acoustic range must transform into the usual anomaly in the group velocity of the phonons at a finite value of the wave vector.

We have considered the Kohn anomalies in the vibration spectra of metals and have established the relation between the singularities in electron and phonon spectra; this opens additional possibilities for research on the topology of the Fermi surface in metals, particularly in alloys, and makes it possible to improve the theoretical conclusions on the basis of experimental information on the fine structure of phonon dispersion curves. In formal terms, the Kohn singularities are described in second-order perturbation theory through the electron interaction and are related to the singularity in $\Gamma^{(2)}$ [see Eq. (27.7)]. It follows from the theory of the non-transition metals [9] that singularities of the type of Kohn singularities and singularities associated with anomalies in $\Gamma^{(3)}$ [see Eq. (27.8)] must also be observed in experiments. However, until recently no experimental studies of the group velocity of phonons have been made for

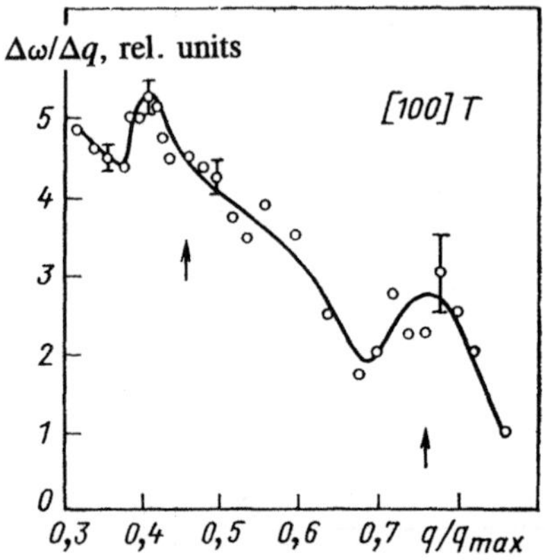

Fig. 97. Appearance of a non-Kohn anomaly ($q/q_{max} \simeq 0.42$) in the group velocity of the transverse phonons of the [100] direction in Al.

finding new three-particle singularities. The authors of [360] observed in their experiments at least two clear singularities which could not be identified as Kohn singularities appearing in addition to the usual Kohn anomalies.

Of greatest interest is the isolated anomaly in $d\omega/d\mathbf{q}$ for the transverse branch in the [100] direction near the phonon wave vector value $q/q_{max} \simeq 0.42$ (Figure 97) because, as put into evidence by a geometrical analysis, conditions for the Kohn effect are missing in the nearest neighborhood of the wave vector values for it. The authors of [9] have shown that the position of the singularity in the phonon spectrum at $q/q_{max} \simeq 0.42$ is close to the calculated position $q/q_{max} = 0.46$ of the anomaly, the anomaly being associated with the appearance of an unpaired interionic interaction in aluminum; the authors of [87] made a detailed geometrical analysis of the conditions for the appearance and the properties of a three-particle anomaly in this particular wave vector range. More specifically, it was established that when the wave vector of the phonon deviates from the [100] direction of high symmetry, the three-particle singularity must be split into four singularities, and the singularities with q_1 and q_4 must appear on the $\mathbf{e}_T \parallel \mathbf{y}$ branch. The Kohn singularity which is observed on the branch under consideration at $q/q_{max} = 0.76$ and which is defined by the vector $\mathbf{b}$ (111) of the reciprocal lattice (Figure 97), must be split into two singularities in an analogous deviation from the direction of symmetry; the conditions of observation and the position of the two anomalies do not depend upon the polarization vector $\mathbf{e}_T$.

The differences in the polarization dependence of these singularities make it possible to confirm by experiments that the anomalies are of the three-article type in the neighborhood $q/q_{max} = 0.42$ of the transverse

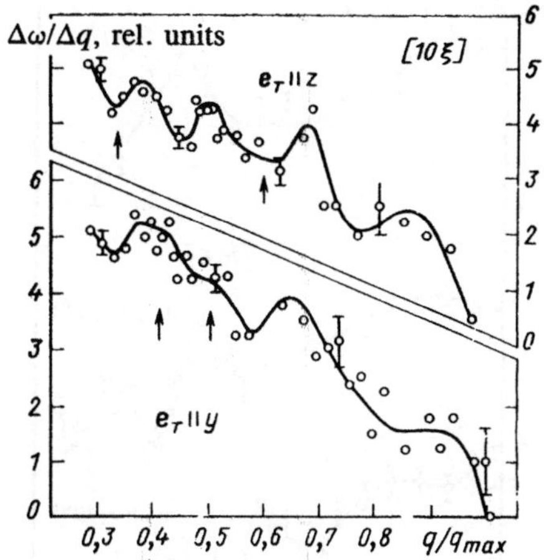

Fig. 98. Group velocity of the transverse phonons in the [10ξ] direction in Al. The arrows denote the position of the three-body singularities for branches with different polarization.

branch in the [100] direction of aluminum. Figure 98 represents the results of $d\omega/d\mathbf{q}$ measurements [87] for the [10ξ] direction, where $\xi = 0.176$ and corresponds to a 10° deviation from the [100] direction. One recognizes that the singularities split, and the form of the splitting is for the second singularity qualitatively the same at $\mathbf{e}_T \parallel \mathbf{y}$ and $\mathbf{e}_T \parallel \mathbf{z}$. By contrast, the form of the splitting of the first singularity depends upon the direction of the polarization vector of the phonons. For $\mathbf{e}_T \parallel \mathbf{z}$, the splitting was much greater than for $\mathbf{e}_T \parallel \mathbf{y}$ when insufficient resolution of the wave vector made it impossible to localize the positions of the singularities (Figure 98). For a detailed investigation of how the splitting of the singularities depends upon the polarization vectors, the authors of [87] made detailed measurements of the phonon velocity around $q/q_{\max} = 0.4$ with improved resolution of the wave vector (improved resolution obtained by reducing the divergence of the neutron beam). Their results (Figure 99) rather convincingly show that the singularity is split in the [10ξ] direction with $\mathbf{e}_T \parallel \mathbf{z}$. At a 7.5° deviation from [100], the splitting is 0.062 in units of $q/q_{\max}$ and amounts to 0.086 in the case of a 10° deviation; for $\mathbf{e}_T \parallel \mathbf{y}$ and a 10° deviation, the splitting could still be estimated as ~0.052 at the limit of the experimental resolution. Thus, in the experiment there appears that type of dependence of the splitting upon the polarization vector, which is typical for a three-particle singularity, and the magnitude is 1.7 times greater for the branch with $\mathbf{e}_T \parallel \mathbf{z}$ (10°) than for the branch with $\mathbf{e}_T \parallel \mathbf{y}$ (10°).

It follows from a comparison of the results obtained for the group velocity of the phonons in the [10ξ] direction and the calculated results

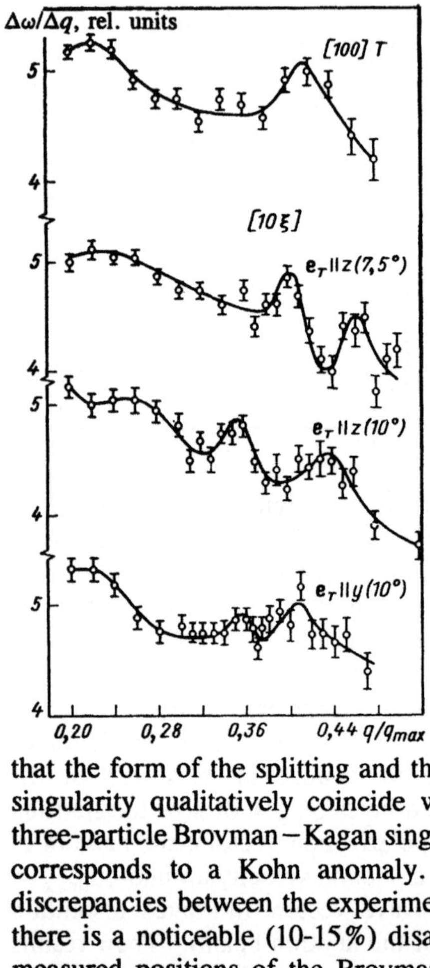

Fig. 99. Dependence of the group velocity of the transverse phonons with different polarization upon the angle of deviation from the [100] direction in Al.

that the form of the splitting and the polarization dependence of the first singularity qualitatively coincide with the properties predicted for the three-particle Brovman–Kagan singularity, whereas the second singularity corresponds to a Kohn anomaly. However, there exist quantitative discrepancies between the experimental results and the calculations, i.e., there is a noticeable (10-15%) disagreement between the calculated and measured positions of the Brovman–Kagan singularities and a slightly smaller disagreement (~5%) in the case of the Kohn anomaly (see Figure 98). The discrepancies may originate from possible experimental errors and also from incorrect theoretical calculations based on the approximation of a spherical Fermi surface of aluminum. The form of the Fermi surface has an increased influence upon the positions of these anomalies can be of the order of magnitude of the energy gap parameter, and this is most likely also observed in experiments.

However, when the real Fermi surface is brought into account, not only the position of the singularities is affected but in the phonon spectrum there appear additional so-called nondiametrical Kohn singularities. These

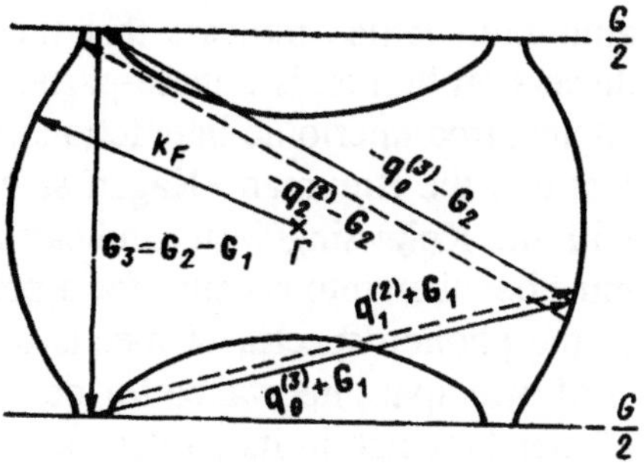

Fig. 100. Scheme of the simplified Fermi surface in a model with two gaps (the dashed line denotes the nondiametrical Kohn transitions; the solid lines correspond to the conditions of three-particle singularities.

singularities are associated with electron transitions between Fermi surface points which are not transformed into each other in an inversion transformation. Of course, for them to appear in the phonon spectrum, the electron velocities at these points usually must be antiparallel. The nondiametrical transitions are indicated by the dashed lines, in Figure 100, which illustrates the simplest Fermi surface in a model with two energy gaps in a scheme of reduced bands. The solid lines of Figure 100 depict the triangle constructed on the arguments of $\Gamma^{(3)}$ (determining the position of the Brovman–Kagan singularity at the phonon wave vector $\mathbf{q}_0^{(3)}$). The positions $\mathbf{q}_i^{(2)}$ of the nondiametrical Kohn singularities (see Figure 100) can be shifted relative to $\mathbf{q}_0^{(3)}$ by the parameter of the energy gap. However, the magnitude of the singularities which are associated with the nondiametrical transitions is given by the region of the strongest coherent conversion of the electron spectrum. Since this region is small (of the order of the gap parameter) in the non-transition metals, the nondiametrical Kohn regions are also small. The authors of [47, 87] have shown that

$$\frac{d\omega}{dq} \sim \left(\frac{V_{\mathbf{q}+\mathbf{b}}}{E_F}\right)(\Delta^{1/2} + 1), \qquad (27.11)$$

where the term with $\Delta^{1/2}$ corresponds to the nondiametrical Kohn singularity accompanying the Brovman–Kagan anomaly so that the ratio of these singularities is proportional to $\Delta^{1/2}$; this accounts for the insignificant influence of the nondiametrical singularities upon the form of $d\omega/dq$ in the majority of the non-transition metals. For example, in the case under consideration, we have $\Delta^{1/2} \simeq 0.1\text{-}0.15$ in aluminum.

Thus, the experimental results which were obtained for the group velocity of the transverse phonons in the [100] and [10ξ] directions of the aluminum lattice indicate that the singularity at $q/q_{max} = 0.42$ on the dispersion branch is a three-particle singularity, i.e., it is associated with an unpaired interionic interaction involving the conduction electrons. The fact that the Brovman–Kagan singularity is in its magnitude comparable with the Kohn singularity indicates not only that the unpaired interaction must be taken into account for a proper interpretation of the fine structure of the phonon spectra of metals but also that introducing this interaction is of principal importance for developing a consistent microscopic theory of metals which in its modern form tries to quantitatively describe all the properties of metals.

28. Anomalies of the Phonon Spectra of Superconductors

Research made with inelastic neutron scattering on superconductors and particularly on compounds with the A-15 structure (V_3Si, Nb_3Sn, Nb_3Al, etc.) is of particular importance in studies of the electron–phonon interaction in metals. In a number of experiments in which the density function of the phonon states of type A-15 compounds was measured [306], it was shown that the phonon spectra of high-temperature superconductors are changed at low temperatures, i.e., the density of the phonon states is noticeably increased at low energies in the compounds V_3Si, V_3Ge, V_3Ga, Nb_3Al, and Nb_3Sn [306] (see e.g., Figure 101). The results show that a softening of the phonon frequencies may occur not only in the long-wave part of the phonon spectrum but may also affect phonons with wave vectors up to the boundary of the Brillouin zone [113, 316].

The acoustic branch of the phonon spectrum could be measured in great detail at various temperatures in the case of Nb_3Sb [113]. It was shown that all measured branches are "softened" when the temperature is reduced from room temperature to 46 K. The strongest temperature dependence is observed in the [110] direction of the transverse acoustic branch polarized in the $[1\bar{1}0]$ direction (Figure 102). Let us note that in the long-wave part of the spectrum, this branch is determined by the shear modulus C' which is practically equal to zero at the temperature of the structural phase transitions ($T = 45.2$ K in the case of Nb_3Sn).

The experimentally observed softening of the phonon frequencies upon decreasing the temperature [113, 316] was qualitatively explained in [22]

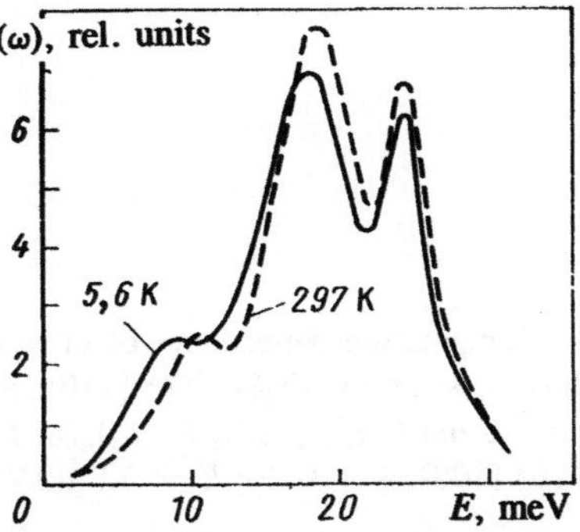

Fig. 101. Density $g(\omega)$ of the phonon states of Nb_3Sn at 5.6 and 297 K.

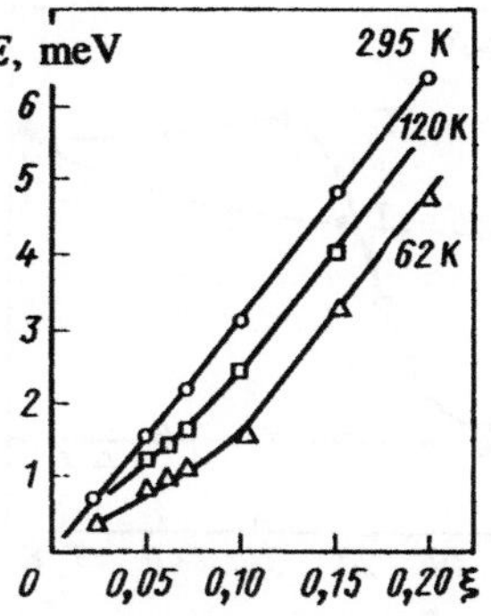

Fig. 102. Temperature dependence of the acoustic phonon branch in Nb_3Sn.

with the model of [21] having two peaks of the electron state density under the assumption that the Fermi level is near one of the peaks. The electron–phonon interaction leads to electron-induced corrections to the phonon frequencies, and this accounts for the temperature-dependent changes in the phonon spectrum. It is an important result of [22] that in systems with a type A-15 structure, the form of the temperature-dependent softening of the phonon branches is substantially determined by the direction of the wave vectors of the phonons. More particularly it was shown that if the latter have some projections upon the directions of the transition metal chains in the A-15 structure, the temperature softening of the phonons with a given wave vector must be minimal in a significant part of the Brillouin zone.

In this connection, research on the temperature dependence of the phonon branches in the [111] direction is of particular interest because as follows from [22], for this direction the softening of the phonons must be less pronounced than for the [110] direction. Measurements [315] of the low-frequency phonons in the Nb_3Sn were made in the [111] direction; it was established that the softening of the long-wave phonons is not as predicted by the theory [22]. Nevertheless, a clear decision on the validity of the theory of [22] can be made only on the basis of detailed studies of the phonon branches in asymmetric directions, which studies must be made for phonon wave vector values up to the boundary of the Brillouin zone.

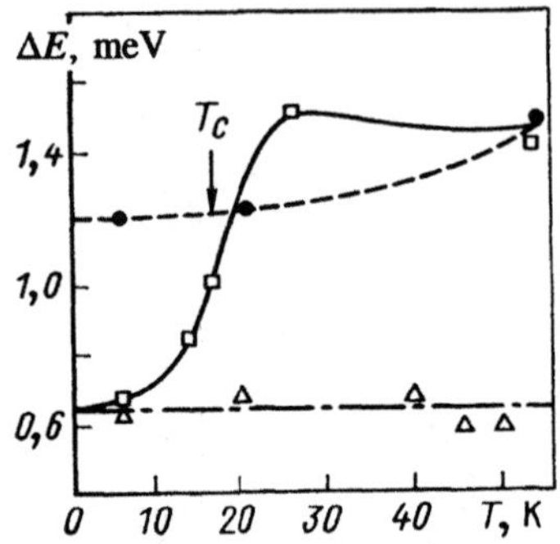

Fig. 103. Temperature dependence of the width of the phonon peaks in Nb_3Sn: ●) [110]; ξ = 0.03; E = 8.0 meV; □) [110]; ξ = 0.18; E = 4.0 meV; Δ) [100]; ξ = 0.20; E = 3.2 meV.

These manifestations of the electron–phonon interaction effects were related to changes in the phonon frequencies during temperature changes; they can be recorded in neutron scattering experiments with high energy resolution and high phonon wave vector resolution. Such experiments are rather laborious but the authors of [113] have presented the results of a unique investigation of the electron–phonon interaction in Nb_3Sn; the interaction is related to changes in the phonon lifetimes and experimentally manifests itself in changes of the width of the coherent maximum corresponding to the phonon. It was noted [113] that in the superconductive state, phonons having an energy smaller than the energy gap cannot excite quasiparticles directly through the gap. Accordingly, the attenuation of such phonons is weak and therefore the corresponding width of the coherent maximum is small.

When the phonon energy increases to the gap value, a threshold excitation process of quasiparticles takes place and the lifetime of the phonons decreases. This sharp increase in phonon attenuation was recorded when the width of the coherent maxima corresponding to transverse phonons of the T_1 branch were measured in the [110] direction of the Brillouin zone of Nb_3Sn [113]. The energy gap was assessed from these data:

$$2\Delta(0) = 7 \pm 1 \text{ meV}.$$

In addition, the authors of [113] studied the attenuation of various phonons in the transition of Nb_3Sn from the superconductive state to the normal state. It follows from the dependences which were obtained for the width of the coherent maxima upon the temperature (Figure 103) that phonons with an energy $\hbar\omega < \Delta(0)$ are weakly attenuated because they cannot excite quasiparticles, whereas phonons with $\hbar\omega > \Delta(0)$ have a shorter lifetime and therefore can decay under excitation of quasiparticles at any temperature. Finally, transverse (T_1) phonons in the [110] direction with $\hbar\omega \simeq \Delta(0)$ begin to have a strong attenuation at the transition of Nb_3Sn from the superconductive state to the normal state; the conclusion is therefore that these phonons are of prime importance for the superconductive properties of Nb_3Sn.

The authors of [319] made similar experimental investigations of the phonon attenuation in Nb, in which the change in the width of the phonons is much smaller than in Nb_3Sn, and also in Pb [320]. In the case of lead, a softening of any branches of the spectra or changes in phonon attenuation during the transition to the superconductive state were not recorded (within the accuracy of the experiments) in the particular directions. This agrees with the existing ideas about the electron–phonon interactions in transition and non-transition metals. The dispersion relations for phonons in superconductive transition metals have been treated in greater detail by the authors of [16].

Chapter 7

THE DYNAMICS OF METAL HYDRIDES

29. General Remarks on the Dynamics of Hydrides

Research on the dynamics of the crystal lattice of hydrides is of interest for a number of reasons relating to principles as well as applications. In particular, since one desires to understand the physical properties of these objects, namely the type of the hydrogen bond in a metal, the influence of hydrogen upon the electron properties and thermodynamic properties of the system, the unusually high mobility of hydrogen diffusion in many metals, isotope effects, etc., it becomes necessary to obtain experimental information on the vibration spectra of hydrides and on the parameters characterizing the hydrogen diffusion in the metal. Besides that, in order to solve applied problems relating to the use of hydrides, one must also know the excitation spectra of these systems, with the phonon spectrum being one of the most important spectra.

Metal hydrides or, as they are termed, hydrogen implantation phases, are of greatest interest. Hydrogen in a metal appears as an implanted impurity which is capable of acquiring order in a hydrogen sublattice at certain concentrations, whereby the structure of the metal is modified. A detailed analysis of the structure of metal hydride phases was made in the review of [90] and in the first volume of the present work (see Vol. 1, Section 21).

In order to describe the vibrations of hydrogen introduced in the crystal lattice of a metal, one can imagine the following clear pattern: during some time τ, the hydrogen is between diffusion jumps trapped at interstitial positions corresponding to the minimum of the potential energy. In the general case, several types of interstitial positions may exist and, from the viewpoint of the potential energy of the interatomic interaction, these interstitial positions differ by the depth of the potential well and the curvature of its walls. A hydrogen atom at an interstitial position vibrates around the equilibrium position with the frequency ω_0. In the limit of an infinitely large mass of the metal atoms and the absence of any interaction between the implanted atoms, a single optical vibration frequency of the hydrogen would exist. When the concentrations of light atoms are high enough, a band of optical vibrations of the crystal develops mainly as a consequence of hydrogen vibrations, whereas the metal atoms vibrate almost exclusively in acoustic modes.

It should be noted that in vibrations with optical frequencies, the light hydrogen atoms have a much greater mean-square deviation from the equilibrium position than the heavy metal atoms. The hydrogen which is present in the metal lattice participates in the acoustic vibrations but the mean square of the displacements is in this case for the hydrogen of the same order of magnitude as the mean square of the displacement of the metal. In vibrations of the metal lattice, the potential relief, i.e., the depth and the curvature of the potential well, are modified; this may facilitate the skipping of hydrogen under the influence of thermal excitations through potential barriers into neighboring interstitial positions in which the hydrogen will be present for a certain time. Thus, hydrogen movement by diffusion occurs in metal lattices, and the activation energy is in a number of cases smaller than, or of the order of magnitude of, the energy of the optical vibrations. Of course, this description is rather crude and schematic but nevertheless renders the basic properties of the behavior of hydrogen in the lattices of the metal hydrides.

The methods of structure analysis, particularly the neutron diffraction method, provide information on the hydrogen equilibrium positions which coincide with the minima of the potential relief. The diffraction methods can also provide information on the mean-square of the displacement of the hydrogen around the equilibrium positions. But the most detailed information on the constants of the interaction forces and the form of the potential barriers can be obtained from research on the dynamics of the lattice of the hydrides.

When the concentration of hydrogen trapped at interstitial positions of a metal lattice is small enough, a hydrogen atom can be considered an isolated atom. One can speak of local modes associated with vibrations of isolated light atoms. In order to describe the dynamics of hydrogen, the model of Einstein oscillators is most appropriate: the light atoms are situated in the field of the forces of heavier metal atoms of lower mobility. If this concept is correct, a sharp maximum must be observed in the spectral distribution of the vibrations; in measurements, the width of the maximum is fully defined by the resolution of the instrument. A broadening of a spectral line can have many reasons, particularly the anharmonicity of vibrations; in the case of the light hydrogen atoms, the anharmonicity can be significant even at low temperatures.

H−H interaction is a possible reason for the broadening. When the interstitial positions are filled in an ordered fashion by implanted atoms, the vibrations of the system can be described by the dispersion relations for the acoustic and optical branches. Since they are collective vibrations of the crystal lattice, the optical branches preserve some features of the local modes in view of the small mass of the hydrogen atoms. In the approximation of an infinite mass of the metal atoms and without long-range forces for the hydrogen atoms, the dispersion curves for the optical vibrations must be flat and the treatment is reduced to the model of non-interacting oscillators. In the general case, a successful description of the dynamics of the crystal lattice of hydrides makes it necessary to include the long-range forces in the considerations; the long-range forces may not be small [48] and will determine the dispersion of the optical branches.

When two sublattices are filled (the hydrogen atoms occupying two types of interstitial sites in the metal), two maxima must appear in the vibration spectrum and two systems of optical branches corresponding to vibrations of the hydrogen at positions of the two types must appear on the dispersion curves. When the system undergoes a structural phase transition which is accompanied by a change in the coordination of the hydrogen atoms, this is immediately manifest in the vibration spectrum which, because of the usually large difference of the vibration energies of the hydrogen at the various interstitial sites, is rather sensitive to a hydrogen redistribution.

The nature of the hydrogen bonding to the metal atoms must manifest itself in the vibration spectra of the hydrides. This can be most clearly observed on the example of the ionic hydrides in which, according to the

Lyddane–Herzfeld rule [236], the transverse and longitudinal optical branches must be split. Such a situation will apparently show up in the case of hydrides with mixed bonds, as assumed for hydrogen compounds of rare-earth metals [239].

The presence of hydrogen in the system must affect the spectrum of the acoustic vibrations; one can imagine at least two mechanisms of this effect:

Firstly, the introduction of hydrogen changes the electron spectrum of the metal and, consequently, the interaction forces, which manifest themselves in the vibration spectrum, must change.

Secondly, the hydrogen proper participates in low-frequency vibrations together with the metal atoms, and this means that new features which are related to vibrations of the hydrogen atoms alone may appear in the spectrum.

The above-described features of the dynamics of the crystal lattice of hydrides mean that investigations of both the dispersion relations of the phonons and the spectral density of the lattice vibrations are of particular interest.

Scattering of thermal neutrons is the research method providing a maximum of information on the dynamics of the crystal lattice of hydrides. Hydrogen has an abnormally large cross section of incoherent scattering of neutrons so that the experimental results of inelastic incoherent scattering of neutrons are relatively easy to interpret. When a sample with coherent scattering is to be preferred, hydrogen can be replaced without difficulties by deuterium which causes mainly coherent scattering of the neutrons.

30. Inelastic Scattering of Neutrons at Metal Hydrides

At least two atoms are present in the unit cell of metal hydrides with order, and therefore the cross section of inelastic incoherent scattering does not have a simple relation to the spectral density of the vibrations as in the case of a cubic crystal with a Bravais lattice. Following [97] we will show which information can be deduced from the double differential cross section of neutron scattering in this case. For the time being we assume that the hydride has a regular crystal lattice of some symmetry. It follows from Eq. (2.24):

$$\frac{d^2\sigma^{+}_{\text{incoh}}}{d\Omega d\varepsilon}=\frac{\hbar k}{2k_0}\sum_i \sigma_i\, e^{-2W_i}\frac{1}{M_i}\sum_{\mathbf{q},J}\frac{|\varkappa \mathbf{e}_J(\mathbf{q},i)|^2}{\omega_J(\mathbf{q})}(n_{J\mathbf{q}}+ \\ +1/2\pm 1/2)\,\delta(\varepsilon\pm\hbar\omega_J(\mathbf{q})). \tag{30.1}$$

The summation over the wave vectors and the branches of the vibrations can be replaced by integration over the energy spectrum $g(\omega)$ with the aid of the expression

$$\sum_{J=1}^{3\sigma}\sum_{\mathbf{q}} F(\mathbf{q},\omega_J(\mathbf{q}))=\int \widetilde{F}(\omega)\, g(\omega)\, d\omega, \tag{30.2}$$

where $F(\mathbf{q}, \omega_J(\mathbf{q}))$ denotes some function of the frequency and the momentum.

By performing the integration, we obtain

$$\frac{d^2\sigma^{+}_{\text{incoh}}}{d\Omega d\varepsilon}=\frac{\hbar k}{2k_0 N}\frac{g(\varepsilon/\hbar)}{\varepsilon}(n(\varepsilon/\hbar)+1/2\pm 1/2)\times \\ \times\sum_i \sigma_i\, e^{-2W_i}\frac{1}{M_i}|\varkappa\varphi_i(\varepsilon/\hbar)|^2, \tag{30.3}$$

where $|\varkappa\varphi_i(\varepsilon/\hbar)|^2$ is defined as

$$\frac{1}{(2\pi)^3}\sum_{J=1}^{3r}\int_{\omega}^{\omega+d\omega}|\varkappa \mathbf{e}_J(\mathbf{q},i)|^2 d^3\mathbf{q}=|\varkappa\varphi_i(\omega)|^2 g(\omega)\,d\omega, \tag{30.4}$$

and $\varphi_i(\omega)$ is the polarization vector of the ith atom averaged over an isoenergetic surface in the case of vibrations with the frequency ω.

In the case of a polycrystalline sample, averaging over the orientation of the vectors $\varphi_i(\omega)$ is possible in Eq. (30.3) and is valid for a polycrystal which is isotropic as a whole. We will assume that the thermal factor is independent of the vibration of the individual crystallites. By using the expression

$$\langle \varphi_i^{\alpha}\varphi_i^{\beta}\rangle = 1/3\,|\varphi_i|^2\,\delta_{\alpha\beta}, \tag{30.5}$$

we obtained

$$\frac{d^2\sigma^{\pm}_{\text{incoh}}}{d\Omega d\varepsilon}=\frac{k\hbar\varkappa^2}{6Nk_0}\frac{(n(\varepsilon/\hbar)+1/2\pm 1/2)}{\varepsilon}g(\varepsilon/\hbar)\sum_i\frac{\sigma_i}{M_i}e^{-2W_i}|\varphi_i(\varepsilon/\hbar)|^2. \tag{30.6}$$

The function $\varphi_i(\varepsilon/\hbar)$ is the square of the absolute value of the polarization vector of the ith atom, averaged over the isoenergetic surface, and is subject to the normalization condition

$$\sum_i |\varphi_i|^2 = 1. \tag{30.7}$$

One can show [24, 97] that for polycrystalline systems which are without order and are composed of atoms of various types, the following expression holds:

$$\frac{d^2\sigma^{\pm}_{\text{incoh}}}{d\Omega d\varepsilon} = \frac{k}{k_0}\,\frac{\hbar\varkappa^2}{2\varepsilon}\,[n(\varepsilon/\hbar) + 1/2 \pm 1/2]\,g(\varepsilon/\hbar)/3N \times$$
$$\times \sum_s f_s \frac{\sigma_s}{M_s} e^{-2W_s} \langle |\varphi_s(\varepsilon/\hbar)|^2 \rangle, \tag{30.8}$$

where s denotes the subscript of a component of the compound; f_s denotes the fraction of atoms of the s component; and $\langle |\varphi_s(\varepsilon/\hbar)|^2 \rangle$ denotes the squares of the absolute values of the polarization vectors of the sth atom, the squares averaged over the distribution of the various atoms at the lattice points and over the isoenergetic surface, wherein

$$\sum_s f_s \langle |\varphi_s(\varepsilon/\hbar)|^2 \rangle = 1. \tag{30.9}$$

Thus, it has been shown that in the case of crystals composed of various atoms, measurements of the double differential cross section of the incoherent scattering of neutrons at a polycrystalline sample can be used to determine the function

$$\Theta(\varepsilon) = \begin{cases} \dfrac{g(\varepsilon/\hbar)}{3N} \displaystyle\sum_s \frac{\sigma_i}{M_i} e^{-2W_i} |\varphi_i(\varepsilon/\hbar)|^2; & (30.10\text{a}) \\ \dfrac{g(\varepsilon/\hbar)}{3N} \displaystyle\sum_s \frac{\sigma_s}{M_s} e^{-2W_s} f_s |\varphi_s(\varepsilon/\hbar)|^2. & (30.10\text{b}) \end{cases}$$

rather than the function $g(\omega)$ of the spectral density of the vibrations.

In the case of Eq. (30.10a), the summation is extended over the atoms of the unit cell of the compound with order; in the case of Eq. (30.10b), the summation is extended over the components of a system without order. the unknown energy dependence of the polarization vectors must be independently determined, for example by using some model for calculations.

Obviously, the function $\Theta(\varepsilon)$ is related to the scattering cross section by the expression

$$\frac{d^2\sigma^{\pm}_{\text{incoh}}}{d\Omega d\varepsilon} = \frac{k}{k_0}\frac{\hbar\varkappa^2}{2\varepsilon}\left(n(\varepsilon/\hbar) + 1/2 \pm 1/2\right)\Theta(\varepsilon). \tag{30.11}$$

In the case of hydrides which are a two-component system, the summation over the atoms in the unit cell can be divided into parts relating to the metal atoms and the hydrogen atoms.

In the simplest case of a diatomic lattice, we have

$$\Theta(\varepsilon) = g(\varepsilon/\hbar)/3N\left(|\varphi_H|^2 e^{-2W_H} + \beta|\varphi_M|^2 e^{-2W_M}\right), \tag{30.12}$$

where $\beta = \sigma_M M_M/\sigma_H M_H$. Since the mass of the hydrogen atoms is small and the cross section of scattering at hydrogen is large, in the majority of metals $\beta \simeq 10^{-3} << 1$ is observed.

The dynamics of the lattice with ionic lithium hydride was calculated [31] with a simple model of the interaction force. It was shown that the square of the absolute value of the polarization vector corresponding to the vibrations of the hydrogen atoms is practically energy-independent in the optical band of the spectrum and close to unity. However, in the range of acoustic vibrations of the lattice, the square of the absolute value of the polarization vector is energy-dependent and remains much smaller than unity. This reflects the fact that the light atom vibrates basically in the optical branches and hardly participates in the acoustic branches. The estimates made by the authors of [31] for the center of the Brillouin zone, i.e., at $\mathbf{q} = 0$, result in

$$|\varphi_H(0)|^2 = \frac{M_H}{M_{Li} + M_H}; \; |\varphi_{Li}(\omega_T^{opt})|^2 = \frac{M_H^2}{M_{Li}^2 + M_H^2}. \tag{30.13}$$

This shows that hydrogen participates in the optical vibrations with a weight factor close to 1 and in the acoustic vibration with the factor $1/M_{Li}$. A similar energy dependence of the polarization vectors occurs also in other hydrides. This has two consequences allowing a simplification of Eq. (30.12). Firstly, since even in the range of the acoustic vibrations, where $|\varphi_H|^2/|\varphi_{Li}|^2 = M_H/M_{Li}$, the second term in the parentheses is smaller than the first term by a factor $\sigma_H/\sigma_{Li} \simeq 10$, the scattering at the metal atoms can be disregarded. There we have

$$\Theta(\varepsilon) = g(\varepsilon/\hbar)/3N|\varphi_H|^2 e^{-2W_H}. \tag{30.14}$$

Secondly, in the band of the optical vibrations of the lattice, we obtain $\Theta(\varepsilon) \sim g(\varepsilon)e^{-2W_H}$ because there is no energy dependence of $|\varphi_H|^2$.

Thus, if our assumption is correct, experiments with inelastic neutron scattering render directly the spectral distribution function of the vibrations in the range of the optical lattice vibrations of the hydrides. In the acoustic part of the spectrum, this function is distorted by the unknown energy dependence of the polarization vectors and only if $|\varphi_H|^2$ depends in a smooth way upon the energy, can the experimental information qualitatively reflect the form of $g(\omega)$.

One should note that disregarding the scattering at the metal atoms is admissible only when the hydrogen concentrations are significant. In the case of systems without order [see Eq. (30.10b)], the partial concentrations f_s of the components appear in the expression which is the analog to Eq. (30.12).

The question of whether the assumed energy dependence of the polarization vectors is valid for any hydride is, generally speaking, without an answer though the calculations which were made for hydrides formed on the basis of a face-centered cubic metal lattice [325, 352] confirm our conclusions. It seems that the influence of the polarization vectors can be brought into account most appropriately through model calculations for each particular case or by direct experimental determination of the polarization vectors with the aid of inelastic neutron scattering.

The problem of the thermal factor, a quantity which is an integral over the spectrum and can be stated in the form [97]

$$W_i = \frac{\hbar\varkappa^2}{4M_i N} \int g(\omega) \frac{|\varphi_i(\omega)|^2}{\omega} \operatorname{cth}\left(\frac{\hbar\omega}{2kT}\right) d\omega. \tag{30.15}$$

was disregarded in all preceding considerations. It follows from a comparison of Eqs. (30.14) and (30.15) that the same product $g(\omega)|\varphi_i|^2$ appears in the function $\Theta(\varepsilon)$ and also in W_i. Thus, assuming neutrons are scattered only at protons, both $\Theta(\varepsilon)$ and W_i can be directly determined from experimental data with an iteration procedure.

A noticeable contribution of multiple and multiphonon scattering of neutrons is an important feature of experiments performed on hydrides because the scattering cross section of protons is large. The contribution of processes in which the neutron is simultaneously or in succession scattered at optical and acoustic vibrations of the hydride lattice is of particular importance for the analysis of experimental data referring to the optical part of the spectrum because these processes can cause distortions in the restored function $\Theta(\varepsilon)$. There exist ways of reducing the influence of these processes. In particular, multiphonon scattering of neutrons, as well as multiple scattering of neutrons, increase with decreasing temperature at a higher rate than single-phonon scattering. Besides that, multiphonon scattering decreases significantly when the momentum transferred in the scattering event decreases, whereas multiple scattering drops off with decreasing sample thickness. Thus, in order to reduce the contribution of multiphonon and multiple processes, it must be possible to make experiments on thin samples at low temperatures, and with a small momentum transfer.

The above-listed conditions are satisfied to the greatest extent in setups in which the neutrons generate phonons while they are scattered. When a sufficiently high energy of the incident neutrons and small angles of scattering are provided for, one can obtain rather small transmitted momenta. A significant flux of epithermal neutrons, which are required for such experiments, can be obtained on pulsed sources with accelerators of charged particles or with stationary reactors having sources of "hot" neutrons. In the first case one uses a time-of-flight spectrometer and in the second case, usually triaxial neutron spectrometers.

In research on hydrides one often uses an other type of setup in which the energy of the incident (in forward geometry) or scattered neutrons is fixed with the aid of a beryllium filter. It is in this case impossible to transfer a momentum of less than 70-80 nm^{-1} for the optical part of the spectrum and multiphonon scattering processes must be numerically brought into account.

The vibration spectrum of hydrides with order is composed of acoustic branches in which mainly the metal atoms vibrate and optical branches which reflect the vibrations of the hydrogen atoms. Incoherent neutron scattering, which with a large cross section occurs at hydrogen, can cause a strong energy-dependent background in experiments designated for determining the dispersion relations for phonons. Obviously, information on the dispersion should be preferably obtained at deuterated targets.

Replacing hydrogen by deuterium usually does not change the structure of the hydride, with vanadium hydride being an exception. But the more general question of whether the model of isotope replacement is applicable in this case has not been definitely answered. Nevertheless, assuming that this model adequately describes the situation, one can expect rather important and general information from the coherent scattering of neutrons at deuterated samples. Unfortunately, to date only a limited number of studies have been made in this way. One must also take into consideration that, in order to obtain the optimal setup for measurements, rather intense beams of monochromatic neutrons with an energy of 100-200 meV are required. Such conditions can be created only on high-flux reactors which are provided with a special source of hot neutrons. In order to obtain adequate intensity of neutron scattering at the optical vibrations, one must reduce the energy and momentum resolution of the spectrometer. Therefore, in the majority of experiments which to date have been made with inelastic coherent neutrons scattering at low energy and momentum resolution, only a low statistical precision of the results was achieved. As to the other aspects, studies of coherent inelastic neutron scattering at hydrides proceed in analogy to investigations on conventional crystals.

31. Hydrides of the Transition Metals of the 5th Group

The transition metals V, Nb, and Ta of the 5th group have a body-centered crystal lattice with tetrahedral (6 per metal atom) and octahedral (3 for each metal atom) interstitial sites. One observed that in practically all cases (an exception is the V−H system in which octahedral positions can be filled at certain concentrations), the hydrogen atoms are located at tetrahedral interstitial sites. It should be noted that the tetrahedra, as well as the octahedra, of the body-centered crystal lattice are not regular polyhedra.

The hydride phases which are formed in the Nb−H and Ta−H systems have a similar crystal lattice, and the structural changes are only small when the hydrogen is replaced by deuterium in these systems. Al hydrides of body-centered cubic metals to the 5th group are systems far from saturation. Indeed, under normal hydrogen pressure, the experimentally obtained maximum hydride composition corresponds to the formula Me−H. Only 1/6 of the tetrahedra are filled. This detail originates from the high diffusion mobility of the hydrogen in hydrides of body-centered

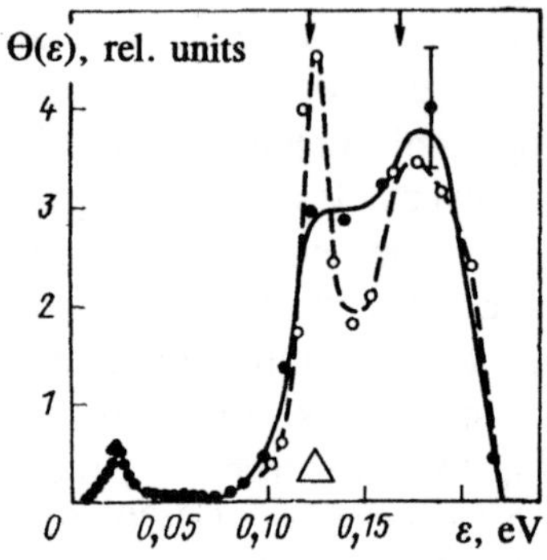

Fig. 104. The function Θ(ε) obtained for $NbH_{0.95}$ with various resolutions of the time-of-flight neutron spectrometer (the solid line refers to 30 μsec/m, and the dashed line, to 15 μsec/m).

cubic transition metals [326]. The extensive defectiveness of the hydrogen sublattice nevertheless does not affect the development of hydride phases with order.

Figure 104 illustrates the energy dependence of the function Θ(ε) obtained in [98] for niobium hydride $NbH_{0.95}$ at room temperature. The structure of this hydride has been considered in Chapter 6 of Vol. 1.

The spectrum is composed of two clearly separated regions; the first ends at 30 meV, whereas the second extends from 90 to 200 meV. We readily conclude that the low-energy part corresponds to the acoustic vibrations and the high-energy part, to the optical vibrations of the lattice. Since in this case the experimental data are given by the scattering of neutrons at hydrogen and since the square of the absolute value of the polarization vector of the hydrogen atoms is $|\varphi_H{}^{ac}|^2 >> 1$ for vibrations in the acoustic branches, the area under the optical part in the restored Θ(ε) is much greater than under the acoustic part. In principle, to Θ(ε) one could attribute a greater similarity to the real phonon spectrum if the acoustic part were represented with a weight factor equal to the mass of Nb, but the unknown energy dependence of $|\varphi_H{}^{ac}|^2$ still does not allow the determination of the precise form of $g(\omega)$. It must be noted that in any case, even if it were possible with, for instance, model calculations to determine $|\varphi_H{}^{ac}(\omega)|^2$, the predominant scattering of neutrons at the hydrogen would render information on "hydrogen acoustics," i.e., on the vibrations of the hydrogen in the acoustic modes. Generally speaking, this may coincide with the true vibration spectrum of the hydride or with the vibrations of the metal atoms. The authors of the recently completed work [233] have carefully studied the acoustic spectrum of $NbH_{0.05}$. The results of the experiment were analyzed in combination with calculated data for a very simple model of forces. It was shown that hydrogen performs resonance vibrations with an energy ~ 16 meV, and the amplitude of these vibrations is several times greater than the amplitude of the corresponding Nb vibrations. This means that at least in some cases, the hydrogen atom does not follow the acoustic vibrations of the metal atom in their phase and amplitude. This problem has not been thoroughly studied from the experimental or theoretical viewpoints but one point is clear: the hydrogen atoms vibrate in the acoustic modes not quite like the metal atoms, and it cannot be ruled out that there exists a relation between the vibrations of the hydrogen and the metal and the diffusion of the light atoms.

In the region of the optical vibrations of the $NbH_{0.95}$ lattice, one observes two well-resolved maxima (see Figure 104) which correspond to the anisotropy of the vibrations of the hydrogen atom trapped in a tetragonally distorted tetrahedral interstitial site of the body-centered crystal lattice of Nb. When the considerations are restricted to the central H–Nb interaction and when it is assumed that the hydrogen atom vibrates in the field generated by the fixed metal atoms, one can show [221] that the vibration spectrum of such an oscillator must comprise two frequencies, one of which is twice degenerate, and that $\omega_1 = \omega_2 = \sqrt{2\omega_3}$.

The ratio of the experimentally obtained vibration energies $\omega_1 = 120$ meV and $\omega_2 = 170$ meV is in good agreement with the calculations. Thus, in the spectrum of the optical vibrations there are two bands and the ratio of the average energies of these bands is well described by the very simple model. But the width of the lines in the optical spectrum cannot be explained with very simple concepts. Indeed, when the model of non-interacting oscillators is assumed to be valid, the line width must be fully given by the resolution of the instrument and $\Delta E/E = M_H/M_M \simeq 1/90$ must hold in the simple model. The width of the lines obtained in experiment substantially exceeds the resolution function and amounts to $\Delta\varepsilon_1 \simeq 29$ meV and $\delta\varepsilon_2 \simeq 50$ meV; when the resolution function is subtracted under the assumption of Gaussian shape of the spectral lines and of the resolution function of the instrument, the width is $\Delta\varepsilon_1' = 11$ meV and $\Delta\varepsilon_2' = 40$ meV. Thus, in the optical spectrum of the Nb hydride, there appears a typical feature of the vibrations of hydrogen which was implanted in a metal lattice: on the one hand, the position of the spectral lines in terms of energy corresponds to the very simple force model of a central interaction of the hydrogen with the nearest metal atoms; on the other hand, the width of the spectral lines is not described by such a simple model.

As indicated above, in principal multiple and multiphonon scattering of neutrons can somewhat contribute to the widths of the spectral lines. Recent investigations of the temperature dependence [82] and of the momentum dependence [245] of the width of the spectral lines in the β phase of the Nb hydride give reason to assume that the contributions of the multiphonon and multiple neutron scattering even at room temperature cannot amount to more than 30% of the experimentally observed width. This means that there must exist physical reasons for a broadening of the spectral lines. A noticeable H–H interaction is a very natural reason [48] leading to the dispersion of the optical branches. Here the width of the spectral lines would have a very simple and logical explanation. It should be noted that $g(\varepsilon)$ or $\Theta(\varepsilon)$ develop in the integration over all momenta in the irreducible part of the Brillouin zone, and therefore the line width of the optical vibration spectrum can be defined as the function $\omega(\mathbf{q})$ in direction of high symmetry and also beyond these directions.

Anharmonicity of the optical vibrations of the hydride lattice can be an other reason, with the anharmonicity associated with the non-parabolic form of the potential well in which the hydrogen atom is located; the non-parabolic form occurs at least in certain directions of the crystal lattice. The authors of [366] have studied the phonon spectrum of Ta–H and found that in the case of the composition $TaH_{0.7}$, the temperature-dependent transition from the ordered β phase (having the same structure as the β phase in the Nb–H system) to the disordered α' is accompanied by a broadening of the spectral lines. The authors have treated the result as a manifestation of the anharmonicity of the vibrations in the disordered α' phase; the mean-square of the displacements of the hydrogen atom becomes large for those directions of the crystal lattice, which correspond to diffusion jumps over neighboring interstitial positions.

It would appear that the broadening related with the H–H interaction must depend upon the concentration of the light component in the hydride. It is therefore interesting to review the results obtained on samples with a low hydrogen concentration, i.e., when the vibrations of the light atoms can be assumed local vibrations. Such experiments were made on the systems α-$NbH_{0.05}$ and α-$VH_{0.04}$ [184, 351]. It was established in both cases that the energies of the optical vibrations change slightly with the concentration, but the form of the spectrum and the line width hardly depend upon the concentration. It seems that the line width can in this case be given by the anharmonicity of the local vibrations, with the anharmonicity appearing more strongly in the phases without order than in the phases with order, as has been shown in [366]. Besides that, it cannot be ruled out that even at very low concentrations, the hydrogen atoms form clusters and, as in the case of a high concentration of light atoms, part of the width can be ascribed to an H−H interaction.

Thus, it was shown that incoherent neutron scattering can provide information on the position of the spectral lines of the optical vibrations of a hydride and on the line width which depends upon the dispersion of the optical branches and the anharmonicity of the vibrations of the light atoms. But in some cases one can obtain information on the position of atoms implanted in a metal lattice. The authors of [298] studied the spectra of the optical vibrations in various phases of the VH system; they could show that the $\alpha \rightarrow \beta$ transition in this system is accompanied by a transition of the hydrogen atoms from tetrahedral interstitial sites to octahedral interstitial sites. This transition was found because hydrogen which is trapped in the octahedral interstitial site of greater freedom vibrates at a lower frequency than the hydrogen trapped in the tetrahedral positions. In particular, for β-$VH_{0.4}$, the vibration energy is 55 meV. It is interesting to note that the tetragonal distortion of an octahedral interstitial position must also cause a splitting of the frequencies of the corresponding vibrations, as in the case of the tetrahedral position of the implant.

The authors of [220] measured the vibration spectrum of β-$VH_{0.5}$ in a wide temperature interval. The 221 meV peak they found is treated as an vibration of hydrogen atoms trapped at octahedral interstitial sites, the vibrations occurring in directions joining the apices of the octahedra. The energy 55 meV corresponds to nonsplit vibrations in the median plane of the octahedron. If this is correct, it turns out that the corresponding force constants differ by more than a factor of 10. Vanadium hydride exhibits still an other important feature. When the hydrogen is replaced by deuterium in $VH_{0.5}$, the vibration frequency of the light atoms at the octahedral interstitial sites changes much less than by a factor of $\sqrt{2}$, predicted by the approximation of isotope replacement [295]. But this approximation works quite well for the vibrations of the hydrogen at tetrahedral sites.

An interesting detail was detected in research on the hydrides of metals of the fifth group by incoherent neutron scattering: despite the considerable changes in the Me−H distance, similar values of the vibration energies of the hydrogen in tetrahedral positions are observed in all hydrides of metals of the fifth group (Figure 105). One can therefore

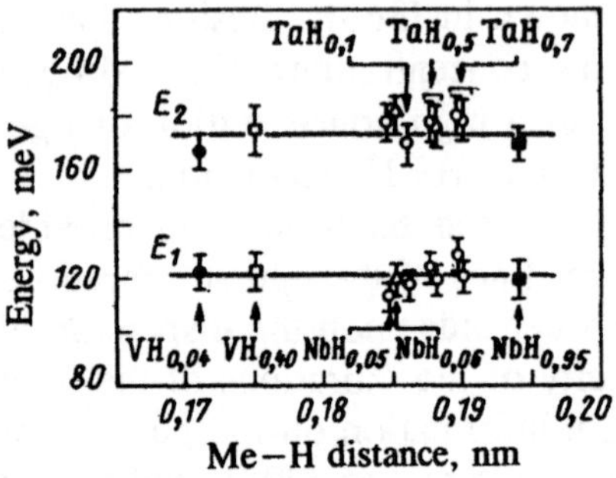

Fig. 105. Energy levels of the optical phonons in hydrides as a function of the metal-hydrogen distance.

conclude that the Me−H bond in this case is given by an interaction which does not depend upon the distance or that there exist other effects compensating for this dependence.

32. Metal Hydrides Based on a Face-Centered Cubic Metal Lattice

The hydrides of the light rare-earth metals. The light rare-earth elements (La, Ce, Pr, Nd) form hydride phases in which the metal atoms have a face-centered cubic lattice, whereas the atoms of the implant can occupy tetrahedral and octahedral interstitial sites. When the composition MeH_2 is reached, almost all the tetrahedral positions turn out to be filled and a structure resembling that of CaF_2 is formed. It should be noted that usually a few percent of the tetrahedral interstitial sites are vacant and that the corresponding hydrogen atoms occupy octahedra [84, 205]. When the concentration of the light atoms increases, the octahedral and the remaining tetrahedral interstitial sites are filled and a structure of the BiF_3 type is obtained at the composition MeH_3.

There are a number of reasons for studying the dynamics of the crystal lattice of these compounds. Firstly, the hydrides of the light rare-earth metals are practically unique systems which absorb hydrogen to complete saturation. Therefore effects related to the mutual interaction of the implanted atoms must manifest themselves to the greatest extent. Secondly, investigations of the vibration spectrum of the hydrides of various rare-earth metals make it possible to observe the dependence of the frequency upon the distance Me−H when one assumes that the type of the metal-hydrogen bond does not change in the entire series of light rare-earth metals. Finally, the possibility of obtaining hydrides of any composition in the concentration range $2 \leq x \leq 3$ of the light component considerably facilitates investigations of the concentration-dependent changes in the vibration spectrum.

The vibration spectra of the di- and trihydrides of cerium were for the first time studied by the authors of [49]. They could show that there exists one maximum in the band of the optical vibrations of the dihydride lattice but that there exist two maxima in the case of the trihydride lattice. This pattern corresponds to the assumption that tetrahedral interstitial sites are filled in the first case and both tetrahedral and octahedral sites are filled in the second case. In a face-centered cubic metal lattice, the positions of the implants have the full symmetry of the cube [90] and therefore symmetric splitting of the vibrational branches does not occur. Nevertheless, a considerable width of the spectral lines was observed in the experiment, with the width significantly exceeding the energy resolution of the instrument [48, 49].

Further research on the vibration spectra of the hydrides of the light rare-earth metals [84, 85, 203, 242, 353] was made on setups with a higher energy resolution than that of [59] and has shown that the optical part of the spectrum has a complicated structure. In particular, it was established that the spectrum of the optical vibrations of the hydrides of the light rare-earth metals cannot be described by two maxima because, when the mass of the rare-earth metals increases, the third maximum becomes more distinct [85]. Besides that, in the hydrides with the composition H/Me ≤ 2, weak maxima corresponding to vibrations of the hydrogen atoms in octahedral sites of the implant were found in the optical vibration spectrum [84, 242]. This means that the octahedral interstitial sites are filled with hydrogen at an earlier stage than the tetrahedral sites. It should be noted that a noticeable isotope effect is observed in this filling of sites. More specifically, at a particular composition of a hydride and a deuteride sample, the fraction of the filled octahedral positions is greater in the first case. This isotope effect is basically related to differences in the vibration energy of the hydride and the deuteride and is consequently given by different energy changes in the transition from the tetrahedron to the octahedron in these two cases [81].

The spectral pattern of the optical vibrations of the trihydrides of the light rare-earth metals is complicated mainly because of the strong interaction of the hydrogen atoms among themselves. This feature is indicated by, for instance, the fact that in the transition from the dihydrides to the trihydrides, a "repulsion" of the lines of the optical

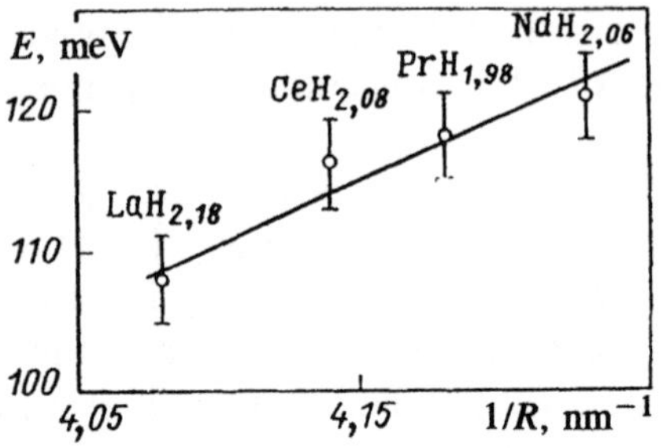

Fig. 106. Dependence of the vibration energy of the hydrogen at tetrahedral sites of the rare-earth metal lattices upon the metal-hydrogen distance.

spectrum occurs. The energy of the vibrations in the octahedral decreases but increases in the tetrahedra [85], and this feature cannot be explained by a simple compression of the lattice. Still another important detail must be noted: in the case of the hydrogen atoms trapped at octahedral interstitial sites, a hydrogen atom in a tetrahedron is the nearest neighbor and a metal atom is only the second neighbor. This leads to the conclusion that the spectrum of the optical vibrations is complicated mainly because the dispersion of the lower optical branches is strong.

The authors of [84, 203] analyzed the dependence of the optical vibration energy of hydrogen atoms trapped at tetrahedral interstitial sites of light rare-earth metals upon the distance Me−H. It was shown that within the experimental error limits, the energies are inversely proportional to the Me−H distance in the entire series of light rare-earth metals (see Figure 106). This feature, on the one hand, indicates that the Me−H bond is the same in all the hydrides considered and points, on the other hand, to differences in the bonding of the hydrogen to the metals in rare-earth metal hydrides and hydrides of metals of the fifth group with a body-centered cubic lattice.

The authors of [352] calculated the dynamics of the lattices of cerium dihydride and trihydride with a simple model of a central force including the forces acting between the second nearest neighbors of the hydrogen. Data on the width and the spectral line positions in terms of energy of the vibration were used as the parameters of the model. Though the model was rather sketchy, the authors could successfully describe the vibration spectrum of Ce dihydride and the temperature dependence of the heat capacity. The results of the calculations point to a significant mutual interaction of the hydrogen atoms.

The dynamics of the trihydride lattice is described with much less success by this model. A comparison of the calculated results and the experimental data shows only qualitative agreement in the range of the optical frequencies. It seems that for properly describing the lattice dynamics of the trihydrides of the light rare-earth metals, a model of greater complexity must be developed, the model taking into account, in particular, possible differences in the Me−H bond at octahedral and tetrahedral positions of the hydrogen.

The specific features of the phase diagram in systems composed of a light rare-earth metal and hydrogen complicate studies of the lattice dynamics of disordered solid solutions of hydrogen because these phases are stable only at temperature $T > 700°C$ [90]. But in the case of cerium hydride, the corresponding measurements were made at room temperature [358]. Using data of inelastic neutron scattering, the authors concluded that a tetrahedral coordination of the hydrogen atoms exists. Besides that, it follows from a comparison of the optical vibration energies $\hbar\omega_H = 96.2$ meV of $CeH_{0.04}$ and $\hbar\omega_H = 106$ meV for $CeH_{1.98}$ and the lattice parameters $a = 0.516$ nm in the first and $a = 0.557$ nm in the second case that the hydrogen is bonded to the metal much more firmly in the dihydride than in a diluted solid solution.

Titanium and zirconium hydrides. Hydride phases are formed in Zr−H and Ti−H on the basis of the face-centered cubic lattice of the metal [90]; in compositions close to H/Me = 2, zirconium hydride has a tetragonally distorted face-centered cubic lattice ($a/c = 1.12$) at a temperature $T < 1000$ K, whereas a tetrahedral lattice distortion arises in titanium dihydride at $T < 310$ K and the tetragonality parameter increases in proportion to the decrease in the temperature, with $a/c = 0.945$ at $T = 80$ K. Tetrahedral coordination of the implanted atoms exists in the two systems, and the available experimental data attest to nonequilibrium of the octahedral interstitial sites in regard to their filling with hydrogen. An explanation of this fact so far has not been found but the authors of [83] have shown that occupied octahedral positions exist in zirconium hydride samples obtained by the nonequilibrium technique of a self-propagating high-temperature synthesis. The investigations involved elastic and inelastic scattering of neutrons. In the band of the optical vibration spectrum restored from data on inelastic neutron scattering, there was found a weak maximum which, as in the case of the dihydrides of the light rare-earth metals [84], corresponds to vibrations of the implanted atoms in octahedral positions.

The spectrum of the optical vibrations of zirconium hydride was most thoroughly studied by the authors of [154] in wide ranges of the

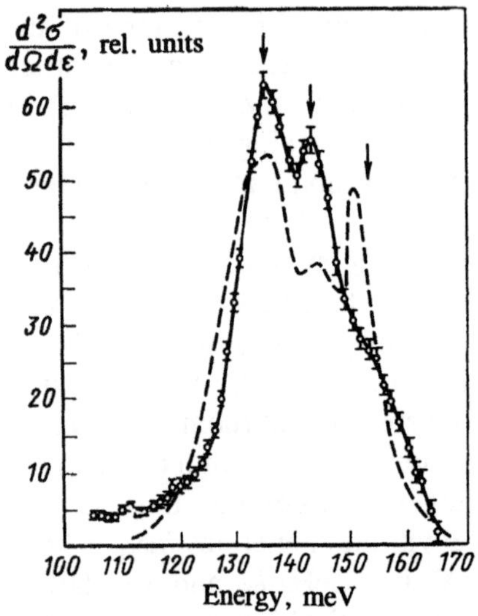

Fig. 107. Spectrum of the optical vibrations in zirconium dihydride $ZrH_{2.00}$; comparison of the calculated results [325] (dashed lines) with the experimental results [154] (dots).

concentration of the light component and of the momenta transmitted in a scattering event. The experiments were made with a spectrometer equipped with a rotating crystal. The authors succeeded in obtaining a monochromatic neutron beam with an energy of 244.8 meV. Taking into account the scattering angles used in the experiment, the average momentum transmitted to the crystal in a single event of scattering at optical phonons was 50-160 nm^{-1}. These experimental capabilities enabled the authors to study the momentum dependence of the width of the optical vibration spectrum and to conclude that multiphonon scattering processes contribute to large transmitted momenta.

The spectrum of the optical vibration of zirconium hydride is a rather complicated curve (Figure 107) with two clear maxima. The complicated form of the spectrum results from the tetragonal distortion of the metal environment of the hydrogen atom, on the one hand, and from the mutual interaction of the implanted atoms, on the other.

When the considerations are restricted to the very simple model of a central-force type interaction of the hydrogen atoms with the nearest metal atoms and when the vibrations of the heavy metal atoms in the acoustic modes are disregarded, one can show that, in the case of a tetragonally distorted tetrahedral interstitial site, the vibration spectrum of the hydrogen atom is split into two bands. One finds a nondegenerate frequency ω_1 corresponding to vibrations in the direction of the c axis and a double-degenerate frequency ω_2 corresponding to vibrations on a plane

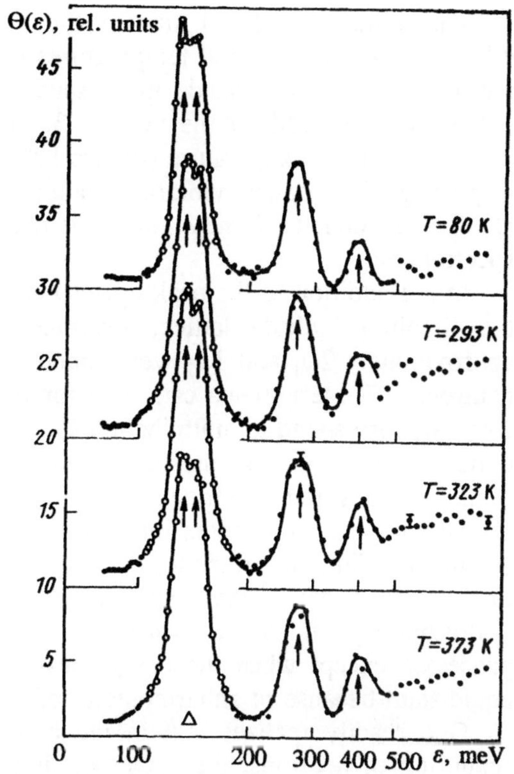

Fig. 108. The optical part of the $TiH_{1.96}$ spectrum in dependence upon the temperature.

perpendicular to the c axis. The ratio of these frequencies is $\omega_2 = (a/c)\omega_1$. Since $a/c = 1.12$ in the case of ZrH_2, one can expect that the spectrum will contain at least two lines separated by an interval of about 17 meV.

The phonon spectrum of zirconium hydride was calculated [33] with a model which is an analog to the earlier described model for the dihydrides of the light rare-earth metals without tetrahedral distortion of the lattice. The resulting spectrum was transformed into $d^2\sigma/d\Omega d\varepsilon$ for a comparison with the experimental results (Figure 107). Qualitative agreement with the experiment was established and it was shown that it is important to bring into account the mutual interaction of the implanted atoms. The model spectrum was used to calculated in [325] the temperature dependence of the heat capacity, which dependence is found to be in good agreement with the experimental results in the entire temperature interval considered.

The authors of [310] studied the vibration spectrum of titanium dihydride and deuteride at temperatures between 80 and 370 K. In the region of the vibration band, in the spectrum there is a peak at the energy 145 meV; the width of the peak is 30 meV and the peak's half-width changes with the temperature (Figure 108). The authors could not detect a splitting of the optical vibrations band; the extent of the splitting should at 80°K amount to 7 meV and is at the limit of the resolution of the spectrometer.

In the addition to the peak corresponding to the optical vibrations of the titanium hydride lattice, maxima at energies corresponding to approximately $2\omega_1$ and $3\omega_1$ were found in the spectrum of the scattered neutrons. These maxima can arise for two reasons. On the one hand, they can correspond to multiphonon and multiple scattering of neutrons at optical vibrations. On the other hand, there may be a relation to an oscillator transition to subsequent excited levels; in a first approximation, a hydrogen atom implanted into the metal lattice can be assumed as an oscillator. Thus, in principle, neutron experiments can provide information on the depth of the potential well for the hydrogen in the metal. Unfortunately, it is experimentally impossible to separate the two processes except when the energy levels of the potential well are not equidistant because of anharmonicity of the well.

One readily realizes that in the case of multiphonon and multiple scattering of neutrons, the corresponding spectral peaks must appear at energies which are multiples of $\hbar\omega_{opt}$, and that in the case of an anharmonic well, these energies cannot be such multiples. Thus, if it were experimentally possible to show that the peaks are not equidistant in the spectrum of scattered neutrons, this would be proof that the potential well is anharmonic and would allow the determination of the well's depth. The authors of [310] could not prove that the peaks are not equidistant because their results had a low statistical accuracy and the errors in the determination of the positions of the peaks on the energy scale were large, owing to the experimental technique used. But a similar principle was employed in [310] in research on an NbD sample for which it could be shown that multiples of the first and second excited vibrational levels do not exist.

Vanadium and niobium dihydrides. As has been shown [90], the dihydride phases of V and Nb are formed on the basis of the face-centered cubic lattice of the metal; the hydrogen atoms occupy tetrahedral interstitial positions. The results of research on the optical vibration spectra of these systems are given in [294]. The data are in good

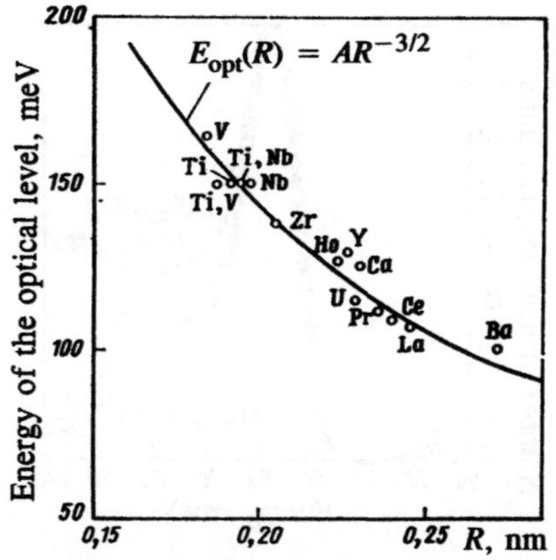

Fig. 109. Relation between the energy of the optical vibrations in hydrides with fluoride structure and the metal-hydrogen distance.

agreement with the results of research on other hydrides having CaF_2 type structure. Comparative considerations of mono- and dihydride phases in these systems are of interest. As indicated above, the absence of a dependence of the optical vibration energy upon the Me–H distance is a distinguishing feature of the lattice dynamics of the niobium and vanadium monohydrides [366]. However, the data of [294] on dihydrides indicate that such a dependence exists. This means that the saturation of Nb and V to the composition MeH_2 implies not only a change in the structure of the metal lattice but also a change in the interaction forces of the metal and hydrogen atoms.

The authors of [294] have analyzed frequency data on the optical vibrations of all hydrides with a CaF_2 type structure. It turned out that the dependence of the average energy of the optical band of the spectrum on the Me–H distance can in all these systems be described by a universal curve $E_{opt}(R) = AR^{-(3/2)}$ (see Figure 109). The authors assume that in such systems, there occurs a type of bonding which is intermediate between the types resulting from the proton and anion models of the electron structure of the hydrides. A fraction of the electrons of the hydrogen occupy levels which are much lower than the Fermi level, and the remaining electrons fill *d* states on the Fermi surface. It has been shown [336] for TiH_2 that in the lower states there are 1.15 of the two additional electrons, whereas the remaining 0.85 are on the Fermi surface.

Palladium hydrides. The hydrogen atoms occupy octahedral positions of implantation in the face-centered cubic lattice of palladium [90]; therefore stoichiometric palladium hydride has NaCl type structure. In compositions with H/Me = 0.63, there exists an ordered β phase in which the hydrogen appears as a lattice liquid. At low concentrations and high

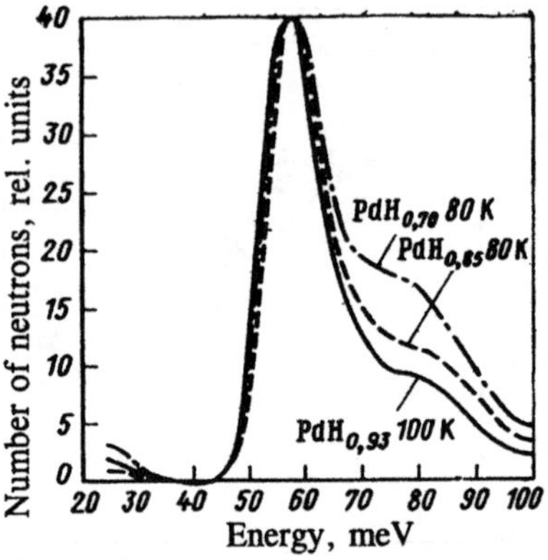

Fig. 110. The optical level for palladium hydride at various temperatures.

temperatures, there exists a region of α solutions which are characterized by disordered arrangements of the hydrogen atoms which in this case can be compared with a lattice gas [90]. Filling of "shallow" octahedral interstitial sites with hydrogen and the lack of order in the hydrides at almost all hydrogen concentrations cause peculiar features in the phonon spectrum of palladium hydride. There is, on the one hand, the (for hydrogen) abnormally low energy of the optical vibrations and, on the other hand, there appears a statistical filling of interstitial sites.

The discovery of superconductivity [324] explains the considerable recent interest in research on the dynamics of the crystal lattice of palladium hydride. It is a well-known fact that pure Pd is not a superconductor. Noteworthy is the fact that an inverse isotope effect is observed at the transition temperature to the superconducting state. This means that for a particular composition, the transition temperature is higher in the deuteride. This leads to the logical assumption that in addition to a restructuring of the electron spectrum by the introduction of hydrogen or deuterium, optical phonons are important for the development of the superconducting state in the hydride.

The authors of [203] have studied the dynamics of nonstoichiometric $PdH_{0.7}$ is in a wide temperature range. The function $\Theta(\varepsilon)$ for this case (Figure 110) is a well-separated peak at 58 meV with a broad "shoulder" on the high-energy side. The authors of [285] have shown that such a form of the phonon spectrum is caused by effects which are directly related to the nonstoichiometric composition of the sample. The calculation was based on a pseudostatistical model in which a cubic

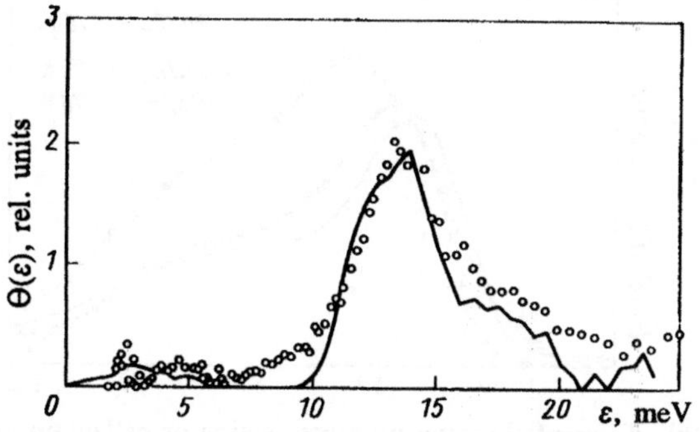

Fig. 111. Comparison of the experimental (dots) and calculated (solid line) functions $\Theta(\varepsilon)$ for $PdH_{0.63}$.

supercell comprising 27 face-centered cubic unit cells is considered. There are 63 hydrogen atoms per 108 metal atoms in such a cell, and the hydrogen atoms are statistically distributed over the 108 octahedral interstitial points. In order to determine the force constants of the interaction of the various atoms in such a supercell, the results of measurements of the dispersion curves in the $PdD_{0.63}$ system were used.

The comparison of the calculated results with the experimental results (Figure 111) seems to indicate that the model must be considered rather adequate. However, it is clear that an increase in the size of a supercell would make it possible to bring into account the nonstoichiometric composition of the sample to an even greater extent. One must recall the result of [294] where inelastic incoherent neutron scattering was used to study PdH_x samples with x equal to 0.7, 0.85, and 0.93. It was found that even with the ratio H/Me = 0.93, a high-energy "shoulder" is observed in the spectrum of the optical vibrations; in this case, the shoulder is hardly explained by the presence of 7% vacancies in the hydrogen sublattice. However, it cannot be ruled out that part of the high-energy "shoulder" is caused by multiple and multiphonon scattering of neutrons.

In the literature [259, 296] there are indications that the vibrations of the hydrogen atoms in palladium hydride are anharmonic. This should manifest itself in the vibration spectra by a temperature dependence of the line width. But since the vibration spectrum is rather broad because of the H−H interaction, the detection of a temperature-dependent contribution is a difficult task. Obviously, experiments in which the temperature

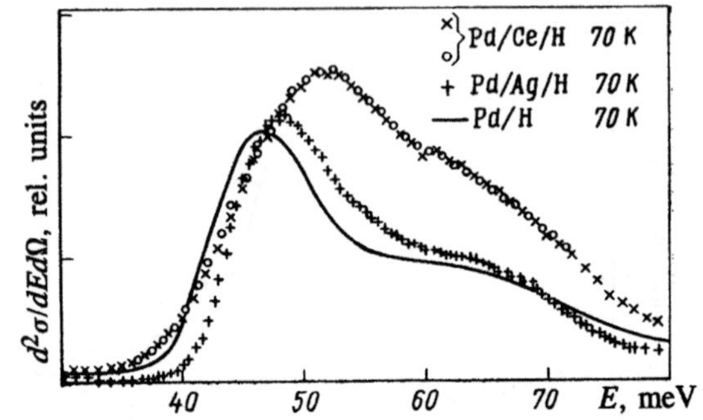

Fig. 112. Double differential scattering cross section of palladium hydride doped with silver and cerium.

dependence of the width of phonon resonances is obtained from inelastic incoherent scattering of neutrons are more appropriate for this purpose.

The discovery that the transition temperature to the superconducting state [137] in the ternary system Pd−Ag−H is higher than in Pd−H has stimulated research on the vibration spectra of the systems $Pd_{0.8}Ag_{0.20}H_{0.4}$ and $Pd_{0.94}Ce_{0.06}H_{0.40}$ [203]. In these compounds, Ag and Ce atoms statistically replace the Pd atoms, the structure of the metal lattice remains unchanged, and the lattice parameter is slightly reduced by the addition of Ag and remains constant when Ce is added. Based on the experimental data obtained at $T = 70°K$ (Figure 112) it was established that the addition of silver strengthens the Me−H bond by 6% on the average and that the H−H bond is slightly weakened in comparison with Pd−H. It turns out that the Pd−H and Ag−H bonds are identical except for the electrostatic repulsion which is partially shielded by the conduction electrons.

A sharper change of the optical vibrations spectrum is observed in the ternary system Pd−Ce−H; this indicates that the binding force of the Me−H bond was enhanced by 20%, on the one hand; and that the degree of disorder in the system was increased, on the other hand. By contrast to the system Pd−Ag−H, the Ce atoms cannot be considered identical with the Pd atoms, and therefore the hydrogen atoms introduced in the lattice have an environment which is distinguished from the neighborhood not only by the light atoms but also by the heavy Pd and Ce atoms. This fact must inevitably broaden the spectrum and smear the details as observed in experiments (see Figure 112).

Inelastic neutron scattering was used [167] to study the local vibrations of the hydrogen and deuterium atoms in solid solutions of the system Pd−H. Such investigations are of interest for explaining the isotope effect in the vibration spectrum and also for clarifying the mutual interaction of the admixed atoms. The experiments were made on single crystal Pd samples in which the atomic hydrogen concentration was varied between 0.2 and 2.72% at 50°C. As a matter of fact, the intensity distribution of neutrons scattered at local vibrations was studied as a function of the temperature, the concentration, the momentum transferred in a scattering event, and various conditions of annealing the samples. The most interesting features which could be observed were: the considerable width of the spectral line corresponding to local vibrations and the change in the position of the line in terms of energy when the hydrogen concentration was increased. In particular, when the atomic hydrogen concentration was changed from 0.2 to 2.72%, the width increased from 20 to 30 meV, and the energy decreased from 66 to 63 meV. Let us note that the energy of the local vibrations turned out to be much higher than the average energy of the optical spectrum in the β phase of the Pd−H system. It cannot be ruled out that this is related to an increased Me−H distance at increasing concentrations of the admixed atoms.

The considerable width of the local vibration levels in the case of the very low concentrations of the implanted atoms cannot be explained by H−H interaction or by fluctuation effects caused by differences in the environment of the implanted atoms. The authors have established that after prolonged sample annealing at 800 K in high vacuum, the band of the local vibrations of the hydride has become very narrow and has a width equal to the width of the resolution function of the spectrometer. It is assumed that this change of the spectrum originates from the annealing of hydrogen clusters which are present even in α solutions and cause local distortions of the crystal lattice. Two additional interesting features of the local vibrations of the hydrogen atoms in the Pd lattice were found. It turned out that the intensity of the peak in the neutron spectrum has an unusual temperature dependence: the intensity decreases with increasing temperature. A dependence of the intensity upon both the magnitude and the direction of the momentum transferred in the scattering event was also observed. In particular, in measurements made at T = 77 K on a $PdH_{0.002}$ sample, the intensity of the spectral lines decreased by a factor of 10 when the direction of the momentum changed from [100] to [110]. These facts assert the conclusion that a strong vibration anisotropy exists in the [100] and [110] directions.

All the experimental data obtained in research on the dynamics of the crystal lattices of palladium hydrides with inelastic incoherent neutron scattering indicate that the statistical distribution effects of the hydrogen atoms are in this system very strong at both low and high concentrations; the interaction between the implanted atoms turns out to be significant at high concentrations. Also the anharmonicity of the vibrations of the hydrogen atoms is significant.

33. Dispersion Curves for Phonons in Hydrides

Palladium hydrides. The technique of inelastic coherent neutron scattering for research on metal hydrides encounters serious difficulties which, on the one hand, are associated with difficulties in the preparation of single-crystal samples, particularly samples with high hydrogen concentration and, on the other hand, involve the optimal experimental conditions required for the observation of high-energy vibrations of light atoms. Still another difficulty results from the use of deuterated samples because, by contrast to hydrogen, deuterium scatters neutrons mainly in a coherent fashion. These difficulties imply restrictions for the use of inelastic coherent neutron scattering in research on the dynamics of the crystal lattices of hydrides and therefore only a limited amount of work has been made with this technique.

Historically, palladium deuteride $PdD_{0.63}$ was the first metal-hydrogen system studied with the technique of coherent inelastic neutron scattering [296]. The acoustic and optical branches of the vibrations were measured in three directions of high lattice symmetry (Figure 113). The solid lines on the figure correspond to results obtained with calculations made with the very simple Born–von Karman model; the dashed lines relate to results for pure Pd. The high dispersion of the optical branches attests to a considerable D–D interaction extending to the second nearest atoms, and this is the principal feature. The experimental data were used to calculate the elastic moduli and the Debye temperature [297]. It was established (see Figure 113) that the spectrum of the acoustic vibrations of the deuteride lattice is much softer than the spectrum of pure Pd. This fact must undoubtedly be brought into account when the thermodynamic characteristics of the deuteride are calculated.

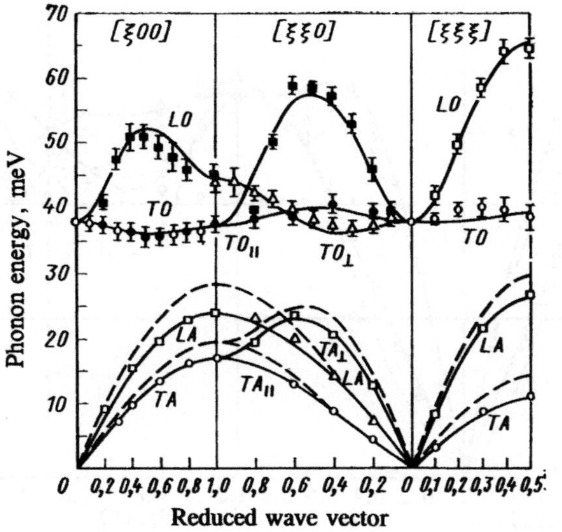

Fig. 113. Dispersion curves for the phonons in $PdD_{0.63}$ in comparison with pure Pd (dashed curves); temperature of the measurements: Pd) 120 K; $PdD_{0.63}$) less than 150 K.

Since the deduction of the acoustic vibration spectrum of the hydride lattice from data obtained in incoherent neutron scattering is complicated, it is impractical to directly compare the experimental information in this frequency range with calculations based on a model the parameters of which were obtained from a best fit to the dispersion curves. However, a comparison is possible in the optical part of the spectrum. It turned out that satisfactory agreement between the calculated $\Theta(\varepsilon)$ and the experimental values for $PdH_{0.7}$ can be obtained only when a quasistatistical model is used and a Pd−H bond 20% stronger than that of Pd−D is assumed [285]. It is interesting to note that this model enables not only the calculation of the function $\Theta(\varepsilon)$ for the vibration band of the hydride lattice but also the calculation of the form of the neutron resonance corresponding to neutron scattering at a particular phonon [297]. It turned out that the calculated results agree qualitatively with the experimental results [297] but there exist considerable quantitative differences in the calculation of the interaction force in a 3 × 3 × 3 supercell. Since such an integral characteristic like the phonon spectrum of the system is rather well-described by the model, one can assume that at least part of the width

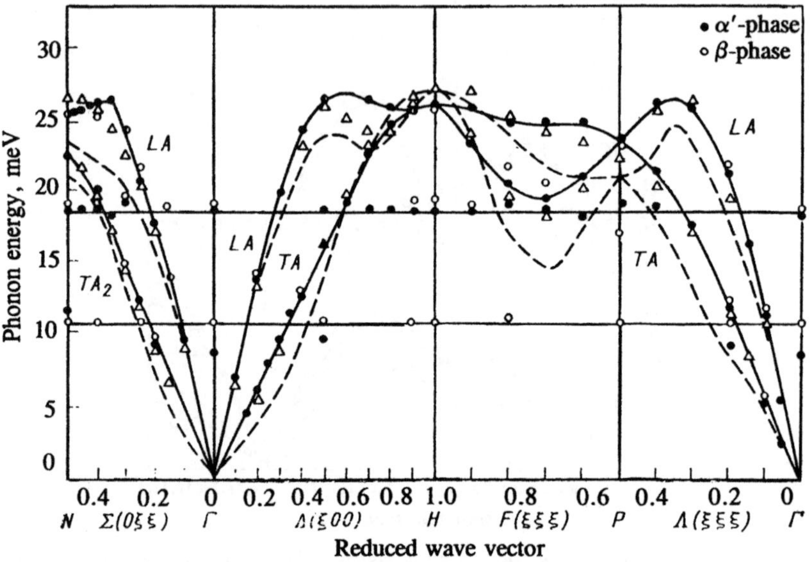

Fig. 114. Comparison of the dispersion for Nb (dashed lines), $NbD_{0.45}$ (Δ, ▲), and $NbD_{0.85}$ (○, ●).

of the neutron resonances originates from a large number of vacancies in the lattice. However, as noted by the authors of [297], there are indications of anharmonicity contributing to the width of the phonon resonances. More specifically, for optical phonons with $\mathbf{q} = 0$ the model predicts a width which depends only on the energy resolution of the spectrometer, whereas a temperature-dependent width is observed in experiments.

We can state in conclusion that the data which were obtained with coherent and incoherent neutron scattering on the dynamics if the crystal lattice of palladium hydrides and deuterides supplement each other.

Hydrides with a body-centered cubic metal lattice. The dispersion of the acoustic branches in the Nb–D system was measured at 473 K [297] and at 380 K [234] (Figure 114). By contrast to the Pd–H system, the deviation of the vibration spectra of the niobium deuterides from the spectrum of the pure metal is irregular. It is a well-known fact that the metal lattice is expanded by the addition of deuterium [90] but a smooth reduction of the frequencies is not observed in the experiments. It was noted that the greatest difference between the spectra occurs in those regions of the Brillouin zone in which anomalies are present in the Nb spectrum. This allowed the authors of [297] to consider the NbD_x and Nb_xMo_{1-x} systems as analogs [282]. When Mo is added to Nb, substitution alloys with a body-centered cubic lattice of the metal, and the lattice parameter is increased by the introduction of D. Despite these differences, the authors of [297] found some common features in the phonon spectra of Nb−Mo and Nb−D. It is assumed that the additional electron from D or Mo enters the conduction band in both cases and, as a consequence, the electron−phonon interaction in the lattice is modified. It is a well-known fact [282] that the dispersion relations in Nb−Mo alloys are modified very strongly when the Mo concentration is increased in the same regions of the Brillouin zone as in the NbD_x system and that the general level of the frequencies rises on this occasion. The entire pattern of the changes of the phonon spectrum apparently can be explained when one uses the information on changes of the Fermi surface in Nb−Mo alloys and the density of the electron states on that surface. A general increase in the frequency can be at least partially explained by a compression of the Nb lattice upon addition of Mo. In the NbD_x system, an overall increase in the frequencies of the acoustic phonons is accompanied by an increase in the lattice parameter. It seems that in this case the expansion of the lattice and the change of the state density on the Fermi surface partially compensate each other [297].

The experimental results reported by the authors of [297] were obtained with a dynamic model in the Born−von Karman scheme with proper regard for the interaction in six coordination spheres. Even in the directions of high symmetry, the dispersion curves of the deuterides could not be precisely described with a multiparameter model. But the general form of the dispersion curves is in good agreement with the experimental results. A comparison of the phonon spectra of pure Nb, of $NbD_{0.115}$, and of $NbD_{0.45}$ shows that at low energies ($E \leq 15$ meV), the density of the states is in the deuterides lower than in Nb and decreases with increasing D concentration, whereas the average energy of the phonons increases. A decrease in the density of the phonon states in the low-energy region of the spectrum is correlated with an increase in the elastic modules obtained with both the ultrasound technique of [243] and neutron scattering [335]. Results which resemble the results for niobium deuteride were obtained in the range of the acoustic lattice vibrations of the systems Ta−D [244] and V−D [299].

Advances in the technique of inelastic neutron scattering facilitated a number of investigations [234, 244, 334] in which the dispersion of the

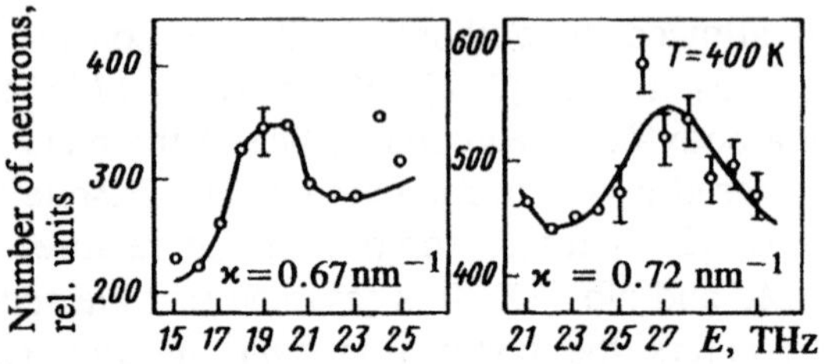

Fig. 115. Examples of experimental phonon peaks in $NbD_{0.6}$ for two values of the transmitted momentum.

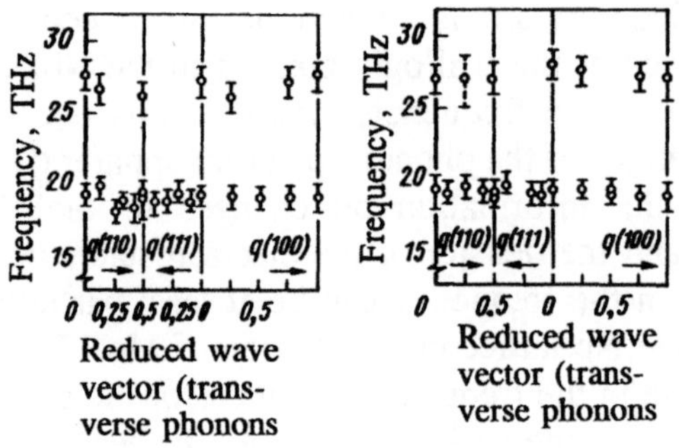

Fig. 116. Dispersion curves for phonons in $NbD_{0.6}$ at 380 K.

optical branches in hydrides of the transition metals of the fifth group were observed. The experiments were made with a triaxial spectrometer mounted on the horizontal channel of a high-flux reactor with a source of "hot" neutrons. The investigations were conducted so that the neutron transmitted part of its energy to the crystal lattice in a scattering event. It must be recalled that even under these conditions, the observation of the optical vibrations is a rather complicated experimental task because the intensity of the phonon resonances is low and the background level is high due to the contribution of incoherent neutron scattering (see Figure 115).

Figure 116 shows the dispersion relations obtained for the optical branches in $NbD_{0.6}$ [334]. The experiments were made at 380 K, which correspond to the cubic α' phase of the Nb−D system. It is generally accepted [90] that the deuterium atoms are in the α' phase statistically distributed over tetrahedral interstitial sites of the body-centered cubic metal lattice but that because of the considerable concentration of the atoms implanted into the system, there exists a short-range order

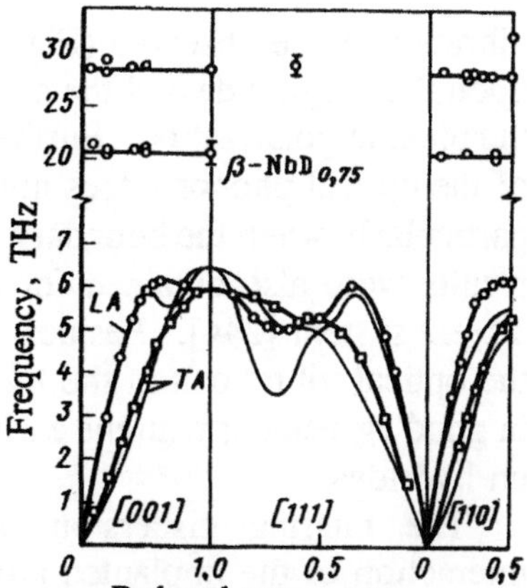

Fig. 117. Dispersion curves for phonons in $NbD_{0.75}$ (the thin solid lines refer to pure Nb).

resembling that of a liquid. Thus, in this case the hydrogen behaves like a lattice liquid: it participates in the diffusion movements and the time during which it remains at interstitial sites varies at the optical frequencies. The missing momentum dependence of the energy of the optical phonons in the three directions of high symmetry of the crystal lattice is the most important feature of the phonon spectrum (Figure 116). It is a well-known fact that experiments with a triaxial spectrometer can be set up so that only neutrons scattered at longitudinal or transverse phonons are recorded. It turned out that in both cases, two identical branches of optical phonons are obtained in the system $\alpha'-NbD_{0.6}$. It was in this way established that the longitudinal and transverse branches are degenerate. Two branches exist because a tetrahedral interstitial site in a body-centered cubic lattice has a tetrahedral distortion. As mentioned above, the observed frequency difference must satisfy the relation $\omega_1 = \omega_2 = \sqrt{2}\omega_3$, which is in rather good agreement with the experimental result. Let us also note that the frequencies are in rather good agreement with the data obtained by incoherent neutron scattering at niobium hydrides [82, 98] when the change in the mass of the isotope was taken into account, i.e., $\omega_H = \sqrt{m_D/m_H m_D} = \sqrt{2 m_D}$.

Analogous experiments have been made on a single crystal of the ordered β phase of $NbD_{0.75}$ in a wide temperature range [234]. The lattice of the β-phase in the Nb–D system has a slight orthorhombic distortion which manifests itself in a broadening of the Bragg reflection but which does not affect the dispersion curves. Accordingly, the entire treatment was made in the approximation of a cubic lattice. As in the case of measurements made on the α' phase, two branches of optical vibrations were observed (Figure 117), and their energies depend neither upon the magnitude and the direction of the phonon momentum nor upon the phonon polarization. Furthermore, as had been established, the energy of the optical phonons does not change when the temperature is reduced, particularly when the boundary of the $\alpha' - \beta$ phases is intersected. Similar results were also obtained for the optical phonons in the α phase of the Ta–D system [244]. Besides that, it turned out that the very energies of the optical vibrations in the two systems are very similar. This result is in good agreement with the earlier evaluated data which had been obtained on hydrides.

The missing dispersion of the optical branches means that the interaction of the implanted atoms is weak. But the experimental results obtained on hydrides with the aid of incoherent neutron scattering show a considerable line width of the optical spectrum. This means either that there exists a significant dispersion of the optical branches in the nonsymmetric directions of a Brillouin zone or that one must look for other reasons for the broadening of the spectral lines. A partial answer to this question could be obtained in an experiment in which the dispersion was measured in nonsymmetric directions of the reciprocal lattice. Let us also note that the difficulties in interpreting the experimental data are enhanced by the peculiar features of the solid solutions of deuterium in Nb and Ta, where there is no ordered sublattice of light atoms and where one can hardly speak of optical vibration branches.

In research on $NbD_{0.85}$ and $NbH_{0.82}$ in the α' and β phases, new low-frequency branches without wave vector dependences resembling those of the optical branches were detected [309]. The energies of these vibrations do not change with the temperature and amount to 18.8 and 10.8 meV when the one isotope is replaced by the other. One readily recognizes that the position of these branches on the energy scale is substantially below the level of the local vibrations and falls into the frequency band of the acoustic spectrum. It was experimentally established that the width of the lines corresponding to these vibrations considerably decreases in the spectrum of the scattered neutrons when the $\alpha' - \beta$ transition takes place and H is replaced by D. When the difference in the cross sections and the volumes of the samples are taken into account, the intensity ratio of the corresponding lines in the spectra of $NbD_{0.85}$ and $NbH_{0.82}$ allows the conclusion that neutron scattering takes place at H or D atoms. It cannot be ruled that these low-frequency excitations reflect the participation of the light atoms in the vibrations of the metal lattice. The vibrations do not

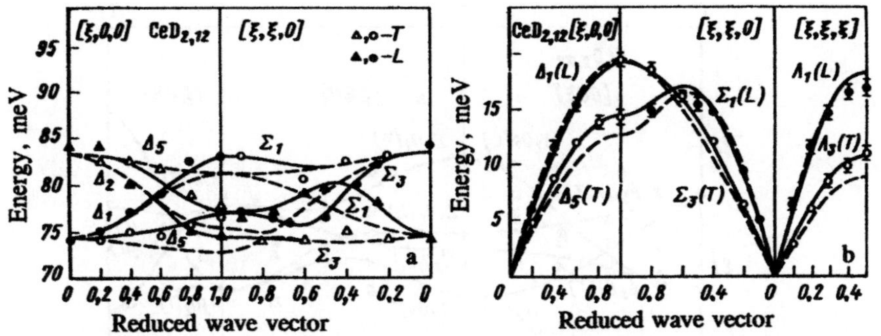

Fig. 118. Dispersion curves for a) the optical and b) the acoustic phonons in $CeD_{2.12}$ (the solid lines correspond to smoothed experimental data).

coincide in phase and amplitude with the vibrations of the metal atoms. The origin of such resonance modes has been earlier discussed in greater detail.

Cerium hydrides. When the results obtained with incoherent neutron scattering at hydrides of the light rare-earth metals were discussed, a considerable width of the spectral levels of the optical vibrations was noted. It was shown that, if the line broadening in the spectrum is caused by H−H interaction, this fact must manifest itself in a substantial momentum dependence of the corresponding phonon branches. The data obtained by coherent scattering of neutrons at $CeD_{2.12}$ [186, 354] immediately confirm this (see Figure 118). The branches of the acoustic vibrations (Figure 118b) have the usual form for face-centered cubic metals and since there exist no data on the dispersion of the phonons in the initial Ce, one cannot draw any conclusions concerning the influence of D atoms upon the acoustic spectrum of the system. The optical branches [354] are more complicated and have a noticeable dispersion, though the dispersion is less pronounced than in $PdD_{0.63}$. The entire band of the optical vibrations extends over about 10 meV. Assuming that the isotope model is valid, one can expect the corresponding band to cover 14 meV in the case of the hydride, and this is still below the value observed in experiments involving incoherent scattering. It seems that a more significant momentum dependence of the energy of the optical phonons exists in directions of low symmetry.

Based on the force model of D-atom interaction up to the second nearest neighbors, both acoustic and optical branches of the spectrum were

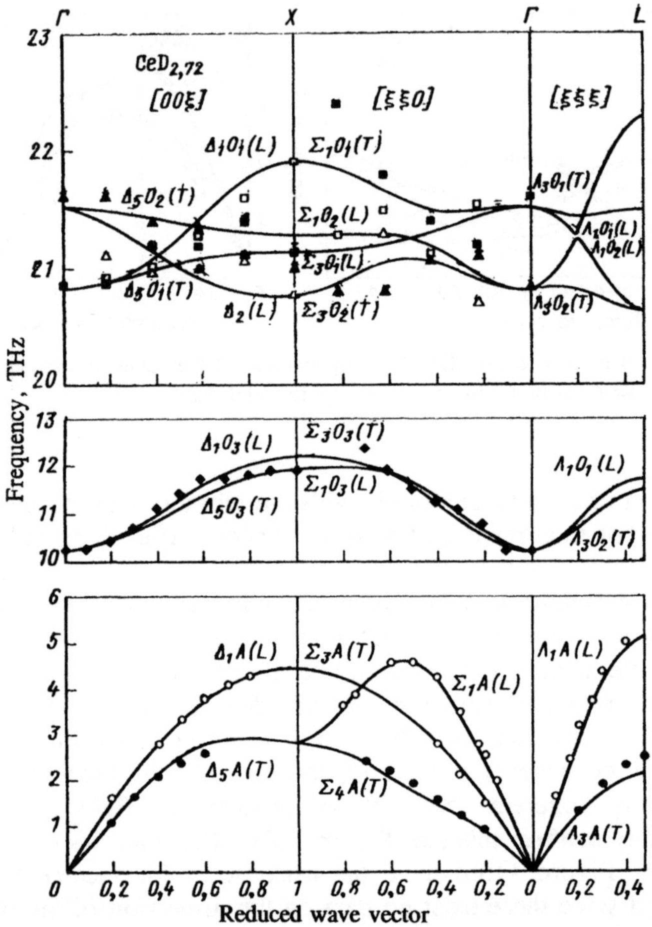

Fig. 119. The dispersion curves for the phonons in $CdD_{2.72}$ (the solid lines represent the results of theoretical calculations which were made in the Born–von Karman model with 28 parameters).

calculated. It turned out that the mutual interaction of the deuterium atoms cannot be described in such a simple way. More particularly, at the boundary of the Brillouin zone there is not even qualitative agreement between the calculated and the experimental results. Apparently a model of greater complexity is required for describing the lattice dynamics of cerium dideuteride. It is interesting to note that the authors of [354] had to introduce the interaction in six coordination spheres for successfully describing the experimental data on the dispersion of phonons in $CeD_{2.12}$.

Measurements of the dispersion of the phonons in $CeD_{2.72}$ were the next step in research on the dynamics of the crystal lattice in the system Ce–D [355]. The following conclusions can be drawn from a comparison of these results (Figure 119) with the data for $CeD_{2.12}$.

1. In the range of the vibration, a slightly greater dispersion for the optical branches $\Sigma_1 A$ and $\Lambda_1 A$ and a slightly smaller dispersion for the transverse branches $\Sigma_4 A$ and $\Lambda_5 A$ are observed.

2. The higher optical branches, which are related to vibrations of the D atoms in tetrahedral interstitial positions, are shifted by about 8 meV toward higher energies.

3. There appeared new optical branches which correspond to octahedral coordination of the D atoms and which have transverse polarization.

The authors note that they could not observe neutron scattering at low-level longitudinal optical phonons. Moreover, it was noted that the width of the neutron resonances, which correspond to neutron scattering at transverse optical phonons due to vibrations of the deuterium atoms at octahedral positions, was much greater than the width of the resolution function of their instrument. A similar situation was observed in $PdD_{0.63}$ in which the deuterium is also trapped in octahedral interstitial positions [285]. More specifically, it was observed that the phonon peaks are considerably broadened; the peaks corresponding to the longitudinal vibrations are noticeably more broadened. The authors of [285] have shown that such a broadening is associated with the statistical distribution of the vacancies and of the implanted atoms over octahedral interstitial sites. It seems that also in $CeD_{2.72}$, which has a noticeable number of vacant octahedra, the broadening of the neutron resonances can be explained in the same way.

The experimental results could be described most appropriately with the 28-parameter Born–von Karman model for a lattice with BiF_3 structure. It turned out that the constants earlier obtained for $CeD_{2.12}$ [354] had to be rather radically modified for an appropriate representation of the entire set of dispersion curves of $CeD_{2.72}$. It was observed that among the interactions involving deuterium atoms which vibrate in octa-cavities, the interaction with the deuterium in the tetra-cavities is strongest, whereas the interaction with the cerium is much weaker and the interaction with the neighbor in the octa-cavity is quite insignificant. This conclusion is in rather good agreement with the experimental data obtained by incoherent neutron scattering [85].

A few words are in place about the diffusion-induced movements in the hydrogen subsystem. Information on the diffusion movement of light implanted atoms can be obtained from data on quasielastic scattering. The intensity and the width of a line of quasielectric scattering are compared with values calculated with some model of diffusion jumps, and in this way all the parameters of interest are determined: the activation energy of diffusion, the diffusion coefficient, and the time dependence of the diffusion coefficient. These problems have been thoroughly treated in [326].

Thus, summarizing the chapter on hydrides, our conclusion is that research made on metal hydrides with neutron scattering to date has provided knowledge about the frequency spectra of the optical vibrations of almost all hydrides, the dispersion relations for the optical and acoustic branches of some deuterides, and the diffusion parameters in cases in which diffusion is not insignificant. A short time ago the authors have started a detailed investigation of the contribution rendered by the light atoms to the spectrum of the acoustic lattice vibrations.

The available experimental information and the theoretical work so far do not allow quantitative conclusions concerning the bonding of the hydrogen to the metal, the form of the potential well for hydrogen, and the influence of any H$-$H interaction upon the properties of the hydrides.

CONCLUSION

Having finished the last volume of the three-volume work "Neutrons and Solids," where the contemporary state of structural and magnetic neutron diffraction work and neutron spectroscopy of crystals have been expounded, it is natural to consider new developments in the use of neutrons in solid state physics. These new developments have been reported in Russian and international conferences of the last four years and relate to the three aspects: the development of neutron sources, the development of experimental techniques, and the directions and the objectives of physics research.

Let us consider each of these aspects.

At the present time, the leading research centers of the world have specialized stationary high-power reactors with fluxes ranging from $5 \cdot 10^{14}$ neutrons/(cm$^2 \cdot$sec) up to $2 \cdot 10^{15}$ neutrons/(cm$^2 \cdot$sec). A large part of the research work on the neutron scattering at solids is performed on these reactors. Stationary reactors with higher fluxes cannot be built, mainly because the capital expenses, as well as the operating costs of such installations, increase excessively. But the development of powerful pulsed neutron sources has been initiated, primarily by making use of high-power proton accelerators of the type of "meson factories" in which pulsed fluxes of about 10^{16} neutrons/(cm$^2 \cdot$sec) can be achieved at an average neutron flux of about $5 \cdot 10^{14}$ neutrons/(cm$^2 \cdot$sec). Such sources, which are characterized by a harder spectrum, open new possibilities for research on neutron scattering in solid state physics, as well as in chemistry, geology, and biology.

As could be expected, the development of experimental methods is, on the one hand, characterized by intensive developments of time-of-flight techniques for measuring elastic and inelastic neutron scattering, mainly on pulsed neutron sources. On the other hand, the traditional improvement of the methods of generating monochromatic neutron beams with defining techniques and the further developments of multi-detector and position-sensitive detector systems will proceed and small-angle scattering research and the three-dimensional analysis of the polarization in elastic and inelastic scattering will become more important.

High-resolution spectrometry will be used on an even larger scale: spin echo, gravitational spectrometers, etc., which allow high-precision measurements of the extremely small energy and momentum transfers.

Elastic and inelastic neutron scattering at crystals has for more than 30 years significantly contributed to solid state physics research and to the development of our ideas of the condensed state of matter; this contribution can hardly be overestimated.

Nowadays, the development of this part of physics cannot be imagined without neutron scattering. We note the following examples among the large number of problems in the solution of which neutron scattering will be of decisive importance.

The improvement of the neutron sources and of experimental techniques will allow investigations of the laws governing the mutual interaction of excitations in a crystal (phonon–phonon, electron–phonon, phonon–magnon, etc. interactions); research on the structure and the excitation spectra of materials under extreme conditions, first of all under high and super-high pressures, will become possible; and the structure and the excitation spectra of systems without order, of the metallic and spun glasses, amorphous semiconductors, etc., will be studied. Investigations of quasi-one-dimensional and two-dimensional structures, of new classes of excitations (phasons and solitons), and of structures and excitations in complicated biological crystals and many other investigations will be made.

Though the stormy development of physics in the last 15 years has provided new powerful tools for solid state research (laser spectroscopy, synchrotron radiation, etc.), neutron scattering techniques will, as in the past, play a decisive role in this field of physics.

REFERENCES

1. K. S. Aleksandrov, *Kristallografiya*, **21**, 245 (1976).

1a. A. M. Al'tshuler, Yu. Kh. Vekilov, and A. D. Izotov, *Fiz. Tverd. Tela*, **18**, 1840 (1975).

2. A. M. Afanas'ev and Yu. Kagan, *Zh. Éksp. Teor. Fiz.*, **43**, 1456 (1962).

3. A. I. Akhiezer, V. G. Bar'yakhtar, and S. V. Peletminskii, *Spin Waves* [in Russian], Nauka, Moscow (1967).

4. V. G. Bar'yakhtar and S. V. Maleev, *Zh. Éksp. Teor. Fiz.*, **39**, 1430 (1960).

5. V. G. Bar'yakhtar and S. V. Maleev, *Fiz. Tverd. Tela*, **5**, 1175 (1963).

5a. O. N. Belenin and V. A. Telezhkin, *Fiz. Tverd. Tela*, **22**, 1988 (1980).

6. R. Blints and B. Zheksh, *Ferroelectric and Antiferroelectric Materials* [Russian translation from English], Mir, Moscow (1975).

7. N. N. Bogolyubov and S. V. Tyablikov, *Dokl. Akad. Nauk SSSR*, **126**, 53 (1959).

8. E. G. Brovman and Yu. Kagan, *Zh. Éksp. Teor. Fiz.*, **63**, 1937 (1972).

9. E. G. Brovman and Yu. Kagan, *Usp. Fiz. Nauk*, **112**, 369 (1974).

10. I. A. Bulat, *Fiz. Tverd. Tela*, **21**, 1001 (1979).

11. V. G. Vaks, S. P. Kravchuk, and A. V. Trefilov, *Fiz. Tverd. Tela*, **21**, 3370 (1979).

12. V. G. Vaks, A. I. Larkin, and S. A. Pikin, *Zh. Éksp. Teor. Fiz.*, **53**, 281; 1089 (1967).

13. V. G. Vaks and A. V. Trefilov, *Fiz. Tverd. Tela*, **19**, 244 (1977).

13a. V. G. Vaks and N. E. Zein, *Fiz. Tverd. Tela*, **23**, 3221 (1981).

13b. S. B. Vakhrushev, Ya. G. Gross, N. M. Okuneva, et al., Preprint of the Physicotechnical Institute (FTI) 585 [in Russian], Leningrad (1978).

13c. B. A. Vindraevskii, S. N. Ishmaev, I. P. Sadikov, et al., Preprint of the

Institute of Atomic Energy (IAE) 3122 [in Russian], Moscow (1971).
14. E. Wigner, *Group Theory* [Russian translation], Nauka, Moscow (1971).
15. K. Wilson and J. Kogut, *The Renormalized Group and the ε Expansion* [Russian Translation], Mir, Moscow (1975).
16. S. V. Vonsovskii, Yu. A. Izyumov, and E. Z. Kumaev, *Superconductivity of the Transition Metals, Their Alloys and Compounds* [in Russian], Nauka, Moscow (1977).
17. V. L. Ginzburg, *Fiz. Tverd. Tela*, **2**, 2031 (1960).
18. V. L. Ginzburg, A. P. Sobyanin, and A. P. Levanyuk, *Usp. Fiz. Nauk*, **130**, 615 (1980).
19. A. E. Golovin, G. I. Zadokhin, M. G. Zemlyanov, et al., *Prib. Tekh. Éksp., No. 4*, **31** (1978).
20. L. P. Gor'kov, *Pis'ma Zh. Éksp. Teor. Fiz.*, **17**, 525 (1973).
21. L. P. Gor'kov, *Pis'ma Zh. Éksp. Teor. Fiz.*, **20**, 571 (1974).
22. L. P. Gor'kov and O. N. Dorokhov, *Pis'ma Zh. Éksp. Teor. Fiz.*, **21**, 656 (1975).
23. V. D. Gorobchenko and E. G. Maksimov, *Usp. Fiz. Nauk*, **130**, 65 (1980).
24. I. I. Gurevich and L. V. Tarasov, *The Physics of Low Energy Neutrons* [in Russian], Nauka, Moscow (1965).
25. A. S. Davydov and N. I. Kislukha, *Zh. Éksp. Teor. Fiz.*, **71**, 1090 (1976).
26. I. E. Dzyaloshinskii, *Zh. Éksp. Teor. Fiz.*, **72**, 1930 (1977).
27. I. E. Dzyaloshinskii and B. A. Ivanov, *Pis'ma Zh. Éksp. Teor. Fiz.* **29**, 592 (1979).
27a. *Dynamics of Solids and Liquids by Neutron Scattering*, S. Lovesay and T. Springer, eds., Springer-Verlag, New York (1977).
28. B. Dorner and R. Comes, in: *Dynamic Properties of Solids and Liquids* [Russian translation], Mir, Moscow (1980).
29. I. P. Eremeev, I. V. Sadikov, and A. A. Chernyshev, *Fiz. Tverd. Tela*, **18**, 1652 (1976).
30. V. V. Eremenko, *Introduction to the Optical Spectroscopy of Magnetic Materials* [in Russian], Naukova Dumka, Kiev (1975).
31. M. G. Zemlyanov, E. G. Brovan, N. A. Chernoplekov, et al., *Inelastic Scattering of Neutrons*, **11**, p. 431, IAEA, Vienna (1965).
32. M. G. Zemlyanov, A. E. Golovin, S. P. Mironov, et al., *Prib. Tekh. Éksp.*, No 5, **34** (1973).
33. M. G. Zemlyanov, Ya. Maier, P. P. Parshan, et al., Preprint Joint Institute for Nuclear Research, 14-11625 [in Russian] (1978).
34. Yu. A. Izyumov, *Fiz. Tverd. Tela*, **21**, 1431 (1979).
35. Yu. A. Izyumov, *Usp. Fiz. Nauk*, **131**, 387 (1980).
36. Yu. A. Izyumov, *Fiz. Tverd. Tela*, **23**, 796 (1981).
37. Yu. A. Izyumov, *Fiz. Tverd. Tela*, **23**, 2666 (1981).
38. Yu. A. Izyumov, *Fiz. Met. Metalloved.*, **53**, 22 (1982).
39. Yu. A. Izyumov, F.A. Kassan-Ogly, and V. E. Naish, *Fiz. Met. Metalloved.*, **51**, 500 (1981).
40. Yu. A. Izyumov and S. V. Maleev, *Zh. Éksp. Teor. Fiz.*, **41**, 1644 (1961).

41. Yu. A. Izyumov and M. V. Medvedev, *Theory of Magnetically Ordered Crystals with Admixtures* [in Russian], Nauka, Moscow (1970).
42. Yu. A. Izyumov, V. E. Naish, and R. P. Ozerov, *Neutron Diffraction of Magnetic Materials*, Consultants Bureau, New York (1991).
43. Yu. A. Izyumov and R. P. Ozerov, *Magnetic Neutron Diffraction* [in Russian], Nauka, Moscow (1966).
44. Yu. A. Izyumov and S. B. Petrov, *Fiz. Met. Metalloved.*, **55**, 24 (1983).
44a. Yu. Kagan, *Zh. Éksp. Teor. Fiz.*, **42**, 1375 (1962).
45. Yu. Kagan and Ya. A. Iosilevskii, *Zh. Éksp. Teor. Fiz.*, **42**, 259 (1962); 44, 1375 (1963).
46. Yu. M. Kagan and Ya. A. Iosilevskii, *Zh. Éksp. Teor. Fiz.*, **45**, 819 (1963).
47. Yu. M. Kagan, V. V. Pushkarev, and A. Kholas, Abstracts of the Reports of the 21st All-Union Conference on Low-Temperature Physics [in Russian], Part 3, p. 43, Khar'kov (1980).
48. I. Kaimov, E. G. Brovman, M. G. Zemlyanov, et al., *Zh. Éksp. Teor. Fiz.*, **9**, 199 (1968).
49. I. Karimov, M. Zemlyanov, M. W. Kost, et al., *Zh. Éksp. Teor. Fiz.*, **9**, 1740 (1967).
50. O. V. Kovalev, *Irreducible Representations of Space Group* [in Russian], Izd. AN Ukr. SSR, Kiev (1961).
51. V. A. Koptsik, *The Shubnikov Groups* [in Russian], Izd. Mosk. Gos. Univ., Moscow (1966).
52. A. M. Kosevich and B. A. Ivanov, *Zh. Éksp. Teor. Fiz.*, **71**, 2000 (1977).
53. A. M. Kosevich, B. A. Ivanov, and A. S. Kovalev, in: *Nonlinear Waves* [in Russian], Nauka, Moscow (1979), p. 45.
54. M. A. Krivoglaz, *Zh. Éksp. Teor. Fiz.*, **40**, 567 (1961).
55. M. A. Krivoglaz, *The Theory of Scattering of X Rays and Thermal Neutrons at Real Crystals* [in Russian], Nauka, Moscow (1967).
56. A. V. Lazuta, S. V. Maleev, and B. P. Toperverg, Preprint of the Leningrad Institute of Nuclear Physics [in Russian], No. 632 (1980).
57. A. V. Lazuta, S. V. Maleev, B. P. Toperverg, et al., Preprint of the Leningrad Institute of Nuclear Physics [in Russian], No. 366 (1977).
58. M. Lines and A. Glass, *Ferroelectric and Similar Materials* [Russian Translation], Mir, Moscow (1981).
59. L. D. Landau and E. M. Lifshits, *Statistical Physics*, Part 1 [in Russian], Nauka, Moscow (1976).
60. A. I. Larkin and D. E. Khmel'nitskii, *Zh. Éksp. Teor. Fiz.*, **56**, 2087 (1969).
61. A. P. Levanyuk, V. V. Osipov, A. S. Sigov, et al., *Zh. Éksp. Teor. Fiz.*, **76**, 345 (1979).
62. A. P. Levanyuk and D. G. Sannikov, *Fiz. Tela*, **12**, 2997 (1970).
63. A. P. Levanyuk and D. G. Sannikov, *Usp. Fiz. Nauk*, **112**, 651 (1974).
64. I. M. Lifshits, *Zh. Éksp. Teor. Fiz.*, **12**, 117; 137; 156 (1942).
65. I. M. Lifshits, S. A. Gredeskul, and L. A. Pastur, *Fiz. Nizk. Temp.*, **2**, 1093 (1976).
66. E. M. Lifshits and L. P. Pitaevskii, *Statistical Physics*, Part 2 [in Russian],

Nauka, Moscow (1978).
67. G. Ya. Lyubarskii, *Group Theory and its Application in Physics* [in Russian], Fizmatgiz, Moscow (1958).
68. S. V. Maleev, *Zh. Éksp. Teor. Fiz.*, **34**, 1518 (1958).
69. S. V. Maleev, *Zh. Éksp. Teor. Fiz.*, **40**, 1224 (1961).
70. S. V. Maleev, *Pis'ma Zh. Éksp. Teor. Fiz.*, **2**, 545 (1965).
71. S. V. Maleev, *Zh. Éksp. Teor. Fiz.*, **66**, 1809 (1974).
72. S. V. Maleev, *Zh. Éksp. Teor. Fiz.*, **73**, 1572 (1977).
73. S. V. Maleev, *Izv. Akad. Nauk SSSR, Ser. Fiz.*, **42**, 1770 (1978).
74. S. V. Maleev, Preprint of the Leningrad Institute of Nuclear Physics [in Russian], No. 392 (1978).
75. S. V. Maleev, V. G. Bar'yakhtar, and R. A. Suris, *Fiz. Tverd. Tela*, **4**, 3461 (1962).
76. A. Maradudin, *Defects and the Vibration Spectrum of Crystals* [Russian Translation], Mir, Moscow (1968).
77. Yu. Z. Nozik, R. P. Ozerov, and K. Hennig, *Structural Neutron Diffractometry* [in Russian], Atomizdat, Moscow (1979).
78. A. N. Okorokov, A. G. Gukasov, V. V. Runov, et al., Preprint of the Leningrad Institute of Nuclear Physics [in Russian], No. 624 (1980).
79. A. I. Okorokov, V. V. Runov, and A. G. Gukasov, *Zh. Éksp. Teor. Fiz.*, **69**, 590 (1975).
80. A. I. Okorokov, V. V. Runov, A. G. Gukasov, et al., *Izv. Akad. Nauk SSSR, Ser. Fiz.*, **442**, 1770 (1978).
80a. V. S. Oskotskii, *Fiz. Tverd. Tela*, **9**, 550 (1967).
81. P. P. Parshin and M. G. Zemlyanov, *Fiz. Tverd. Tela*, **23**, 2204 (1981).
82. P. P. Parshin, M. G. Zemlyanov, A. S. Ivanov, et al., *Fiz. Tverd. Tela*, **23**, 2316 (1981).
83. V. F. Petrunin, S. K. Dolukhanov, M. G. Zemlyanov, et al., *Fiz. Tverd. Tela*, **23**, 1926 (1981).
84. P. P. Parshin, M. G. Zemlyanov, M. E. Kost, et al., *Izv. Akad. Nauk SSSR, Inorg. Mater.*, **9**, 1653 (1978).
85. P. P. Parshan, M. G. Zemlyanov, M. E. Kost, et al., *Fiz. Tverd. Tela*, **22**, 472 (1980).
86. A. Z. Patashinskii and V. L. Pokrovskii, *The Fluctuation Theory of Phase Transitions* [in Russian], Nauka, Moscow (1979).
87. A. Yu. Rumyantsev, V. V. Pushkarev, M. G. Zemlyanov, et al., *Zh. Éksp. Teor. Fiz.*, **75**, 712 (1978).
88. V. V. Runov, A. N. Okorokov, and A. G. Gukasov, Preprint of the Leningrad Institute of Nuclear Physics [in Russian] No. 507 (1979).
89. G. A. Smolenskii, V. A. Bokov, V. A. Isupov, et al., *Ferroelectric and Antiferro-electric Materials* [in Russian], Nauka, Moscow (1971).
90. V. A. Somenkov and S. Sh. Shil'shtein, *Phase Transitions of Hydrogen in Metals* [in Russian], Preprint of the Institute of Atomic Energy, Moscow (1978).
90a. G. F. Syrykh, A. P. Zhernov, M. G. Zemlyanov, et al., *Zh. Éksp. Teor. Fiz.*, **70**, 353 (1976).

91. L. Testardi, M. Weger, and N. Goldberg, *Superconducting Compounds with the Structure of β-Tungsten* [Russian translation], Mir, Moscow (1977).
91b. K. B. Tolpygo, *Fiz. Tverd. Tela*, **3**, 685 (1961).
92. E. A. Turov and Yu.P. Irkhin, *Fiz. Met. Metalloved.*, **3**, 15 (1956).
93. S. V. Tyablikov, *Methods of the Quantum Theory of Magnetism* [in Russian], Nauka, Moscow (1965).
94. W. Harrison, *Pseudopotential in the Theory of Metals*, Benjamin, New York (1966).
95. V. Heine, M. Kohn, and D. Weir, *The Theory of the Pseudopotential* [Russian translation], Mir, Moscow (1973).
96. Yu. N. Tsarev, V. K. Nikulin, and A. D. Chevychelov, *Fiz. Met. Metalloved.*, **33**, 939 (1972).
97. N. A. Chernoplekov, M. G. Zemlyanov, E. G. Brovman, et al., *Fiz. Tverd. Tela*, **5**, 112 (1963).
97a. N. A. Chernoplekov, M. G. Zemlyanov, and A. G. Chicherin, *Zh. Éksp. Teor. Fiz.*, **43**, 2080 (1962).
97b. N. A. Chernoplekov, M. G. Zemlyanov, A. G. Chicherin, et al., *Zh. Éksp. Teor. Fiz.*, **44**, 858 (1963).
98. N. A. Chernoplekov, M. G. Zemlyanov, V. A. Somenkov, et al., *Fiz. Tverd. Tela*, **11**, 2894 (1969).
99. A. A. Chernyshev, P. P. Parshin, A. Yu. Rumyantsev, et al., *Zh. Éksp. Teor. Fiz.*, **68**, 347 (1975).
100. Yu. L. Shitikov, M. G. Zemlyanov, and S. P. Mironov, *Prib. Tekh. Éksp.*, No. 4, 35 (1977).
101. M. S. Shur, *Kristallografiya*, **11**, 448 (1966); **12**, 215 (1967).
102. I. R. Entin, V. P. Glazkov, V. P. Moryakov, et al., *Prib. Tekh. Éksp.*, No. 5, 56 (1976).
103. A. Aharony and M. E. Fisher, *Phys. Rev. B*, **8**, 3323 (1973).
104. K. Aizu, *J. Phys. Soc. Jpn.*, **31**, 802 (1971).
105. K. S. Alexandrov, *Ferroelectrics*, **14**, 801 (1976).
106. K. S. Alexandrov, *Ferroelectrics*, **20**, 61 (1978).
107. L. Almqvist and R. Stedman, *J. Phys. F.*, 1, 785 (1971).
108. J. Als-Nielsen, *Phys. Rev. Lett.*, **11**, 730 (1970).
109. J. Als-Nielsen, O. W. Dietrich, and L. Passel, *Phys. Rev.*, **B14**, 4908 (1976).
110. P. W. Anderson and E. I. Blount, *Phys. Rev. Lett.*, **14**, 217 (1965).
110a. A. O. E. Animalu, *Phys. Rev. B*, **8**, 3542 (1973).
111. J. D. Axe, B. Dorner and G. Shirane, *Phys. Rev. Lett.*, **26**, 519 (1971).
112. J. Axe, J. Harada, and G. Shirane, *Phys. Rev. B*, 1, 1227 (1970).
113. J. D. Axe and G. Shirane, *Phys. Rev. B*, **8**, 1965 (1973).
113a. J. D. Axe, M. Iizumi, and G. Shirane, *Phys. Rev. B*, **22**, 3408 (1980).
114. G. E. Bacon and R. S. Pease, *Proc. Soc.*, **230**, 359 (1955); **A220**, 397 (1953).
115. P. Bak, S. Krinsky, and D. Mukamel, *Phys. Rev. Lett.*, **36**, 52 (1976).
116. P. Bak and D. Mukamel, *Phys. Rev. B*, **13**, 5086 (1976).
116a. P. Bak, *Rep. Progr. Phys.* **45**, 587 (1982).
117. C. M. Bertoni, O. Bisi, and F. Nizzoli, *Phys. Lett. A*, **47**, 466 (1974).

118. R. N. Bhatt and W. L. McMillan, *Phys. Rev. B*, **14**, 1007 (1976).
118a. H. Bilz, B. Gliss, and W. Hanke, in: *Dynamic Properties of Solids*, Vol. 1, G. K. Horton and A. A. Maradudin, eds., North-Holland Publishing Company (1974).
118b. M. Birr, A. Hiedemann, and B. Alefeld, *Nucl. Instrum. Methods*, **95**, 435 (1971).
119. R. J. Birgenau, M. T. Hutchings, et al., *Phys. Rev. Lett.*, **26**, 718 (1971).
120. R. J. Birgenau, J. R. Kjems, G. Shirane, et al., *Phys. Rev. B*, **10**, 2512 (1974).
121. R. J. Birgenau and G. Shirane, *Phys. Today*, **31**, 32 (1978).
122. R. J. Birgenau, W.B. Yelon, E. Cohen, et al., *Phys. Rev. B*, **5**, 2607 (1972).
123. J. L. Birman, *Phys. Rev. Lett.*, **17**, 1216 (1966).
124. R. Blinc and D. Phadzi, *Mol. Phys.*, **1**, 391 (1958).
125. M. Blume, *Phys. Rev.*, **130**, 1670 (1963).
126. G. Bogonovi, G. Caglioti, and M. Antonini, *Inelastic Scattering of Neutrons*, Vol. 1, IAEA, Vienna (1965), p. 117.
127. S. Boronkay and M. F. Collins, *J. Magn. Magn. Mater.*, **4**, 205 (1973).
128. V. Bortolani and G. Giunch, *J. Phys. F.*, **6**, 1269 (1976).
129. J. P. Boucher, *Solid State Commun.*, **33**, 1025 (1980).
130. J. P. Boucher, L. P. Regnault, J. Rossat-Mignot, et al., *Solid State Commun.*, **33**, 171 (1980).
131. C. J. Bradley and B. L. Davies, *Rev. Mod. Phys.*, **40**, 359 (1968).
132. B. N. Brockhouse, *Inelastic Scattering of Neutrons in Solids and Liquids*, IAEA, Vienna (1961), p. 113.
133. B. N. Brockhouse, K. R. Rao, and A. D. B. Woods, *Phys. Rev. Lett.*, **7**, 93 (1961).
133a. B. N. Brockhouse and P. K. Iyengar, *Phys. Rev.*, **III**, 747 (1958).
134. A. D. Bruce, *Adv. Phys.* **29**, 111 (1980).
135. A. D. Bruce and A. Aharony, *Phys. Rev. B*, **10**, 2078 (1974).
136. A. D. Bruce and R. A. Cowley, *Adv. Phys.* **29**, (1980).
137. W. Buckee, *Phys. Lett.*, **A42**, 40 (1973).
138. R. S. Carter, H. Palevsky, and D. J. Hughes, *Phys. Rev.*, **106**, 1168 (1957).
139. P. H. Catler, R. Day, and W. F. King, *J. Phys. F.*, **5**, 1801 (1975)
140. A. A. Chernyschov, V. V. Pushkarev, A. Yu. Rumyantsev, et al., *J. Phys. F.*, **9**, 1983 (1979).
140a. S. T. Chui, *Phys. Rev. B*, **9**, 3300 (1974).
141. R. Clarke and A. M. Glaser, *J. Phys. C*, **7**, 2147 (1974).
142. J. Cloizeaux and J. J. Pearson, *Phys. Rev.*, **128**, 2131 (1962).
143. W. Cochran, *Adv. Phys.* **9**, 387 (1960).
144. W. Cochran, *Adv. Phys.* **10**, 401 (1961).
145. W. Cochran, *Adv. Phys.* **72**, 157 (1969).
146. M. F. Collins, V. J. Mankiewicz, R. Nathans, et al., *Phys. Rev.*, **179**, 417 (1969).
147. J. F. Cooke, *Phys. Rev. B*, **7**, 1108 (1973).
148. J. F. Cooke, Proc. Conf. on Neutron Scattering, Gatlinburg, Tennessee (1976).
149. J. F. Cooke, J. W. Lynn and H. J. Davis, *Solid State Commun.*, **20**, 799

(1976).
150. M. J. Cooper and R. Nathans, *Acta Crystallogr.*, **23**, 357 (1967).
151. A. P. Cracknell, *Rep. Progress Physics*, **32**, 633 (1969).
152. A. P. Cracknell, *J. Phys. C.*, **6**, 1054 (1973).
153. A. P. Cracknell, *J. Phys. F.*, **4**, 456 (1974).
154. J. G. Couch, O. K. Harling, and L. C. Clune, *Phys. Rev.B*, **4**, 2675 (1971).
155. R. A. Cowley, *Phys. Rev. A*, **134**, 981 (1964).
156. R. A. Cowley, *J. Phys. Soc. Jpn.*, **28S**, 239 (1970).
157. R. A. Cowley, *Adv. Phys.* **29**, 1 (1980).
157a. E. R. Cowley, *Can. J. Phys.*, **51**, 843 (1973).
158. R. A. Cowley, W. J. L. Buyers, and G. Dolling, *Solid State Commun.*, **7**, 181 (1969).
159. R. A. Cowley, A. D. B. Woods, and G. Dolling, *Phys. Rev.*, **150**, 487 (1966).
160. L. Dagens, M. Rasolt, and R. Taylor, *Phys. Rev. B*, **11**, 2726 (1975).
161. O. W. Dietrich, J. Als-Nielson, and L. Passel, *Phys. Rev. B*, **14**, 4923 (1976).
162. G. Dolling, in: *Lattice Dynamics*, A. A. Maradudin and G. K. Horton, eds., North-Holland Publishing Company, Amsterdam (1974), Vol. 1, p. 541.
163. B. Dorner, *Acta Crystallogr. A* **28**, 319 (1972).
164. B. Dorner, J. D. Axe, and G. Shirane, *Phys. Rev. B*, **6**, 1950 (1972).
165. B. Dorner, A. A. Chernyshev, V. V. Pushkarev, et al., *J. Phys. F.*, **11**, 365 (1981).
166. W. Drexel, *Z. Phys.*, **255**, 281 (1972).
167. W. Drexel, *J. Phys. Chem. Solids*, **37**, 1135 (1976).
168. L. Dumoulin, P. Nedellec, J. Arzoumanian, et al., *Phys. Status Solidi (B)*, **90**, 207 (1978).
169. D. H. Dutton, B. N. Brockhouse, and A. P. Miller, *Can. J. Phys.*, **50**, 2915 (1972).
170. V. Dvorak, *Ferroelectrics*, **7**, 1 (1974).
171. R. Elliott and R. Lowde, *Proc. R. Soc. London A*, **230**, 46 (1955).
172. R. J. Elliott and A. A. Maradudin, Proc. Intern. Conf. Inelastic Neutron Scattering from Solids and Liquids, 1964; Vienna, p. 231 (1965).
173. R. J. Elliott and M. F. Thorpe, *Proc. Phys. Soc.*, **91**, 903 (1967).
174. R. J. Elliott and D. W. Taylor, *Proc. Phys. Soc.*, **83**, 189 (1964).
175. C. P. Enz, *Solid State Commun.*, **15**, 459 (1974).
176. J. Feder and E. Pyette, *Phys. Rev. B*, **1**, 4803 (1970).
177. V. K. Fedyanin and V. Yu. Yushankhai, *Phys. Lett.*, **A70**, 459 (1979).
177a. B. L. Fielek, *J. Phys. F*, **10**, 1665 (1980).
178. M. E. Fisher, *Am. J. Phys.*, **32**, 343 (1964).
179. M. E. Fisher and R. J. Burford, *Phys. Rev.*, **156**, 583 (1967).
180. P. A. Fleury, J. F. Scott, and J. M. Worlock, *Phys. Rev. Lett.*, **21**, 16 (1968).
181. B. C. Frazer and R. Pepinsky, *Acta Crystallogr.* **6**, 273 (1953).
182. P. De Gennes and J. Villian, *J. Phys. Chem. Solids*, **13**, 10 (1960).
183. V. L. Ginzburg, A. P. Levanyuk, and A. A. Sobyanin, *Phys. Rep.* **57**, 153 (1980).
184. W. Gissler, G. Alefeld, and T. Springer, *J. Phys. Chem. Solids*, **31**, 2361

(1970).

185. A. M. Glazer, *Acta Crystallogr. B* **28**, 3384 (1972).

186. C. J. Glinka, *Solid State Commun.*, **22**, 541 (1977).

187. C. J. Glinka, V. J. Mankiewicz, and L. Passel, *Phys. Rev. B*, **16**, 4084 (1977).

187a. H. N. Gupta and R. S. Upadhyaya, *Phys. Status Solidi (B)*, **93**, 781 (1979).

187b. H. N. Gupta and R. S. Upadhyaya, *Phys. Status Solidi (B)*, **102**, 143 (1980).

188. J. Hafner and P. Schmuck, *Phys. Rev. B*, **9**, 4138 (1974).

189. O. Halpern and M. Johnson, *Phys. Rev.*, **55**, 898 (1939).

190. B. I. Halperin and P. C. Hohenberg, *Phys. Rev.*, **177**, 952 (1969).

191. B. I. Halpern and C.M. Varma, *Phys. Rev. B*, **14**, 4030 (1976).

192. J. Harada, J. D. Axe, and G. Shirane, *Phys. Rev. B*, **4**, 155 (1971).

193. H. Hasegawa, *J. Phys. Soc. Jpn.*, **46**, 1504 (1979).

194. J. B. Hastings, S. M. Shapiro, and B. C. Frazer, *Phys. Rev. Lett.*, **40**, 237 (1978).

195. I. Hatta, *J. Phys. Soc. Jpn.*, **28**, 1266 (1970).

196. P. C. Hohenberg and B. I. Halperin, *Rev. Mod. Phys.*, **49**, 435 (1977).

197. T. Holstein and H. Primakoff, *Phys. Rev.*, **58**, 1098 (1940).

198. J. G. Houmann, M. Chapellier, A. R. Macintoch, et al., *Phys. Rev. Lett.*, **34**, 587 (1975).

199. J. G. Houmann and B. Moller, in: Proc. of the Conf. on Neutron Scattering, Gatlinburg, Tennessee (1976).

200. J. Hubbard, *Proc. R. Soc. London A*, **276**, 238 (1963).

201. J. Hubbard, *J. Phys. C.*, **4**, 53 (1971).

202. J. Hubbard, *Phys. Rev. B*, **19**, 2626; **20**, 4584 (1979).

203. D. J. Hunt, *J. Less-Common Met.*, **48**, 169 (1976).

204. M. T. Hutchings, G. Shirane, R. J. Birgeneau, et al., *Phys. Rev. B*, **6**, 1999 (1972).

205. Instrumentation for Neutron Inelastic Scattering Research, Panel Proceedings Series, IAEA, Vienna (1970).

206. P. K. Iyengar, G. Venkataramann, Y.H. Gameel, et al., *Inelastic Scattering of Neutrons*, IAEA, Vienna (1968), Vol. 1, p. 195.

207. P. K. Iyengar, G. Venkataramann, P. R. Vijiayaghaven, et al., *Inelastic Scattering of Neutrons*, IAEA, Vienna (1965), Vol. 1, p. 15.

208. T. Izuyama, D. Kim, and R. Kubo, *J. Phys. Soc. Jpn.*, **18**, 1025 (1963).

209. Yu. A. Izyumov, *J. Magn. Magn. Mater.* **21**, 33 (1980).

210. Yu. A. Izyumov, O. V. Gurin, and S. B. Petrov, *J. Magn. Magn. Mater.* **24**, 291 (1981).

211. Yu. A. Izymov and and V. E. Naish, *J. Magn. Magn. Mater.* **12**, 239 (1979).

212. J. K. Jacobs, C.R. Brown, and F.D. Manchester, Intern. Conf. on Hydrogen in Metals, Julich, (1972), Vol. 2, p. 601.

213. W. Jeitschko, *Naturwissenschaften*, **57**, 544 (1970).

214. L. J. Jongh and A. R. Miedema, *Adv. Phys.* **23**, 1 (1974).

215. L. P. Kadanoff, *Physics*, **2**, 263 (1966).

216. W. A. Kamitakahara, *Ferroelectrics*, **16**, 111 (1977).

217. W. A. Kamitakara and B. N. Brockhouse, *Phys. Lett. A*, **29**, 639 (1969).

218. K. Kawasaki, *Progr. Theor. Phys.* **39**, 1133 (1968); **40**, 930 (1968).
219. J. K. Kjems and M. Steiner, *Phys. Rev. Lett.*, **41**, 1137 (1978).
219a. E. W. Kellermann, *Philos. Trans. A*, **238**, 513 (1978).
220. D. Klauder, V. Lottner, and H. Sehener, *Solid State Commun.*, **12**, 617 (1972).
221. W. Kley, J. Peretti, R. Rubin, et al., Inelastic Scattering of Neutrons by Condensed Systems, Brookhaven National Laboratory-940, (1965), p. 105.
222. W. Kohn, *Phys. Rev. Lett.*, **2**, 393 (1959).
223. V. Koreman, J. L. Murray, and R. E. Prange, *Phys. Rev. B*, **16**, 4032; 4048; 4058 (1977).
224. V. Koreman and R. E. Prange, *Solid State Commun.*, **31**, 909 (1979).
225. V. Koreman and R. E. Prange, *J. Magn. Magn. Mater.*, **15-18**, Part 1, 355 (1980).
226. J. A. Krumhansl and J. R. Schrieffer, *Phys. Rev. B*, **11**, 3535 (1975).
227. A. Larous and B. N. Brockhouse, *Can. J. Phys.*, **54**, 1990 (1976).
228. K. M. Leung, D. N. Hone, D. L. Mills, et al., *Phys. Rev. B*, **21**, 4017 (1980).
229. K. M. Leung and D. L. Huber, *Solid State Commun.*, **32**, 127 (1979).
230. G. G. Libowitz, *The Solid State Chemistry of Binary Metal Hydrides*, Benjamin Inc., New York (1965).
231. E. Lieb, T. Schultz, and D. Mattis, *Ann. Phys.*, **16**, 407 (1961).
232. P. A. Lindgard and J. G. Houseman, in: *Rare Earth and Actinides*, Institute of Physics, London (1971).
233. S. G. Lonie and M. L. Cohen, *Phys. Rev. B*, **17**, 3174 (1978).
234. V. Lottner, J. M. Rowey, Glinka, et al., Proc. Conf. on Lattice Dynamics, M. Balkanski, ed., Paris, (1977), p. 247.
235. S. W. Lovesey, *J. Phys. C.*, **5**, 2769 (1972).
236. R. H. Lyddane and K. F. Herzfeld, *Phys. Rev.*, **54**, 846 (1938).
237. J. W. Lynn, *Phys. Rev. B*, **11**, 2624 (1975).
238. J. W. Lynn, H.G. Smith, and R. M. Nicklow, *Phys. Rev. B*, **8**, 3493 (1973).
239. A. R. Mackintosh and H. Moller, in: *Magnetic Properties of Rare-Earth Metals*, R. J. Elliott, ed., Plenum Press, New York (1972).
240. A. R. Mackintosh, *Phys. Today*, **30**, 23 (1977).
241. A. R. Mackintosh, *J. Magn. Magn. Mater.* **15-18**, 326 (1980).
242. A. Maelend, *J. Chem. Phys.*, **54**, 39 (1971).
243. A. Magerl, B. Berre, and G. Alefeld, *Phys. Status Solidi (A)*, **36**, 161 (1976).
244. A. Magerl, *J. Phys. C*, **10**, 2783 (1977).
245. A. Magerl, and N. Stump, *Solid State Commun.*, **33**, 627 (1980).
246. A. A. Maradudin and S. H. Vosko, *Rev. Mod. Phys.*, **40**, 1 (1968).
247. W. Marshall and S. W. Lovesey, *Theory of Thermal Neutron Scattering*, Clarendon Press, Oxford (1971).
247a. B. M. Martin, *Phys. Rev.*, **186**, 871 (1969).
248. K. A. McEven, W.G. Stirling, and C. Vettier, *Phys. Rev. Lett.*, **41**, 343 (1978).
249. H. Meister, J. Skalyo, B. C. Frazer, et al., *Phys. Rev.*, **184**, 550 (1969).
250. F. Mezei, *Z. Phys.*, **225**, 146 (1972).
251. H. J. Mikeska, *J. Phys. C*, **11**, L29 (1978).

252. V. J. Minkiewicz, M. F. Collins, R. Nathans, et al., *Phys. Rev.*, **182**, 624 (1969).
253. V. J. Minkiewicz, K. Gesi, and E. Hirakawa, *J. Appl. Phys.*, **42**, 1374 (1971).
254. H. Montgomery and A. P. Cracknell, *J. Phys. C*, **6**, 3156 (1973).
255. H.A. Mook, J. W. Lynn, and R. M. Nicklow, *Phys. Rev. Lett.*, **30**, 556 (1973).
256. H. Mori and K. Kawasaki, *Progr. Theor. Phys.*, **27**, 529 (1962).
257. J. A. Moriarty, *Phys. Rev. B*, **13**, 2337 (1976).
258. J. A. Moriarty, *Phys. Rev. B*, **16**, 2537 (1977).
259. J. A. Moriarty, *Phys. Rev. B*, **19**, 609 (1979).
260. T. Moriya, *J. Magn. Magn. Mater.* **14**, 1 (1979).
261. D. Mukamel, *Phys. Rev. Lett.*, **34**, 481 (1975).
262. D. Mukamel and S. Krinsky, *J. Phys. C*, **8**, L496 (1975).
263. D. Mukamel and S. Krinsky, *Phys. Rev. B*, **13**, 5065; 5078 (1976).
264. K. A. Muller and W. Berlinger, *Phys. Rev.*, **26**, 13 (1975).
265. K. A. Muller, W. Berlinger, and J. C. Slonszewski, *Phys. Rev. Lett.*, **25**, 734 (1970).
266. K. A. Muller, W. Berlinger, and F. Waldner, *Phys. Rev. Lett.*, **21**, 814 (1970).
267. V. E. Naish, *J. Magn. Magn. Mater.* **24**, 299 (1981).
268. Y. Nakagawa and A. D. B. Woods, *Phys. Rev. Lett.*, **11**, 271 (1963).
269. S. Naud, B. B. Tripathi, and H.C. Gupta, *Nuovo Cimento B*, **41**, 7 (1977).
270. M. Nielsen, *Phys. Rev. B*, **7**, 1626 (1973).
271. G. Nilsson, *Inelastic Scattering of Neutrons*, IAEA, Vienna (1968), Vol. 1, p. 187.
271a. G. Nilsson and G. Nelin, *Phys. Rev. B*, **3**, 364 (1971).
271b. G. Nillson and G. Nelin, *Phys. Rev. B*, **10**, 612 (1974).
272. G. Nilsson and S. Rolandson, *Phys. Rev. B*, **7**, 2393 (1973).
273. G. Nilsson and S. Rolandson, *Phys. Rev. B*, **9**, 3278 (1974).
274. A. C. Nunes, J. D. Axe, and G. Shirane, *Phys. Rev. B*, **4**, 2472 (1971).
275. A. I. Okorokow, V. V. Runov, and A. G. Gukasov, *Nucl. Instrum. Methods*, **157**, 487 (1978).
276. A. A. Oles, F. Kajzer, M. Kucab, and W. Sikora, *Magnetic Structures Determined by Neutron Diffraction*, Warsaw-Krakow (1976).
277. B. A. Oli and A. O. E. Animalu, *Phys. Rev. B*, **13**, 2398 (1976).
278. H. R. Oft, J. K. Kjems, and F. Hullinger, *Phys. Rev. Lett.*, **24**, 100 (1970).
278a. K. Parlinski, *Research Applications of Nuclear Pulsed Systems*, Vienna (1976), p. 179.
279. G. Parette and R. Kohn, *J. Phys.*, **32**, 447 (1971).
280. G. E. Peckham, D. H. Sannderson, and R. J. Sharp, *Brit. J. Appl. Phys.*, **18**, 73 (1967).
280a. R. M. Pick, M. N. Cohen, and R. M. Martin, *Phys. Rev. B*, **1**, 910 (1970).
280b. A. S. Pine and P. Dresselhouse, *Phys. Rev. B*, **4**, 356 (1971).
280c. J. Petzelt, in: *Phase Transitions*, Gordon and Breach Science Publishers, Great Britain (1981), Vol. 2, p. 155.
281. V. Plakhty and W. Cochran, *Phys. Status Solidi*, **29**, K81 (1968).

282. B. M. Powell, P. Martel, and A. D. B. Woods, *Phys. Rev.*, **171**, 727 (1968).

282a. B. M. Powell, P. Martel, and A. D. B. Woods, *Can. J. Phys.*, **55**, 1601 (1977).

283. R. E. Prange and V. Koreman, *Phys. Rev. B*, **19**, 4691; 4698 (1979).

284. R. Pynn and G. L. Squires, *Proc. R. Soc. London A*, **326**, 347 (1972).

285. A. Rahman, *Phys. Rev. B*, **14**, 3630 (1976).

286. B. D. Rainford, J. G. Houmann, and H.J. Guggenheim, *Neutron Inelastic Scattering*, IAEA, Grenoble (1972).

287. M. Rasolt and R. Taylor, *Phys. Rev. B*, **11**, 2717 (1975).

288. G. Reiter, *Phys. Rev. Lett.*, **46**, 202 (1981).

289. P. Resibois and C. Piette, *Phys. Rev. Lett.*, **24**, 514 (1970).

290. B. Renker, H. Rietschel, W. Glaser, et al., *Phys. Rev. Lett.*, **30**, 1144 (1973).

291. J. J. Phyne, N. C. Koon, J. B. Milstein, et al., Proc. Conf. Neutron Scattering, Gatlinburg, Tennessee, (1976), p. 783.

292. T. Riste, E. J. Samuelson, K. Ofnes, et al., *Solid State Commun.*, **9**, (1971).

293. D. S. Ritchie and M. E. Fisher, *Phys. Rev. B*, **5**, 2668 (1972).

294. D. K. Ross, P. F. Martin, W.A. Oates, et al., *Z. Phys., Chem., New Ser.*, **114**, 221 (1979).

295. J. M. Rowe, *Solid State Commun.*, **11**, 1299 (1972).

296. J. M. Rowe, *Phys. Rev. Lett.*, **33**, 1297 (1974).

297. J. M. Lowe, in: *Neutron Inelastic Scattering*, IAEA, Vienna (1978), Vol. 1, p. 151.

298. J. J. Rush and H.E. Flotow, *J. Chem. Phys.*, **48**, 3795 (1968).

299. J. J. Rush, J. M. Rowe, C.J. Glinka, et al., Proc. Conf. Neutron Scattering, Gatlinburg, Tennessee, June 1976-CONF-760601-PI.

300. V. C. Sahni and G. Venkataraman, *Adv. Phys.* **23**, 547 (1974).

301. M. Sakamoto, J. Chihara, H. Kadotani, et al., *Bibliography for Thermal Neutron Scattering*, 6th Ed., JAERI-M (1979), p. 8417.

302. J. Sakuri, R. A. Cowley, and G. Dolling, *J. Phys. Soc. Jpn.*, **28**, 1426 (1970).

303. E. J. Samuelsen, R. Silberglitt, G. Shirane, et al., *Phys. Rev. B*, **3**, 157 (1971).

304. H. Schmidt and F. Schabl, *Phys. Lett.*, **61**, 476 (1977).

305. D. S. Schreiber and R. M. Gotts, *Phys. Rev.*, **131**, 1148 (1963).

306. B. P. Schweiss, B. Renker, E. Schneider, et al., in: *Superconductivity of D- and F-Band Metals*, New York-London, p. 189.

307. J. F. Scott, *Rev. Mod. Phys.* **46**, 83 (1974).

308. S. M. Shapiro, J. F. Axe, G. Shirane, and T. Riste, *Phys. Rev. B*, **6**, 4332 (1972).

308a. L. I. Sham, *Phys. Rev.*, **188**, 1431 (1969).

308b. L.I. Sham, in: *Dynamical Properties of Solids*, G.K. Horton and A.A. Maradudin, eds., North-Holland Publ. Company, Amsterdam (1974), Vol. 1.

309. S. M. Sahpiro, Y. Noda, T. O. Brun, et al., *Phys. Rev. Lett.*, **41**, 1051 (1978).

310. S. M. Shapiro, D. Richter, Y. Noda, et al., *Phys. Rev. B*, **23**, 1594 (1980).

311. R. I. Sharp, *J. Phys. C*, **2**, 432 (1969).

312. W. M. Shaw and L. D. Muhlestein, *Phys. Rev. B*, **4**, 969 (1971).

313. S. R. Shenoy and N. C. Halder, *Phys. Rev. B*, **11**, 690 (1975).

314. G. Shirane, *Rev. Mod. Phys.* **46**, 437 (1974).
315. G. Shirane and J. D. Axe, *Phys. Rev. B*, **18**, 3742 (1978).
316. G. Shirane, J. D. Axe, and R. J. Birgeneau, *Solid State Commun.*, **9**, 397 (1971).
317. G. Shirane, J. D. Axe, J. Harada, et al., *Phys. Rev. B*, **2**, 155 (1970).
318. G. Shirane, S. M. Shapiro, and R. Comes, *Phys. Rev. B*, **14**, 2329 (1976).
319. J. Singh and S. Prakash, *Nuovo Cimento B*, **37**, 131 (1977).
320. H. P. Sinha and J. C. Upadhyaya, *J. Phys. Chem. Solids*, **38**, 41 (1977).
321. S. P. Singh and S. S. Kushwaha, *J. Phys. B*, **39**, 1317 (1978).
322. N. Singh and S. Prakash, *Phys. Rev. B*, **8**, 5532 (1973).
323. J. Skalyo, B. C. Frazer, and G. Shirane, *Phys. Rev. B*, **1**, 278 (1970).
324. T. Skoskiewicz, *Phys. Status Solidi*, **11**, K123 (1972).
325. E. L. Slaggie, *J. Phys. Chem. Solids*, **29**, 923 (1968).
326. T. Springer, in: *Dynamic of Solids and Liquids by Neutron Scattering*, Springer Verlag, New York (1977).
327. R. Stedman, Z. Amilius, R. Pauli, et al., *J. Phys. F*, **6**, 157 (1976).
327a. C. Stassis, I. Zarestky, and B. Harmon, *Phys. Rev. B*, **18**, 2632 (1978).
328. R. Stedman, L. Almqvist, G. Nilsson, et al., *Phys. Rev.*, **162**, 545 (1967).
329. R. Stedman and G. Nilsson, *Phys. Rev. Lett.*, **15**, 634 (1965).
330. R. Stedman and G. Nilsson, *Phys. Rev.*, **145**, 492 (1966).
331. M. Steiner, *J. Appl. Phys.*, **50**, 7395 (1979).
332. M. Steiner, B. Dorner, and J. Villian, *J. Phys. C*, **8**, 165 (1975).
333. M. Steiner, J. Villian, and C. G. Windsor, *Adv. Phys.* **25**, 87 (1976).
333a. A. Steyerl, *Z. Phys. B*, **30**, 231 (1978).
334. N. Stump, *Solid State Commun.*, **19**, 805 (1976).
335. N. Stump, *J. Phys. F*, **7**, LI (1977).
336. A. C. Switendick, *J. Less-Common Met.* **49**, 283 (1978).
337. K. Tajima, Y. Ishikawa, and H. Obera, *J. Magn. Magn. Mater.* **15-18**, Part 1, 373 (1980).
338. R. Taylor and A. H. McDonald, *J. Phys. F*, **10**, 2387 (1980).
339. W.D. Teuchert and R. Geick, *Phys. Status Solidi (B)*, **61**, 123 (1974).
340. H. Thomas and K. A. Muller, *Phys. Rev. Lett.*, **21**, 1256 (1968).
341. M. P. H. Thurlings, E. Frikkee, and H. W. De Wijn, *J. Magn. Magn. Mater.* **15-18**, Part 1, 369 (1980).
342. B. P. Toperverg, V. V. Runov, A. I. Chernov, et al., *Phys. Lett. A*, **71**, 289 (1979).
343. A. Tucciarone, H. Y. Lau, L. M. Corliss, et al., *Phys. Rev. B*, **4**, 3206 (1971).
344. I. C. Upadhyaya and L. Dagens, *J. Phys. F*, **9**, 2117 (1979).
345. V. G. Vaks, S. P. Kravchuk, and A. V. Tref, *J. Phys. F*, **10**, 2105 (1980).
346. V. G. Vaks, S. P. Kravchuk, and A. V. Trefilov, *J. Phys. F*, **10**, 2172 (1980).
347. V. G. Vaks, N. E. Zein, S. P. Kravchuk, et al., *Phys. Status Solidi (B)*, **96**, 857 (1979).
348. L. Van Hove, *Phys. Rev.*, **95**, 249 (1954).
349. L. Van Hove, *Phys. Rev.*, **95**, 1374 (1954).
349a. P. E. Van Camp and V. E. Van Doren, *Phys. Status Solidi (B)*, **93**, 483 (1979).

349b. P. E. Van Camp, V. E. Van Doren, and I. T. Devreese, *J. Phys. C*, **19**, 1421 (1981).
350. C. Vettier, D. B. McWhan, E. I. Blount, G. Shirane, *Phys. Rev. Lett.*, **39**, 1028 (1977).
351. G. Verdau, R. Rubin, and W. Kley, *Inelastic Scattering of Neutrons*, Vienna (1968), Vol. 1, p. 223.
352. P. Vorderwish, *Phys. Status Solidi (B)*, **65**, 171 (1974).
353. P. Vorderwish, *Phys. Status Solidi (B)*, **64**, 495 (1974).
354. P. Vonderwish, *Solid State Commun.*, **25**, 213 (1978).
355. P. Vorderwish, *Solid State Commun.*, **34**, 853 (1980).
356. W. B. Waeber, *J. Phys. C*, **2**, 903 (1969).
357. T. Waldkirch, K. A. Muller, and W. Berlinger, *Phys. Rev.*, **B7**, 1052 (1973).
357a. W. Weber and C. M. Varma, *Phys. Rev. Lett.*, **39**, 1094 (1977).
358. W. Wegener, P. Vorderwish, and S. Hauteckler, *Phys. Status Solidi (B)*, **98**, K171 (1980).
359. D. G. Westlake and J. F. Miller, *J. Phys. F, Met. Phys.*, **10**, 853 (1980).
360. J. W. Weymouth and R. Stedman, *Phys. Rev.*, **B2**, 474 (1970).
361. K. G. Wilson, *Phys. Rev.*, **B4**, 3174, 3184 (1971).
362. C. G. Windsor, R. K. Heenan, B. Boland, et al., *Nucl. Instrum. Methods*, **151**, 477 (1978).
363. E. J. Woll and W. Kohn, *Phys. Rev.*, **126**, 1693 (1962).
364. A. D. B. Woods and S. H. Chen, *Solid State Comm.*, **2**, 233 (1964).
365. Y. Yamada, Y. Fujii, and I. Hatta, *J. Phys. Soc. Jpn.*, **24**, 1503 (1968).
366. R. Yamaba, N. Watanabe, K. Sato, et al., *J. Phys. Soc. Jpn.*, **211**, 85 (1976).
367. A. Yoshimori, *J. Phys. Soc. Jpn.*, **14**, 807 (1959).
368. K. R. A. Ziebeck and J. G. Houman, in: Proc. Conf. on Neutron Scattering, Gatlinburg, Tennessee (1976).
369. A. Zungar and M. L. Cohen, *Phys. Rev. B*, **19**, 568 (1979).
370. A. Zungar and M. L. Cohen, *Phys. Rev. B*, **20**, 4082 (1979).

GPSR Compliance

The European Union's (EU) General Product Safety Regulation (GPSR) is a set of rules that requires consumer products to be safe and our obligations to ensure this.

If you have any concerns about our products, you can contact us on ProductSafety@springernature.com

In case Publisher is established outside the EU, the EU authorized representative is:

Springer Nature Customer Service Center GmbH
Europaplatz 3
69115 Heidelberg, Germany

Batch number: 10755284

Printed by Printforce, the Netherlands